the information store

☎ 01603 773114
email: tis@ccn.ac.uk

Atkins' M

7 DAY LOAN ITEM

This is a bran
beautiful chen
responsible fo
rics, drugs, pla
With engaging
an incredible r
expected conn
world can be u
which it is bui
completely nev
and enthrallin

Peter Atkins

Atkins' Molecules

Second edition

CAMBRIDGE
UNIVERSITY PRESS

PUBLISHED BY THE PRESS SYNDICATE OF THE UNIVERSITY OF CAMBRIDGE

The Pitt Building, Trumpington Street, Cambridge, United Kingdom

CAMBRIDGE UNIVERSITY PRESS

The Edinburgh Building, Cambridge CB2 2RU, UK

40 West 20th Street, New York, NY 10011–4211, USA

477 Williamstown Road, Port Melbourne, VIC 3207, Australia

Ruiz de Alarcón 13, 28014 Madrid, Spain

Dock House, The Waterfront, Cape Town 8001, South Africa

http://www.cambridge.org

First published 2003
First edition published by W. H. Freeman and Company, 1987

Printed in the United Kingdom at the University Press, Cambridge

Typeface Columbus 11.25/14 pt.   *System* LATEX 2_ε [TB]

A catalogue record for this book is available from the British Library

Library of Congress Cataloguing in Publication data
Atkins, P. W. (Peter William), 1940–
    Atkins' molecules/Peter Atkins – 2nd ed.
        p.   cm.
    Rev. ed. of: Molecules. 1987.
    includes bibliographical references and index
    ISBN 0 521 82397 9 – ISBN 0 521 53536 0 (pbk.)
        1. Molecular structure.   2. Molecules.   I. Atkins, P. W. (Peter William), 1940–
    Molecules.   II. Titles.
QD461.A83   2003
547'.122–dc21   2003048571

ISBN 0 521 82397 8 hardback
ISBN 0 521 53536 0 paperback

Contents

Preface

...from the first edition

Joy may be inarticulate, but reflection is empty without understanding. There is delight to be had merely by looking at the world, but that delight can be deepened when the mind's eye can penetrate the surface of things to see the connections within.

The following pages are intended to augment our delight when looking at the world. They introduce one facet of chemistry – its portrayal of the structure of substances – and they aim to show what makes up the things that make up the everyday world. These pages are an introduction to *molecules*. They are meant, among other things, to show the shapes of molecules and the connections between them, to show why some molecules behave as they do, and to reveal the economy of nature. In short, this book aims to make the molecular familiar.

I have purposely included molecules that I find interesting. Even so, I could have chosen any of a thousand others and still remained with the familiar. However, limitations of space dictate the brevity of the selection, and I must ask readers to quell the irritation they will certainly feel when they look in the index for a substance and find it ignored. There are several million known compounds, and manufacturers of pills, potions, and detergents – and nature especially – have at their disposal a vast chemical organ on which they can conjure symphonies of mixtures. It is inevitable that a book such as this will be incomplete. Its purpose is only to open an eye, not to show the world.

There is no particular order in which the book should be read; indeed, it is not necessarily meant to be read in any order: it is a book for occasional delectation. I wrote it, however, with a particular flow of thought in mind, so it is not completely amorphous and can profitably be read from front to back.

I tried to avoid technical terms throughout, but some inevitably (or at least unintentionally) crept in and are explained in the Glossary. Where possible, I also tried to explain; but do not expect too much fulfilment here, for some explanations are not yet known to anyone, and others require too much technical background. Moreover, I did not want to diminish delight by overloading the text with too much explanation: this is only an *introduction* to understanding. I particularly wanted to show that *some* appreciation of the features of molecules can be achieved without a college degree (or even a freshman course) in chemistry. Most of the information presented here has been culled from a dozen or so books that I have come to respect. I hope that, if the authors find their thoughts on these pages, they will regard that as a tribute. Many of the points I mention are discussed in more detail in those books, and readers would be well advised to check them before using the information I provide – for I have also cut corners in my wish to simplify and render palatable a sometimes tough and complex dish.

. . .and for the second edition

My, what a change there has been in the fifteen years since the first edition! Most obviously, there is the change in graphical representation of molecules. For the first edition, I had to build physical models of the molecules, photograph them, process the film, then trace the images and colour them by hand. Now, of course, computer software does the whole business far more quickly and far more realistically than in those bygone days. I have used *WebLab Pro* to construct all the images. You will also find them on the website for the book, where you can fiddle with them, rotate them, and so on.

I have culled some of the less interesting molecules from the first edition and added new ones: there are about 50 new molecules in this edition. Even the descriptions of most of the original molecules have been completely revised in the light of new knowledge or because I wanted to say something different or illuminate a new feature that has come to light.

What has not changed between the editions, I hope, is the sense of the sheer joy of seeing interconnections and explanations, the sense of delight at seeing why nature is the way it is, and the sense of understanding why a little change can have profound consequences. That is the deep pleasure of chemistry, that it opens a third eye onto the inner nature of things.

Introduction

When you hold this book you are holding molecules. When you drink coffee you are ingesting molecules. As you sit in a room you are bombarded by a continuous storm of molecules. When you appreciate the colour of an orchid and the textures of a landscape you are admiring molecules. When you savour food and drink you are enjoying molecules. When you sense decay you are smelling molecules. You are clothed in molecules, you eat molecules, and you excrete molecules. In fact, you are made of molecules.

A molecule is a characteristic grouping of atoms of the kind shown in the illustrations throughout this book. Until the beginning of the twentieth century, molecules were regarded as little more than abstract accounting symbols used by chemists to describe their reactions. However, in an extraordinary collaboration, physicists and chemists have confirmed the reality of molecules. First, they used indirect methods to infer the existence of these tiny particles of matter. Later, they used more sophisticated techniques to obtain what had long been sought – compelling images of individual molecules and atoms.

The following pages are intended to show a little of what has been found. They show the molecules we breathe, wear, eat, burn, and see all around us. The primary purpose of the illustrations is to acquaint you with their variety with a minimum of technical prerequisites. The molecules described here range from the simplest possible to the highly complex. Some do ostensibly humdrum things, such as methane (15), which is merely burned.[1] Others are included because they act as molecular building blocks, or they happen to typify a taste or an odour, or they are responsible for a colour. Some molecules do very grand things and are included here because of their importance. Among these is the most ubiquitous chemical in the world, cellulose (93), which grows as great forests and softens the face of the Earth, and deoxyribonucleic acid, DNA (203), which encodes the generations. We shall see how the replacement of one or two

---

[1] Numbers in parentheses refer to molecules described in the text.

atoms can convert a fuel into a poison, change a colour, render an inedible substance edible, or replace a pungent odour with a fragrant one. That changing a single atom can have such consequences is the wonder of the chemical world.

The illustrations alone will tell you much about the compositions and appearances of the molecules, and your perusal could, with profit, stop there. However, the drawings are enriched by knowing what they represent and understanding how a molecule performs its function. The remaining paragraphs of this introduction explain some of the background to the illustrations, suggest how to think about them, and sketch a few of the arguments that lead from atoms to molecules to properties. Chemistry provides a bridge between the familiar and the fundamental. These pages will give but a mere glimpse of this bridge but, with luck, you will see a little through the mist and understand how a chemist thinks.

A single atom can make a considerable difference to the properties of a molecule. The molecules responsible for the blue of a cornflower (*Centaurea cyanus*) and the red of a poppy (*Papaver orientale*) differ by only one hydrogen atom. This is explained in more detail on page 175.

ELEMENTS AND ATOMS

One of the great achievements of chemistry has been to show that all the matter in the world, be it a lump of rock, a glass of water, an ostrich feather, or a tree, is built from no more than a hundred or so simple substances called *chemical elements*. The elements include hydrogen, carbon, oxygen, and copper, and are so called because they cannot be broken down into simpler substances by heating, roasting, boiling, treatment with acid, or any of the other techniques that chemists use for changing matter. Physicists, of course, have developed more aggressive techniques, and they can smash elements apart into electrons, protons, and the other fundamental particles of nature using their particle accelerators. However, for our purpose, which is to explore our surroundings, we can stay with the hundred or so elements and marvel that the rich tapestry of the world can be stitched from so meagre a selection of thread.

The smallest particle of an element that can exist is an *atom* (from the Greek *atomos*, 'uncuttable'). A lump of a pure element, such as a lump of pure gold, is a collection of identical atoms; a lump of carbon is also a collection of identical atoms, but each atom is different from an atom of gold. Atoms are very small: the diameter of a carbon atom is only about 0.15 billionths of a metre (0.000 000 000 15 metres, 1.5×10^{-10} m), so a 1.5-centimetre line of carbon (about this long: ————) is a hundred million carbon atoms from end to end and about a million atoms across. Any visible lump of matter – even the merest speck – contains more atoms than there are stars in our galaxy. When we lift an apple we feel the total weight of a colossal number of almost weightless atoms. When we hear the ripple of water we are hearing shockwaves as a myriad of almost imperceptible molecules crash down and collide with other molecules. When we dress we pull across our bodies a great web spun from almost infinitesimal dots and held together by the forces acting between them. When we see a flame we are seeing the release of an almost negligible droplet of energy, but in such a Niagara that the heat sears and consumes.

Each atom consists of a very tiny central *nucleus* with a positive electric charge. That nucleus is surrounded by a sufficient number of negatively charged electrons to cancel out its charge, so the atom as a whole is electrically neutral. The electrons form a series of concentric shell-like clouds round the nucleus, so it is convenient to think of atoms as minute spheres. In this book, we represent atoms by spheres magnified up to 50 million times, so a carbon atom is represented by a sphere nearly 1 centimetre in diameter. Oxygen and nitrogen atoms have about the same number of electrons as carbon (eight and seven, respectively, in place of carbon's six) and are almost the same size as carbon atoms. A hydrogen atom is appreciably smaller because in place of carbon's six electrons it has only one. Most of the other atoms that we shall meet are appreciably larger than carbon. Phosphorus, sulfur, and chlorine atoms all have more than twice as many electrons as carbon (15, 16, and 17, respectively) and we represent them by spheres of correspondingly larger diameters.

We consider fewer than a dozen elements in this book. Each one is denoted by a chemical symbol, which is commonly the first letter of its name (with occasionally the inclusion of a later letter):

H hydrogen   C carbon   N nitrogen   O oxygen
F fluorine   P phosphorus S sulfur   Cl chlorine

Remnants of Latin derivatives sometimes raise their head, as in Na for sodium (*natrium*), K for potassium (*kalium*), and Fe for iron (*ferrum*).

Very occasionally we shall meet the concept of an 'ion'. An *ion* is an atom that has lost or gained one or more (negatively charged) electrons and hence has acquired an electric charge. When an atom loses electrons it becomes positively charged and is called a *cation*. An example is the sodium ion, Na^+, which is a sodium atom that has lost one electron. Potassium, K, sodium's neighbour in the periodic table, can also lose an electron to form a singly charged potassium cation, K^+. A magnesium cation, Mg^{2+}, is a magnesium atom that has lost two electrons and become doubly positively charged. When an atom gains electrons it becomes negatively charged and is called an *anion*. A chlorine atom forms a singly charged chloride anion, Cl^-, by gaining one electron. An oxygen atom forms the doubly charged oxide anion, O^{2-}, by gaining two electrons. Once again, the precise number of electrons that an atom can gain or lose depends on the internal structure of its atoms, and the ions mentioned in this paragraph include most of those we need consider.

COMPOUNDS

A *compound* is a definite, fixed combination of elements. Thus, water (5) is a combination of hydrogen and oxygen, and aspirin (170) is a combination of carbon, hydrogen, and oxygen.

Many compounds consist of *molecules*, our subject. A molecule, as mentioned before, is a specific, discrete grouping of atoms in a definite geometrical arrangement. Numerous illustrations of models of molecules appear later in the book, so you can turn to almost any page to see an example. Almost all the molecules we describe consist of atoms of no more than half a dozen elements, and in many cases just two or three. That is one of the wonders of the world, that so much can be spun from so little, just as the world's literature can be spun from two dozen letters. Because we need to depict so few elements, we can distinguish between them by using spheres of different colours and will use the convention shown overleaf. These colours are commonly used, and in some cases have been chosen to allude to typical properties of the elements themselves. Thus, hydrogen is shown as white (pale grey in practice) because it is the simplest atom; carbon is as black as soot; and oxygen, the life-giver, is red. Chlorine is a greenish-yellow gas (hence, its name,

The atoms of the elements shown in the illustrations later in the book are distinguished by this colour code.

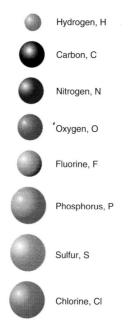

Hydrogen, H

Carbon, C

Nitrogen, N

Oxygen, O

Fluorine, F

Phosphorus, P

Sulfur, S

Chlorine, Cl

from the Greek *khloros*, 'green') and sulfur is a yellow solid. Just occasionally, we shall need to invoke another element and introduce it as required. Finally, we use purple spheres to represent the metal ions (specifically, sodium and potassium ions). The molecular structures based on spheres shown in the following pages are called 'space-filling models' because they give a reasonably accurate impression of the bulk of the molecule.

The composition of a molecule is denoted by listing the chemical symbols of the elements it contains and denoting the number of atoms of each element by a subscript (with 1 omitted). Thus, H_2O tells us that a molecule of water consists of two hydrogen atoms and one oxygen atom. The chemical formulas of the male and female sex hormones, testosterone, $C_{19}H_{28}O_2$ (195), and oestradiol, $C_{18}H_{24}O_2$ (196), show that a great deal of strife and joy, not to mention literature and warfare, stems from the difference amounting to one carbon atom and four hydrogen atoms.

BONDS BETWEEN ATOMS

The links between atoms that hold them in specific geometrical arrangements are called *bonds*. For our purposes all it is necessary to know is that a chemical bond is a shared pair of electrons. This idea,

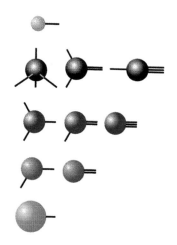

The atoms of the elements form characteristic numbers of bonds as shown here.

which was first proposed by the American chemist G. N. Lewis in the opening decades of the twentieth century, has survived the rigours of quantum-mechanical scrutiny with only minor changes of detail. We can picture a bond as two electrons that hover between the nuclei of the atoms they join and act as a kind of electrostatic glue.

The number of bonds that a given atom can form is a reflection of the number of electrons it can share with its neighbours. The rules governing this ability, which can be explained by going further into atomic structure, are as follows:

- a hydrogen atom usually forms only one bond
- a carbon atom usually forms four bonds
- a nitrogen atom usually forms three bonds
- an oxygen atom usually forms two bonds
- a chlorine atom usually forms one bond.

In writing the *structural formula* of a molecule, a depiction of the bonding pattern in a molecule, a bond is represented by a short single line (–) between the chemical symbols of the atoms it joins. The bond between a hydrogen atom and a chlorine atom in hydrogen chloride, HCl, is therefore represented as H–Cl. Because the arrangement of atoms in models of complex molecules is often difficult to make out, we shall also include *tube structures* or 'stick' structures showing the bonding pattern alone, with the tubes representing the bonds coloured at each end to indicate the identity of the element using the same code as for the 'space-filling' spheres. Thus, an ethanol molecule can be represented as

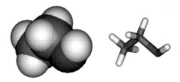

The illustration on the left is a space-filling model and that on the right is a tube structure.

Some atoms can form more than one bond to another atom. When a carbon atom shares two pairs of electrons with a neighbouring oxygen atom, for example, there is a *double bond* between them. This double bond is denoted C=O, and it can be seen in many of the structures shown later, including acetic acid (31) and testosterone (195). Similarly, two atoms can share three pairs of electrons, in which case the atoms are joined by a *triple bond*, as in the hydrogen cyanide molecule H–C≡N (115).

Single bonds act like hinges, because one end of the molecule can be twisted relative to the other end of the molecule, so molecules can coil into many different shapes. Although we might show a chain of carbon atoms laid out in a straight line, as in

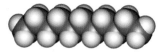

we should think of the chain as ceaselessly writhing and wriggling and adopting shapes like

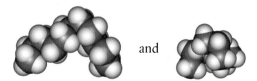   and

Where the two parts of a molecule are joined by a double bond that ability to coil is lost because one end of the molecule cannot rotate relative to the other around the torsionally rigid double bond.

ORGANIC COMPOUNDS

Most of the compounds shown in the following pages are *organic.* That is, they are compounds containing carbon and (usually) hydrogen. Compounds that are not organic are called *inorganic.* Some very simple carbon compounds, particularly those not containing hydrogen (carbon dioxide, chalk, and other carbonates, for example) are honorary inorganic compounds.

The term 'organic' does not mean that the compounds are necessarily made by biological organisms, although that was once thought to be the case and is the origin of the name. It was once believed that organic compounds contained some kind of 'vital force' that led to life. That view was overturned in the nineteenth century, when it was shown that a typical organic compound, urea (147), a component of urine, could be made from inorganic starting materials.

Organic compounds are prominent in these pages because they are so important and interesting. They are responsible for the colours and odours of flowers and vegetation and for the taste of food. Indeed, virtually the whole of the natural world, apart from the rocks and the oceans, consists of organic compounds. Many of the newer construction materials, notably plastics, are also organic, as are almost all pharmaceuticals.

Carbon, through the organic compounds it forms, plays a special role in the world because it has a unique ability to form bonds with itself. A glance at the following pages will show many examples of molecules that consist of chains and rings of black carbon spheres. A few other elements can link to themselves (sulfur among them), but none so extensively as carbon, and none gives so many stable structures. Carbon has this unique ability because it is rather mediocre and undemanding as an element. It has a middling ability to attract electrons from other atoms, and its own electrons can be removed fairly easily. In other words, it is promiscuous in its ability to share electrons, and can therefore form a huge number of liaisons.

When looking at the organic molecules pictured in the following pages, it is often helpful to think of them as chains or rings of carbon atoms that form an underlying framework. With the framework in mind, the structure of the molecule can be identified by noting the other groups of atoms that are attached to it. For example, a common structural motif is the hexagonal benzene ring:

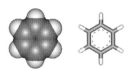

The dashed line inside the hexagon on the right indicates that the double-bond character of the bonding between three pairs of carbon atoms is shared equally around the ring. A molecule derived from benzene is benzoic acid, a benzene ring to which the 'carboxyl group', –COOH, is attached:

The commonly occurring groups, such as –COOH, that decorate basic frameworks are called *functional groups* and are often the chemically active parts of organic molecules. Examples of functional groups that occur frequently in the molecules we consider are the hydroxyl group, –OH, the carbonyl group (>C=O), and the carboxyl group (–COOH).

Another feature to which we must be alert is the fact that the same molecular formula may apply to two or more different substances because the same atoms can be linked in a variety of ways, each bonding pattern corresponding to a different compound. An example is the molecular formula C_2H_6O, which applies both to ethanol and to dimethyl ether:

Ethanol, C_2H_6O Dimethyl ether, C_2H_6O

Different compounds with the same molecular formula are called *isomers*, from the Greek for 'equal parts', as the different molecules can be thought of as built from the same kit of atoms. These two isomers, ethanol and dimethyl ether, differ in the 'connectivity' of the molecules, which atom is joined to which. Another kind of isomerism occurs when two molecules have the same atoms joined together (that is, have the same connectivity) but differ in the arrangement of those atoms in space. This *geometrical isomerism* is illustrated by the following two compounds of composition C_4H_8:

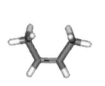 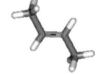

cis-2-Butene   *trans*-2-Butene

(Note the double bond in the centre of the molecule; the 2 in the name tells us that the double bond starts at carbon atom number 2.) These are distinct compounds with different properties, because the $C{=}C$ double bond is rigid and one end of the molecule cannot be rotated relative to the other. The two C_4H_{10} molecules

in which there is only a single bond between the two ends are not isomers, because one end can be rotated relative to the other, so one arrangement can be rotated into the other without breaking any bonds. In the text we shall refer to the isomer on the left as the *cis isomer* because the two $-CH_3$ groups (which are called 'methyl groups') are on the same side of the double bond (and *cis* is Latin for 'on this side'). The isomer pictured on the right is the *trans isomer*, because the $-CH_3$ groups are on opposite sides of the double bond (and *trans* is Latin for 'across'). At this stage, isomerism may seem of minor technical importance. However, in due course (see molecule 175), we will see that it can have horrendous consequences.

FORCES BETWEEN MOLECULES

The fact that individual molecules form liquids and solids rather than always flying freely as in a gas is an indication that there are interactions between them. Thus, the fact that water is a liquid at room temperature is a sign that the H_2O molecules adhere to each other. To boil water, we have to supply sufficient energy to rip the molecules away from each other. The forces responsible for holding molecules together are called *van der Waals forces*, after the nineteenth-century Dutch scientist who first studied them. Here we shall consider only three contributions to the attractions between molecules.

Although molecules are electrically neutral, because electrons pile up in some regions and are depleted in others, the electric charge varies over the surface of the molecule. This variation can be calculated and depicted by colours applied to a surface bounding the molecule. For instance, the distribution of charge in an ethanol molecule is

(The transparent version on the right allows us to see the location of the underlying atoms.) In this illustration, and in others like it later in the book, regions of slight negative charge are depicted red (on the electron-attracting oxygen atom, in this case) and regions of slight positive charge are depicted blue, with intermediate colours (yellow and green) indicating intermediate distributions of charge. The separation of positive and negative regions of charge endows the molecule with an electric dipole, and dipoles on different molecules can attract each other by a *dipole–dipole interaction*. That is, the region of slight excess negative charge (red) is attracted to the region of slight positive charge (blue) in a neighbouring molecule, and the molecules stick together. Only molecules with this imbalance of charge distribution – so-called *polar molecules* – interact in this way. Hydrocarbons are normally considered not to be polar, so they do not adhere by this dipole–dipole interaction.

All molecules, regardless of whether or not they are polar, interact with one another through the so-called *dispersion interaction*. As we have seen, molecules are composed of atoms that consist of a tiny central positively charged nucleus surrounded by a cloud of negatively charged electrons. We should not think of that cloud as frozen in time. Instead, it is like a swirling mist, thin here one instant, thick here the next. Where the cloud briefly clears, the positive charge of the nucleus shines through. Where the cloud briefly thickens, the negative charge of the electrons outweighs the positive charge of the nucleus. When two molecules lie close to one another, the charges associated with the swirling clouds interact, with the partially revealed positive charge of a nucleus attracted to the partially accumulated negative charge of the denser electron cloud. Thus, the two molecules adhere together. All molecules interact in this way, but the strength of the interaction is greater for molecules built from atoms that have large numbers of electrons, such as chlorine and sulfur.

The third type of interaction we consider is called a *hydrogen bond*. This very important interaction, which is denoted by a dotted line,

A· · ·B, in place of a single dash, consists of two atoms with a hydrogen atom lying between them:

The bond, which has about one-tenth the strength of a normal chemical bond, is formed because the oxygen atom in –N· · ·H–O– attracts electrons so strongly that it draws the shared electron pair in the H–O– bond towards itself, leaving the positive charge of hydrogen's nucleus almost fully exposed. That partially exposed positive charge is strongly attracted to other electrons nearby, particularly those of another nitrogen or oxygen atom. For reasons related to the ability of atomic nuclei to pull electrons towards themselves, a hydrogen bond can occur only between atoms that can attract electrons strongly, which for our limited palette of elements are

–O–H· · ·O–   –O–H· · ·N–   –N–H· · ·O–   –N–H· · ·N–

Fluorine, F, is the only other element for which hydrogen bonding is important. As we shall see, hydrogen bonding is of enormous importance in the world, for among numerous other things, it accounts for the existence of oceans and the rigidity of wood. Without hydrogen bonds, the oceans would be vapour and trees would lie flat like moss.

A final point to note is that one substance is likely to dissolve in another if they have similar types of interactions. Thus, molecules that interact with each other by forming hydrogen bonds are likely to dissolve in a liquid that can form hydrogen bonds. In short, 'like dissolves like'.

MIXTURES

Mixtures consist of substances, such as molecular compounds, simply mingled together. Food and the living things that use landscapes as a stage, and which are often destined to become food themselves, are extremely complex mixtures of organic compounds, for they consist of varieties of biological cells, with all the paraphernalia of life packed inside. The flavour of an orange, for example, is due to an unconscious conspiracy of hundreds of different compounds.

The task of presenting the molecules of our natural and synthetic world might therefore seem hopeless. However, simplifications are

Oranges have an odour that arises in part from a terpene (page 149) molecule (limonene) that is the mirror image of the molecule that contributes to the odour of lemons. The colour of an orange is due largely to anthocyanin molecules (162). Their tartness is due to citric acid (102).

possible. Some substances are single compounds and we can depict them precisely, as in the case of aspirin (170) and sugar (89). For others, we can display the molecules *typical* of their composition. This is the case with gasoline (20), a complex mixture dominated by several characteristic types of molecules. For still other substances, we can show molecules that *typify* their characteristics. Thus, although we cannot show all the known components of an orange, we can show the typical molecules that contribute to its flavour or its texture. In this way we can at least begin to distinguish an orange molecularly from a banana.

This account will show that the world is simultaneously both simple and complex. It is complex because even a leaf is composed of myriad types of molecules with a composition that changes with the seasons. It is simple because scientists can unravel the leaf, distinguish its components, and see how one component may dominate in the summer but that another will dominate in the autumn. The tale of unravelling is far from complete, and it may be an age before we are certain that we know every detail about the differences between an orange and a banana, let alone a man and a woman. Nevertheless, these pages will show that the horde of molecules that make up the world is less an amorphous swarm and more a group of individual, understandable personalities.

1

Simple substances

Simple these molecules may be, but unimportant they are certainly not, for they include the gases of the air and the water of the oceans. They are the enablers, if not the components, of life. They are also agents of havoc, for they include the substances that pollute and destroy where nature, left alone, might flourish.

Unless you are currently working as a deep-sea diver, you are immersed in a gas that is about four-fifths nitrogen and one-fifth oxygen. As in any gas, its molecules are in continuous chaotic motion (indeed, the word 'gas' comes from the same Greek root as 'chaos'). The molecules hurtle through space at about the speed of sound (about 340 metres per second at sea level), collide with each other, and go hurtling off in other directions until they collide again a fraction of a second later. The incessant impact of this storm of molecules on the surface of a container – including your container, your skin – is experienced as a virtually constant pressure, a pressure equivalent to a lump of lead weighing 14 pounds applied to every square inch of your body. We stagger under this unnoticed but immense load throughout our lives. Even on a still, warm summer's day, or in a quiet room such as you may now be in, you are in fact at the centre of an unseen storm of molecules. When the wind blows, the molecules stream predominantly in one

An early version of our atmosphere emerged from the Earth beneath our feet, as at the outgassing that accompanied this lava flow from the Kivu lava lake situated in East Africa between Zaire and Rwanda.

direction and strike that side of your face. Sometimes this unseen stream of molecules can be strong enough to fell trees and destroy buildings. The modulated puffs of air we call speech can be even more powerful, for they can fell empires.

AIR

It is not at all clear where our atmosphere came from or how it has changed. There is general agreement that the Earth acquired an early version of its atmosphere as a result of the outgassing of the rocks and planetesimals that aggregated to form our initially primitive planet. A similar kind of outgassing occurs today at volcanoes, and it is surmised that the gases they release – largely water vapour, hydrogen, hydrogen chloride, carbon monoxide, carbon dioxide, nitrogen, and molecules containing sulfur – were abundant in the first atmosphere. Of these substances, only nitrogen is abundant in our present atmosphere. Hence there is a question about where the rest have gone and where other gases have come from. Some answers – which are little more than wise but perhaps erroneous speculations – are presented here and in the next section. One substance can be dealt with at once: hydrogen molecules, being very light and moving very fast, escaped the gravitational pull of the planet and disappeared into space, as any that are newly formed would soon do today.

1 OXYGEN O$_2$

Oxygen accounts for about 20 per cent of the volume of the atmosphere. It is the most abundant element in the Earth's crust, accounting for almost half its total mass, and is present there in combination with other elements in the form of water, carbonates, silicates, and oxides. It is also abundant on the Moon, but not in the free state; there, oxygen is trapped in compounds in the lunar rocks. When space travel becomes more commonplace, we may find it economical to mine the lunar surface for its oxygen; here on Earth we mine for metals and discard the unwanted oxygen from their ores.

Although oxygen is abundant in the atmosphere and is obtained industrially from this rich source by distillation of liquid air, it is a relative latecomer. The atmosphere of the newly formed Earth contained very little oxygen. Some of it arrived when water molecules (5) that had outgassed from the rocks were broken apart by the intense ultraviolet radiation from the Sun. The bulk of our oxygen arrived when the first photosynthesizing cells evolved about 2.7 billion years ago: the prokaryotes we know as cyanobacteria (blue–green bacteria). These single-celled, nucleus-free life forms acquired hydrogen from water (discarding the oxygen) and carbon and oxygen from carbon dioxide, in the process of building their own carbohydrates (93). Thus, the oxygen that we now prize so highly, that is essential to most animal life, and which must be carried whenever we explore

The colour of these piglets is due to haemoglobin molecules to which oxygen molecules have attached by linking to the iron atoms they contain. 'White' humans are pink for the same reason.

alien environments, was originally a pollutant in an atmosphere that favoured a different form of life. That great pollution left its imprint on the Earth, for the surge of oxygen that accompanied the emergence of photosynthesis oxidized the iron dissolved in the seas. The Earth rusted, and the great deposits of red iron ore record the moment.

The history of the atmosphere, and in particular the abundance of oxygen in it, is still the subject of much controversy. Evidence is accumulating that the concentration of oxygen has fluctuated widely, perhaps reaching 35 per cent during the Carboniferous (300 million years ago), and that the effect of the asteroid impact that eliminated the dinosaurs was intensified by a conflagration that the impact ignited in the oxygen-rich air of the time. The current concentration of oxygen appears to be related to the balance of carbon-bearing material available as biosphere and secured out of conflagration's way in the rocks beneath the biosphere's feet and roots. It is estimated that there is over 20 000 times as much carbon trapped underground, including the tiny particles trapped in sandstone, as there is in the current biosphere.

Oxygen is an extraordinarily dangerous material, especially to life, and its use by living things is almost literally playing with fire. Breathing oxygen has much the same effect as exposure to radiation and the damage caused is chemically much the same. Thus, in each case, damage is the result of the formation of *radicals*, which are molecules with an odd number of electrons. In the case of oxygen, the radical formed by the acquisition of one electron to give the species denoted O_2^- is called the superoxide ion, which goes on to produce the virulent hydrogen peroxide molecule, H_2O_2 (11), and hydroxyl radicals, OH, which are also responsible for radiation damage. Estimates suggest that between one and two per cent of the oxygen used by cells is converted into superoxide ions, and up to 10 per cent during vigorous exercise. That might not sound much, but it amounts to nearly 2 kilograms of superoxide ions produced each year in our body even if we take no exercise. This danger can be seen from another perspective: the damage done by breathing for one year is equivalent to about 10 000 chest X-rays. The only way to avoid death, it appears, is to stop breathing. More positively, we can now begin to see how organisms may have been able, unconsciously, to prepare themselves for dealing with the toxicity of oxygen by learning to cope with exposure to intense solar radiation, for the problem of tackling the same radicals is common to both.

Oxygen itself is an odourless, colourless, tasteless gas that condenses to a pale blue liquid. The colour change comes about when pairs of molecules cooperate in the absorption of light, a phenomenon that is possible only when the molecules spend a lot of time close together in a liquid. Oxygen also has the unusual property of being magnetic. This is most clearly shown by the ability of a magnet to attract liquid oxygen, but the gas is also magnetic. To understand this magnetism, we need to know that an electron spins either clockwise or counterclockwise, and in doing so it acts as a tiny magnet. In most molecules there are as many clockwise-spinning electrons as there counterclockwise-spinning electrons, so the magnetic effects of the electrons cancel out and the molecules are not magnetic. An oxygen molecule is peculiar: two of its electrons spin in the same direction (for instance, both clockwise), so their magnetic effects do not cancel out. One application of this property is to the measurement of concentrations of oxygen in artificial atmospheres, such in incubators for premature babies: the magnetism of the gas is monitored and then interpreted in terms of the concentration of oxygen molecules.

Huge quantities of oxygen gas are used in steelmaking, for which about 1 tonne of oxygen is needed in order to prepare 1 tonne of metal. It is blown through the impure molten iron, combines with the impurities present (particularly carbon), and carries most of them away as gas. Oxygen is better for this purpose than air, which is mainly nitrogen (2) and which carries away too much heat when it is blown through the molten metal.

2 NITROGEN N_2

Nitrogen is the most abundant gas in the atmosphere, accounting for 78 per cent of its volume: over three-quarters of every breath we inhale is nitrogen. This abundance of nitrogen probably outgassed from the rocks, just as did the other gases of the early atmosphere. However, nitrogen molecules are too unreactive to have combined with other substances to as great an extent as oxygen has done; what was abundant has remained abundant and much of the nitrogen we breathe has been around for billions of years.

Like oxygen, nitrogen atoms go round in pairs. However, the atoms in the N_2 molecule are bound together by a triple bond: $N{\equiv}N$. This results in one of the most strongly bonded molecules known, one that can survive collisions with other molecules. Nitrogen's relative lack of reactivity allows it to dilute the dangerous oxygen of the air.

Many of the molecules in living cells, particularly the proteins, contain nitrogen atoms. Nitrogen is therefore essential to the growth of plants and crops, and somehow the biosphere must get its hands on these unreactive molecules. A typical human body contains about 1.7 kilograms of nitrogen. Its incorporation into life, first into plants and then into animals, begins with its conversion into nitrogen oxides (9) by lightning flashing through the air and by solar radiation acting on the upper atmosphere. These more reactive compounds are then washed into the soil by rain.

A major highway for the migration of nitrogen from the atmosphere to the soil is *nitrogen fixation*, the incorporation of atmospheric nitrogen into compounds. Nitrogen fixation may be natural or it may be contrived by industry. Biological nitrogen fixation is caused only by certain prokaryotes, including bacteria, particularly cyanobacteria and actinomycetes (branching, multicellular mould-like organisms). Some of these bacteria (particularly *Azotobacter* and *Clostridium*) can exist and operate individually but the most important, *Rhizobium*, form symbiotic associations with higher plants, particularly the legumes (clover, pulses such as peas and beans, alfalfa, acacia), the roots of which they colonize. The bacteria invade the plant and cause the formation of a nodule by inducing localized proliferation of the plant host cells. In all cases the agent responsible for the fixation is the enzyme *nitrogenase*, which consists of two protein molecules. One of these molecules is based on two molybdenum atoms, 32 iron atoms, and between 25 and 30 sulfur atoms (as well as the usual extensive scaffolding of carbon atoms). The other protein is based on iron. Industrial chemists are intensely interested in the mode of action of nitrogenase, for, if it could be mimicked, then the world would have a ready means for harvesting the nitrogen of the skies and redistributing it as fertilizer or, more elegantly, for genetically engineering nitrogenase into a crop that would generate its own fertilizer as it grew.

Nitrogenase is inactivated when it is exposed to oxygen, which reacts with the iron component of the proteins. Cyanobacteria (which generate oxygen during photosynthesis) and the free-living aerobic bacteria of soils, such as *Azotobacter* and *Beijerinckia*, have various methods to overcome the problem. Thus, *Azotobacter* has the highest known rate of respiratory metabolism of any organism, so it protects the enzyme by maintaining a very low level of oxygen in its cells. *Azotobacter* also produces copious amounts of extracellular polysaccharide slime, which inhibits the access of oxygen into the cell. *Rhizobium* contains haemoglobin-like oxygen-scavenging molecules that are seen as a pink colour when the active nitrogen-fixing nodules of legume roots are cut open.

Cyanobacteria, which obtain their energy from sunlight, their carbon from carbon dioxide, and their nitrogen directly from the atmosphere, are believed to have been the first microorganisms to colonize land. This hypothesis is supported by the observation that they were the first to re-establish themselves after an eruption of the volcano on Krakatoa in 1883 had completely destroyed all life in an extensive surrounding area. Indeed, some pioneering algae, specifically the red cyanobacteria that occur in huge colonies in the Red Sea, have their settlements named after them.

Biological nitrogen fixation is too slow and too local to support the burden of production we currently impose on the land, so additional nitrogen must be reaped from the air in huge amounts and applied as fertilizer. This is done on an industrial scale by finding ways to help nitrogen molecules to combine with hydrogen to form ammonia (6), which can be absorbed by the soil or converted into other fertilizers. Nitrogen lost from the atmosphere is replenished by the decomposition of vegetation and flesh, for, when protein molecules decompose, the nitrogen atoms are released as ammonia molecules, which in due course degrade into molecular gaseous nitrogen.

3 CARBON DIOXIDE CO_2

Carbon dioxide is the gas we exhale, for it is a product of the consumption of the organic compounds we ingest as food. When an organic compound burns (and here I include that extremely sophisticated type of burning in living cells we call metabolism), each carbon atom is removed from its molecule by two oxygen atoms and carried away as carbon dioxide. In a flame, the disruption of the molecules of the fuel (such as methane, 15) and the formation of strong carbon–oxygen bonds is accompanied by the release of energy as heat. Carbon dioxide is the end of the road for the combination of carbon with oxygen and its formation corresponds to the maximum release of energy. In that sense, carbon dioxide is a dead form of carbon, the end of its road energetically. However, it is not inert, for green

Much of the original carbon dioxide of the planet has been trapped as carbonate rock, the compressed remains of shell and skeletons. These shells are coloured by the impurities, particularly iron ions, that they have incorporated.

vegetation uses the energy of sunlight to restore it to life, plucking carbon dioxide from the skies, combining it with hydrogen obtained from water, and building carbohydrates (93) in the process known as *photosynthesis* which is under the control of chlorophyll (157).

Photosynthesis and the dissolving of carbon dioxide in the oceans maintain the balance of carbon dioxide in the atmosphere (or so we hope). Carbon, as carbon dioxide, enters the surface layer of the ocean at a net rate of about 10 billion tonnes (10^{13} kg) per year. Once there, it is taken up by photosynthetic organisms in the surface layers of the ocean and transformed into organic molecules, or into the deep ocean as particulate organic matter. In the deep ocean the organic matter is broken down by bacteria into dissolved inorganic compounds in a process of re-mineralization. That inorganic matter is returned slowly to the surface as the deep water mixes with the surface water. The overall process takes about a century.

It is now common knowledge, however, that the carbon dioxide burden of the atmosphere is being affected by human activity, not least humans' commitment to the internal combustion engine. A car produces 2.5 kilograms of carbon dioxide for every litre of fuel it burns, so a typical year's driving contributes about 5 tonnes of the gas to the atmosphere. Your garden should contain about 1000 mature trees to clean up the mess you make and absorb that amount of carbon dioxide by photosynthesis.

In a muscle or a brain, the energy released when carbon dioxide is formed in a metabolic process may be used to raise a weight or produce an idea. Carbon dioxide is also the end product of the partial consumption of carbohydrates during fermentation, an incomplete form of respiration that forms alcohol (26) as another principal product. Hence, carbon dioxide is the gas in the head of beer and the bubbles in champagne, for it comes out of solution when the bottle is opened and the pressure is released. Carbon dioxide in water is common as soda water or seltzer, and, with added flavours, as beverages of various kinds. In water it forms the very weak acid *carbonic acid*, which tingles the tongue, is a taste enhancer, and acts as a mild bactericide. Carbonic acid is also said to encourage flow from the stomach to the intestine, which perhaps accounts for the rapid inebriating effect of champagne.

Carbon dioxide is the fourth most abundant component of the dry atmosphere on Earth and the most abundant gas in the atmospheres of Mars and Venus. A great deal of carbon dioxide was removed from the early atmosphere as the oceans fell from the skies, for the gas was dissolved in them. Now most of the carbon dioxide of the early planet lies beneath our feet, in the form of carbonate rock, as chalk and limestone. No similar precipitation of water occurred on the hot surface of Venus, and that on Mars (if there was any) was insufficient, so on those planets the carbon dioxide remains in the atmosphere. It has been calculated that the mass of carbonate rock on Earth, plus the amount of carbon dioxide in the atmosphere and dissolved in the oceans, is approximately the same as the mass of carbon dioxide that now hangs in the skies of Venus. Had the Earth been only 10 million kilometres closer to the Sun than its present 140 million kilometres, the temperature of its surface would have been too high for the oceans to form, and Earth would have evolved into a planet like Venus.

Carbon dioxide in the atmosphere acts partly to trap the infrared radiation emitted from the warm surface of the Earth. Because carbon dioxide is transparent to the Sun's visible light, that light can penetrate through the atmosphere to the Earth's surface. As the surface warms, it emits infrared radiation that cannot escape back into space because carbon dioxide molecules absorb it. This trapped energy warms the atmosphere in a process known as the *greenhouse effect.*

In a real greenhouse the build-up of heat is due more to the glass preventing a convective mixing of the warm air inside with the cold air outside than to the absorption of infrared radiation.

Carbon dioxide is used as a leavening agent in baking. Typical *baking powders* consist of sodium bicarbonate ($NaHCO_3$), an acid (or, typically, two acids, such as tartaric acid and the acidic salt sodium aluminium sulfate), and starch, which acts as a filler and helps to separate the acid and bicarbonate particles and keeps them from reacting prematurely. However, even so mundane a product as baking powder has an interesting technology, because it has to release carbon dioxide in two separate bursts. The first burst occurs at room temperature as a result of the action of the moistened tartaric acid, and it produces many tiny cavities in the batter. The second burst of activity is due to the action of the aluminium salt, and it occurs at high temperature. This second surge of carbon dioxide swells the cavities to give the desirable final light texture.

The carbon dioxide used in bread-making is usually formed by the action of yeast on sugar or other similar carbohydrate molecules. Such yeast particles are present in the air, but, to achieve more uniform characteristics in baking, a particular strain, *Saccharomyces cerevisiae*, is normally cultured in dilute molasses and then used.

Carbon dioxide does not exist as a liquid at atmospheric pressure at any temperature, and to form the liquid at room temperature requires a pressure of at least 30 atmospheres, which is called the 'critical pressure'. *Supercritical carbon dioxide* is carbon dioxide at greater than its critical pressure. This fluid is currently of great interest as a solvent, partly because it is not toxic but also because the solubility of a substance in it increases with pressure. Thus, it has been developed as a solvent to remove caffeine from coffee (179), to extract the flavour components of hops (104), and to replace the possibly carcinogenic chlorinated hydrocarbons that hitherto have been used for dry cleaning. Supercritical carbon dioxide has other advantages over chlorinated hydrocarbons, for it can be used to clean items that cannot be dry cleaned with chlorinated hydrocarbons, such as leather, fur, and some synthetics.

4 OZONE O_3

Ozone is present in the upper atmosphere in the *ozone layer*, a band about 20 kilometres thick centred between 25 and 35 kilometres above the surface of the Earth. If all of it were collected and compressed to the atmospheric pressure characteristic of the Earth's surface, it would form a layer about 3 millimetres thick. The ozone is formed when the Sun's ultraviolet radiation is absorbed by molecules containing oxygen: oxygen atoms are driven out of those molecules and subsequently bond to O_2 molecules that they strike. Once it has formed, the ozone molecule absorbs more ultraviolet radiation at a different wavelength and is blasted apart. Both processes, formation and decomposition of ozone, absorb radiation and hence help to protect the living organisms on the surface below. The absorption of ultraviolet radiation by the gas is so efficient that, at wavelengths near 250 nanometres, in the ultraviolet, only one part in 10^{30} of the incident solar radiation penetrates the ozone layer. A being with eyes able to see only in 250-nanometre light would see the sky pitch black at noon.

The *ozone hole* is a region of depletion in the concentration of ozone. It opens over Antarctica each southern spring and incipient holes are starting to form over the Arctic too. Each year the Antarctic hole forms earlier and grows bigger, and now extends to an area greater than that of continental North America. Culprits include highly reactive chlorine atoms that are ejected from chlorofluoro-carbons (13) when they are high in the atmosphere and exposed to intense solar radiation. These atoms attack ozone molecules and strip out an oxygen atom, leaving ordinary oxygen. The resulting ClO molecules later collide with oxygen atoms produced by the impact of solar radiation on ozone molecules, and give up their oxygen atoms to form O_2 molecules. This step releases the chlorine atom, which can continue to cause its destructive havoc. Even a single chorine atom can go on to kill thousands of ozone molecules.

Depletion of ozone occurs mainly over the poles, and especially the South Pole, because the reactions responsible proceed most rapidly on clouds of ice particles in the stratosphere. There is also a link with global warming caused by the accumulation of carbon dioxide in the atmosphere (3). Thus, whereas an accumulation of carbon dioxide warms the lower atmosphere, it acts in an opposite way in the stratosphere, causing it to radiate more heat into space and to grow colder. Cooling results in more ice clouds, and therefore a more rapid depletion of ozone.

Ozone is a blue pungent gas (*ozein* is the Greek for 'to smell') which condenses to an inky blue–black explosive liquid. Its smell can be detected near electrical equipment and after lightning, since it is also formed by an electric discharge through oxygen. Because ozone would be encountered in aircraft cabins on commercial flight at altitudes of about 15 kilometres and would cause coughing and chest pains, the incoming air is passed through filters that catalytically decompose ozone into ordinary oxygen. Atmospheric ozone attacks the carbon–carbon double bonds in rubber (68) and contributes to its weathering.

WATER AND AMMONIA

Some important molecules consist of a single central atom surrounded by enough hydrogen atoms to satisfy its tendency to form bonds. Each of these molecules, which include the two described here and methane (15), can be regarded as the simplest member of a series of increasingly complex compounds in which the hydrogen atoms are replaced by other atoms or by groups of atoms. In water the central atom is oxygen, and in ammonia it is nitrogen. In each case the central element is temporarily protected from reaction by a shell of hydrogen atoms.

5 WATER H_2O

Water occurs in huge abundance on the Earth, where most of it lies in the great pools we call oceans that cover 71 per cent of the Earth's surface to an average depth of 6 kilometres. This water dropped from clouds formed when the heat in the interior of the young Earth drove oxygen and hydrogen out of chemical combination in rocks built of compounds like *mica*, a form of potassium aluminium silicate. The newly formed molecules were brought to the surface in streams of lava and then released as water vapour, to form great clouds that could rain once the Earth had cooled. Hence our oceans were once our rocks. All three forms of water – ice, liquid, and vapour – are abundant on Earth, but very little is in a form suitable for human consumption: 97 per cent of it is too salty, and 75 per cent of the Earth's fresh water is solidified at the poles. The remaining 1 per cent of the total water is drinkable, but most of that is inaccessibly deep groundwater. Thus, only 0.05 per cent of the total, the water that runs through lakes and streams, is readily available. Some ancient groundwater is mined in deep wells, but that water recedes further from the surface as we use it but do not replace it.

Water and ice in bulk are pale blue, as in this ice tunnel near Spitsbergen in Norway. The colour arises from the presence of hydrogen bonds between the water molecules.

The oddest property of water is that it is a liquid at room temperature. This is surprising because so small a molecule would be expected to be a gas, like ammonia (6), methane (15), and its even closer relative hydrogen sulfide (H$_2$S, 112). That water is a liquid stems from the ability of its molecules to link together by forming networks of hydrogen bonds. One molecule can form these weak but important bonds to four others, and all four neighbours link to their neighbours, and so on. As a result, the molecules cluster together as a mobile liquid, rather than moving independently as a gas.

Most solids expand when they melt, but ice expands: ice at 0 °C is *less* dense than water at 0 °C. As a result, ice floats on water, giving us icebergs and a solid skin on frozen ponds. This skin insulates the liquid water beneath, protects it from the cooling winds above, and can keep it from freezing during the winter. Thus, aquatic life can survive in the liquid, even though the roof of its world is frozen. This quirk of density is again due to the presence of hydrogen bonds, for, when water freezes, its molecules are held apart, as well as held together, by the hydrogen bonds between them: each molecule grips its neighbours firmly, but holds them at arm's length. When the solid melts, this open framework-like structure partially collapses, the molecules lie closer together, and the liquid is more dense. Floating icebergs are thus a sign of the strength of hydrogen bonds; indeed, it was hydrogen bonds that brought death to the *Titanic.*

Water is also an excellent solvent. It mixes readily with alcohol (26) as in wine, beer, and spirits, because it can form hydrogen bonds with alcohol molecules. Sugar (89) dissolves readily in water for similar reasons. Many ionic solids dissolve in water, as salt and other minerals dissolve to yield the brine of the seas. This ability to dissolve ionic solids stems from the fact that electrons accumulate slightly on the oxygen atom, giving it a partial negative charge, as indicated by the red region in the structure in the margin, and denude the hydrogen nuclei slightly, giving them a partial positive charge (depicted by the blue region). As a result, the oxygen atom can emulate an anion and the hydrogen atoms can emulate a cation, with the result that the actual cations and anions of an ionic solid are seduced into spreading into the surrounding water molecules.

Water is a perfect medium for such processes as the transport of nutrients into cells and the carving of landscapes by flushing minerals out of rocks. Water appears to be essential for life, since it can provide a fluid environment within cells through which other molecules can migrate. Water can transport molecules up to cells, give them mobility within cells, and transport molecules away from cells to other locations, including transferring molecules to the outside environment as waste. It can transport organic molecules like glucose (87) and the ions of such elements as sodium, potassium, and calcium that are so essential to an organism's functioning. Moreover, water, when it is a liquid, can do all this at body temperature. Fortunately, it cannot dissolve the calcium phosphate of our bones, so our skeletons do not dissolve in our own fluids and we do not wilt.

Natural waters have contributions to their colour from microscopic green organisms, scattering from microscopic dissolved particles, and reflections of the sky and vegetation, but pure water and pure ice are both very pale blue–green. The colour is due to absorption of infrared and red light from white light (so leaving it with a tinge of blue) by the excitation of the stretching vibration of the O–H bonds in the molecule.

Ice, it might come as a surprise, is black. To be precise, if water is frozen in a tube with opaque walls, there are so many bubbles and cracks in the ice that any light entering from one end of the tube is scattered and doesn't travel through to the other end of the tube, even for quite thin samples. If the opaque walls are removed, light enters from the side, scatters around inside the sample, and leaves in all directions; as a result, the ice looks its normal translucent grey. Frost and snow look white – neither black nor pale blue–green –

because the incident sunlight is reflected from very small ice grains close to the surface and very little penetrates into the bulk.

Under some circumstances water is red. Superheated steam (steam heated to well over 100 °C), for instance, glows pale red. Very hot individual water molecules luminesce as the vigorously vibrating molecules discard their energy as photons of red light.

6 AMMONIA NH$_3$

The pungent gas ammonia takes its name from the Ammonians – worshippers of the Egyptian god Amun who, like the fainthearted of later years, used *sal volatile* (ammonium carbonate, $(NH_4)_2CO_3$) in their rites. Smelling salts releases ammonia, which irritates the linings of the nose and lungs, triggering a reflex that increases the rate of breathing and as a result enhances alertness.

Ammonia is one of the most important industrial chemicals, for it begins the chain of industrial food production. More molecules of ammonia are manufactured each year than of any other industrial chemical. Almost all these new ammonia molecules are made by the *Haber process*, in which atmospheric nitrogen (2) and hydrogen are forced to combine under high pressure (about 100 times atmospheric pressure) and temperature (about 700 °C) in the presence of a catalyst. This process for harvesting the air was invented by Fritz Haber during World War I when Germany was cut off from its normal supply of nitrates from Chile.

Although the conversion of atmospheric nitrogen to ammonia is called nitrogen 'fixation' (2), it might more accurately be called nitrogen mobilization. In the process, the tightly held nitrogen atoms of N_2 are ripped apart, and each one is surrounded by less tightly held hydrogen atoms. The nitrogen is thereby made more susceptible to attack, attack that may result in its incorporation into amino acids and thence into the proteins (80) of organisms. Ammonia is manufactured in such mountainous abundance because it is a source of usable nitrogen. Its major use is in fertilizers, either as ammonia or after conversion first into nitric acid (10) and then into nitrates. Ammonia is also the starting point for the incorporation of nitrogen atoms into industrially produced molecules, including nylon (79) and pharmaceuticals.

Ammonia is a colourless, flammable gas under normal conditions. It dissolves very readily in water because it can form hydrogen bonds with the water molecules. This high solubility contributes to the perceived odour of the gas, because ammonia is able to dissolve readily in the aqueous mucus that coats the olfactory epithelium of the nose (see Chapter 4). Water would probably smell just as pungent if our sensors were not constantly saturated by it. An overripe Camembert or Brie also smells of ammonia, for the gas is formed as the nitrogen-containing protein molecules decompose. The same process takes place in stale urine (147). The smell of ammonia can often be found near compost heaps, especially when the composted material is rich in nitrogen, such as grass clippings and manure.

Although livestock waste is an important source of urban, airborne ammonia, another major source is the internal combustion engine – or, more strictly, the catalytic converters intended to reduce polluting emissions – some of which can produce a tenth of a gram of ammonia per kilometre. Once ammonia is formed, it reacts with other nitrogen oxides (9) present in the exhaust stream and forms small particles of solid ammonium nitrate. These particles pick up water and contribute to a smog-like haze. The particles can also irritate the lungs and cause lung and heart disease.

Reports are current that the addition of ammonia to cigarettes increases the impact of the nicotine (184) they contain up to a hundred-fold. The process is called 'free basing', and is similar to the process used to heighten the effects of cocaine (183). When ammonia is added, the nicotine donates a hydrogen ion to the ammonia molecule and converts it from its acid form, an ionic form that binds strongly with other charged ions present, to its base form, an electrically neutral, more weakly interacting form. The latter evaporates more easily from the smoke particles, deposits directly on the lung tissue, and diffuses immediately throughout the body.

Ammonia can be regarded as the parent of the compounds called *amines*. Amines often retain the pungency of ammonia but frequently in a modified and more disagreeable form. To see the kind of change that occurs, you should turn to putrescine (150) and its partner cadaverine (151), as one day you certainly will.

SMOG, POLLUTION, AND ACID RAIN

We described air as perhaps it ought to be, with innocuous, essential components. However, air as it truly exists is quite a different thing, for it contains alien molecules that stem from natural and 'civilized' sources that use the sky as a sewer. Some of the more common pollutants that often inhabit our skies are described in this section. Although chemistry undeniably contributes to pollution (particularly in the hands of the economists, politicians, physicists, farmers, travellers, homeowners, and others who make use of the benefits it brings), the deleterious properties of the molecules described here – and those discussed in the section entitled 'Nasty compounds' in Chapter 6 – should be weighed against the colossal beneficial contributions of the other molecules described in this book.

7 SULFUR OXIDES SO_2, SO_3

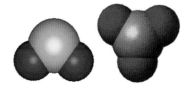

Just as carbon dioxide is formed when oxygen carries off carbon atoms from burning carbon and organic compounds, so sulfur dioxide is formed as a colourless, dense, toxic, non-flammable gas with a suffocating odour when the yellow solid non-metallic element sulfur burns in air. The same end product is obtained from the combustion of compounds containing sulfur atoms, including some of the components of oil and coal. Sulfur dioxide also belches out of the ground at volcanoes and emerges from industrial plants where iron and copper ores (pyrite, FeS_2, and copper sulfide, CuS) are heated during the extraction of the metal. It is also formed where volatile sulfur compounds produced by the decay of plant and animal matter are oxidized in the air. About 200 billion kilograms of sulfur reach the sky each year from industrial sources, joining there the 300 billion kilograms that nature donates.

Sulfur dioxide (SO_2) is not particularly soluble in water, and gives rise to a weak acid, sulfurous acid. It is not *directly* responsible for the formation of acid rain. Indeed, a lot of it is deposited directly onto solid surfaces, and conversion to sulfur trioxide and sulfuric acid (8) occurs on these surfaces. Some, however, is converted into sulfuric acid over the course of hours, and this strong acid falls to the Earth as acid rain. (Acid rain also contains nitric acid, 10.)

Sulfur dioxide can profoundly affect weather patterns on Earth by reflecting sunlight and cooling the atmosphere. Sulfur dioxide is converted into a sulfuric acid aerosol by the action of solar radiation in the lower atmosphere and reaction with stratospheric water vapour. This aerosol remains in suspension long after solid volcanic ash particles have fallen to Earth (the short lifetime aloft of these particles has little long-term effect on the weather) and forms a layer of sulfuric acid droplets at an altitude of between 15 and 25 kilometres.

Sulfur dioxide is used as a preservative in a number of foods and beverages, for it is able to combine with the oxygen that would otherwise attack the commodity being preserved. The molecules responsible for the colours and flavours of dried fruits and fruit juices survive longer if sulfur dioxide is present. Sulfur dioxide is used in wine-making, partly to suppress the growth of wild yeasts and bacteria that would sour the grape juice into vinegar (31) and partly to prevent oxidation. More is used with white wine than with red so that oxidation of the pale yellow pigment is prevented.

Huge numbers of sulfur dioxide molecules are manufactured deliberately, by burning sulfur, in the production of sulfuric acid (8). They are then encouraged by a catalyst to combine with another oxygen atom so that most of the manufactured sulfur dioxide is converted into sulfur trioxide (SO_3). That conversion takes place only slowly without a catalyst, as when atmospheric sulfur dioxide forms the trioxide in water droplets. The oxygen-rich trioxide reacts vigorously with many substances and is usually converted into sulfuric acid as soon as it appears.

8 SULFURIC ACID H_2SO_4

Sulfuric acid, which when pure is a viscous, oily liquid, is the world's most widely used industrial chemical, and no other chemical is manufactured in greater tonnage. (More *molecules* of ammonia are produced, but each one is lighter.) Almost every manufactured item comes into contact with the acid at some stage. So important is sulfuric acid that its annual production was once taken as a measure of the degree of industrialization of a country and of commercial activity in the world. However, there has been a shift in its use: most of it is now going into the production of phosphate fertilizers, and it is increasingly being replaced by hydrochloric acid in the surface treatment of iron. Hence, it is now probably more useful as an indicator of agricultural – rather than industrial – activity.

Sulfuric acid also falls from the skies, for it is one of the acids of *acid rain.* It is formed there when sulfur dioxide (7) is converted into sulfur trioxide in airborne droplets of water.

Nudibranchs (sea slugs) are marine gastropods that adopt a number of defence mechanisms and sometimes warn off potential predators by their flamboyant colouration. One defence mechanism is the secretion of acids, including sulfuric acid, by cells in their outer layers. The yellow outgrowths of these slugs are respiratory organs.

9 NITROGEN OXIDES NO, NO₂, N₂O

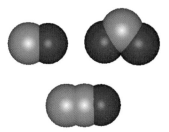

The first two of these oxides, nitric oxide (NO) and nitrogen dioxide (NO$_2$), are formed in the combustion chambers and hot exhausts of automobile and aircraft engines and are collectively known as NO$_x$. Nitric oxide is formed first, and concentrations of it and its reaction product, nitrogen dioxide, build up before dawn and in heavy traffic. At dawn, when the Sun's radiation begins to stimulate chemical reactions, ozone (4) is produced. Nitrogen dioxide is a brown gas that, together with smoke-like particles, contributes to the colour of smog.

In the first edition of this book, there was little to say about nitric oxide other than its role in pollution. In the intervening years, the molecule has grown enormously in stature, for we now know that it plays a role in a wide range of biological phenomena. It is produced by enzymatic action on amino acids and, being small, can readily diffuse through cell membranes and take part in reactions elsewhere in the cell.

The flashing of fireflies (*Photinus* and *Photuris*) is regulated by nitric oxide, which inhibits the use of oxygen molecules that diffuse into the lantern area of the firefly and normally participate in the metabolic processes taking place in the mitochondria (the energy-producing organelles) of the photocytes (the cells where light is produced). As a result, the oxygen is free to attach itself to the light-producing molecules, and a flash of light is produced. In mammals, NO helps to maintain blood pressure by dilating blood vessels, assists the immune system in killing invaders, and is a major factor in the control of penile erection (197). In the brain, it plays a role in development, signalling, and the laying down of memories. Nitric oxide is a subtly useful messenger, partly because it is so mobile but also because, being a reactive radical, it is rapidly destroyed. With these central physiological roles, it is not surprising that nitric oxide is now a prime target for drug therapies. Nitric oxide gas itself is used in the treatment of premature babies with persistent pulmonary hypertension, where it dilates the blood vessels in their immature lungs. Other potential therapies include the possibility of reducing cell death and brain damage when the brain releases an excess of NO after it is injured or starved of oxygen in a stroke. All these therapies are echoes of an early, once little understood treatment for heart disease and *angina pectoris*, the administration of nitrates, such as amyl nitrate and nitroglycerine (189). We now know that their function is to enhance the availability of nitric oxide, thereby dilating blood vessels, and enhancing the flow of oxygenated blood.

Nitrous oxide (N_2O) is laughing gas. When it is inhaled, it produces insensibility to pain, preceded by mild hysteria. It is used for surgical operations of short duration, such as in dentistry. Because the gas is almost inert and not toxic, it is also used as a propellant for foodstuffs, such as whipped cream. Nitrous oxide is a source of oxygen when it is exposed to a flame; indeed, as N_2O it can be thought of as a molecular analogue of air, with a higher proportion of oxygen (one atom in three) than there is in actual air (about one atom in five); as such it is injected into internal combustion engines as a chemical supercharger to achieve, as in the previously impotent, spurts of higher performance.

10 NITRIC ACID HNO₃

Nitric acid is formed in the atmosphere when nitrogen dioxide dissolves in water, and it falls from the skies as a component of acid rain. Pure nitric acid is a colourless liquid; the straw colour familiar from school-laboratory bottles of dilute acid is due to nitrogen dioxide formed by the acid's partial decomposition. It is manufactured by dissolving in water the nitrogen dioxide produced by the oxidation of ammonia.

The acid is the parent of the *nitrates*, which are ionic solids that contain the nitrate ion (NO_3^-). Among these is ammonium nitrate (NH_4NO_3), which is widely used as a fertilizer because it contains a high proportion of nitrogen and (like almost all nitrates) is soluble in water. The solubility of nitrates accounts for their rarity as minerals, because they simply get washed away. However, *Chile saltpetre* (impure sodium nitrate), which is found in large deposits in the arid regions of Chile, is an exception. Its origin is uncertain, but it may be the result of weathering on animal or vegetable remains. *Guano*, a phosphate deposit that also contains nitrates, was once used as a fertilizer; it is the excrement of fish-eating sea birds and is found in large quantities on the dry, rocky islands off the coast of Peru.

When the hydrogen atom of nitric acid is replaced by a carbon atom and the atoms the carbon atom carries, the properties of the molecule often change dramatically. To see what I mean, jump to TNT (188).

11 HYDROGEN PEROXIDE H₂O₂

The additional oxygen atom has little effect on the physical properties of hydrogen peroxide in comparison with water, but it profoundly changes the chemical properties. Hydrogen peroxide is a powerful oxidizing agent, destroying organic compounds that come into contact with it. That is partly its role in pollution, for fragments of the peroxide (such as HO_2 and OH) occur in photochemical smog, where they attack unburned fuel molecules and convert them into lachrymators such as PAN (12). This chemical activity is put to use as a bleach for hair and teeth, and on a larger scale, for paper pulp. The action of hydrogen peroxide as a bleach stems from its ability to oxidize and destroy pigments, including the melanin (165) responsible for the colour of black, brown, and fair hair. Many such colours are due to molecules with strings of carbon atoms joined by alternating single and double bonds. The hydrogen peroxide molecule attacks the double bonds: after breaking one of the bonds open, it dumps its excess oxygen between the fragments, forming a three-membered ring called an *epoxide:*

With the alternating sequence of single and double bonds disrupted, the colour disappears. An advantage of hydrogen peroxide over some other bleaching agents, such as chlorine gas, is that its decomposition products, oxygen and water, are not pollutants.

Hydrogen peroxide is present in trace amounts in honey, where it is the product of an enzyme that oxidizes glucose (87). Its presence accounts for the antibiotic action that once led to honey being used for dressing wounds. There are, however, better and more hygienic sources of hydrogen peroxide than bees. Our immune system uses hydrogen peroxide as a weapon. Of the white blood cells that are used to oppose infection and invasion, the granulocytes are most common, and one of their versions is the neutrophils. Neutrophils engulf bacteria and other microorganisms and microscopic particles, and discharge hydrogen peroxide into the cavity containing their victim.

12 PEROXYACETYL NITRATE $C_2H_3O_5N$

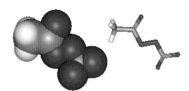

Peroxyacetyl nitrate (PAN) is a rubbish dump for the molecules found in the exhausts of motor traffic, and indeed it occurs in photochemical smog. Incompletely burned hydrocarbon fuels are introduced into the atmosphere as partially oxidized fragments that include the radical CH_3CO (you can pick this fragment out on the left of the model). This radical attacks the abundant oxygen molecules that surround it, and forms CH_3CO-OO, a modification of the hydrogen peroxide molecule (11). During morning smogs, when there is a lot of nitric oxide and sunlight around, the CH_3CO-OO oxidizes the NO to NO_2. During the afternoon, when the concentration of NO has been depleted and NO_2 is more abundant, the CH_3CO-OO attaches to NO_2 molecules to form PAN itself.

PAN is the powerful lachrymator responsible for much of the eye irritation caused by smog. It brings tears to the eyes because the tear ducts respond to invasion by secreting a saline fluid (which also includes various antibodies in case the invasion is bacterial) in an attempt to wash the invader away. The molecule is also largely responsible for the damage that smog does to vegetation; its very high oxygen content causes the oxidation (in effect, partial combustion) of any organic matter it touches. The molecule damages plants by causing collapse of tissue on the lower surface of leaves.

13 DICHLORODIFLUOROMETHANE CF_2Cl_2

14 TETRAFLUOROETHANE $C_2H_2F_4$

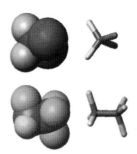

Dichlorodifluoromethane is an example of a *chlorofluorocarbon* (CFC), and is specifically CFC-12. These compounds, which are sold under the DuPont tradename 'Freon', include CH_3CFCl_2 and CH_2FCF_3. The code number can be interpreted as a structure by adding 90, to give a number of the form *xyz* (102 in the case of CFC-12), then the first digit is the number of carbon atoms, the second the number of hydrogen atoms, and the third the number of fluorine atoms. The number of chlorine atoms is then worked out by ensuring that each carbon atom is attached to four atoms. Thus, $CFCl_3$ is CFC-11.

The CFCs are non-toxic, non-flammable, unreactive, and volatile compounds. They have been widely used as coolants in refrigerators because they are readily liquefied. The CFCs have also been used to foam plastics for insulation material and as propellants in aerosol spray cans. The problem, though, is that CFCs that are released to the atmosphere rise up into the stratosphere where, under the impact of solar radiation, most notably high-energy UV-C radiation (166), they dissociate into radicals, particularly free chlorine atoms. These chlorine atoms then go on to destroy ozone molecules (4). In the process, in which O_3 molecules are converted into O_2 molecules, the chlorine atoms are regenerated. As a result, they go on with their mischief, and a single chlorine atom can kill hundreds of thousands of ozone molecules.

Chlorofluorocarbons that contain no hydrogen atoms (such as CFC-11 and CFC-12) are insoluble in water and so are not removed from the lower atmosphere by rain. They survive the relatively low-energy ultraviolet solar radiation that reaches the lower atmosphere (UV-A). The replacement compounds for these dangerously persistent molecules do contain hydrogen, and are called *hydrochlorofluorocarbons* (HCFCs). They include CHF_2Cl (CFC-22, or HCFC-22), and have the advantage of being degradable at low altitudes. One property that has to be monitored is that, as the proportion of hydrogen atoms in the molecule increases, it becomes more flammable: so a balance must be struck between the degradability of the compound and its flammability.

Because chlorine atoms are the rogues in CFCs and HCFCs, the obvious next step is to eliminate them completely, and to use *hydrofluorocarbons* (HFCs) such as tetrafluoroethane (14). This compound survives in the atmosphere until it gives in to attack by hydroxyl (OH) radicals. The resulting HF molecules survive and do not contribute to the destruction of ozone.

2

Fuels, fats, and soaps

Now we come to the principal types of molecules, all of them organic, that will be the focus of our attention in the remainder of the book. With these molecules we begin to see the web of intricacy that carbon spins, largely by linking to itself. The structural prototype of all organic compounds is methane (15). This little molecule can be extended into chains, rings, and networks of linked carbon atoms on which atoms of other elements, most notably hydrogen, oxygen, and nitrogen, can hang. Many of the molecules that spring structurally from methane have an enormous impact on the world, being deployed by nature, modified and synthesized by industry, and used by everyone. Some are merely burned as fuels. Two common but important types of molecules — alcohols and carboxylic acids — are used by nature and by chemists to forge links between chains of carbon atoms. You will read about their combinations, which are called *esters*, in many places throughout the book. There are esters in the fats and oils that act as food reserves in our bodies, which we consume by the tonne during a lifetime and which we do not always entirely discard. Some fats we spill, and for those you will see how chemists engineer molecules that can infiltrate grease yet dissolve in water and hence can act as soap.

HYDROCARBON FUELS

The replacement of gas manufactured from coal by natural gas, the gas trapped in porous rocks, transformed the pattern of energy usage during the 1950s, and now natural gas is pumped through pipelines that cross countries and continents. The dominant producers are Russia and the USA, and the propane and butane also present in the raw gas are removed to leave a gas that is predominantly methane (15). Much of that gas is used for domestic and industrial heating, but some is also used as the starting point for the synthesis of organic compounds. This section describes some of the typical components of natural gas and of the *liquefied petroleum gas* (LPG) used for mobile applications, including camping. All the molecules discussed are *hydrocarbons*, compounds containing only carbon and hydrogen.

Natural gas, like petroleum oil, originated from decayed organic matter, probably as the result of bacteria (typically cyanobacteria) scavenging for oxygen atoms and leaving strings of hydrocarbon molecules as residue. For such bacterial action to dominate oxidation, there must be a high productivity of organic matter and a low concentration of competing oxygen. These conditions are satisfied in sedimentary layers in coastal waters, where marine organisms thrive. The hydrocarbons that are produced are eventually squeezed out of the compacted muddy sediments into neighbouring porous rocks – most often sandstone (particles of quartz and silicates bound together by a cement of silicate or carbonate) or limestone (calcium carbonate). Loss of the hydrocarbons by means of upward migration is prevented if the rock formation is roofed over by an impermeable layer. A typical hydrocarbon trap consists of an *anticline* (a convex upward fold) in which sandstone is capped by an impermeable shale (a fine-grained sedimentary rock composed of compressed silt and clay). The gaseous hydrocarbons rise highest in the porous rock, lying above the petroleum that fills the remaining pores and floats on the ground water in the pores below.

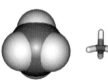

15 METHANE CH$_4$

A single central atom of an element surrounded by hydrogen atoms constitutes a primitive but portentous molecule. In no case is this better demonstrated than when the central atom is carbon, as in

methane (for the origin of the name, see methanol, 25). Carbon's tendency to form four bonds is satisfied if four hydrogen atoms are attached, as in this tetrahedral molecule. However, as we shall see, methane is the precursor of all organic chemicals in which these hydrogen atoms are successively replaced by other atoms and by groups of atoms. Primitive, simple methane is the most pregnant of molecules.

Methane is an odourless, non-toxic, flammable gas. It is a gas of carbon atoms in flight, but each carbon atom is prevented, by its casing of hydrogen atoms, from immediately reacting and from sticking together with others to form a solid block of carbon. Methane is formed naturally when bacteria release single carbon atoms from digested organic material, and it occurs as marsh gas when it is not trapped by a suitable rock formation. It is the main component of natural gas, accounting for about 95 per cent of the gas from midcontinental-US sources, 75 per cent of the gas from the Texas Panhandle, and nearly 70 per cent of the gas from Pennsylvania.

In a gas flame in which methane combines with oxygen, the hydrogen atoms are ripped off the tetrahedral methane molecules by the oxygen to form water molecules (5), and the remaining carbon atom picks up oxygen atoms to form carbon dioxide (3). The blue and green light of the flame is emitted by energetic C_2 and CH molecules formed briefly in the course of the combustion. In a limited supply of air, the carbon may be oxidized incompletely, forming carbon monoxide (CO) and a smoke of unburned carbon particles in which billions of atoms have stuck together to give soot. Then the flame is yellow with the glow of light emitted by the incandescent carbon particles.

The blue and green colours of the methane flame are due to the excitation of an electron to a higher energy location in the C_2 and CH molecules. The electron falls back to its original location almost immediately and, in doing so, discards the excess energy as light.

16 BUCKMINSTERFULLERENE C_{60}

Soot is a remarkably interesting commodity, not least because it turned out to be the home of a wholly unexpected and possibly highly important form of the element carbon. The buckminsterfullerene molecule is a sixty-atom spherical molecule of carbon with an alternation of five- and six-membered rings of atoms, just like the pentagons and hexagons on a modern football. It is one of a number of fullerenes, which are analogous molecules with various numbers of carbon atoms. The molecule was discovered in space in 1987 but has since been prepared in abundance on Earth. Buckminsterfullerene, despite its late discovery, might be one of the most abundant forms of carbon, for it is present not only in soot but also in the atmospheres of red giant stars. The unwieldy name is a tribute to the geodesic domes produced by the American architect R. Buckminster Fuller, who was driven by an intense desire to do more with less.

Like the laser, which was a laboratory curiosity for a decade or so after its invention, the fullerenes (as they are settling into being called) are profoundly interesting, but as yet with promise largely unfulfilled. One intriguing property is the possibility of inserting a single atom of any element into the interior of the football, to give a shrink-wrapped version of the periodic table.

The fullerenes have open-ended cousins, the carbon nanotubes:

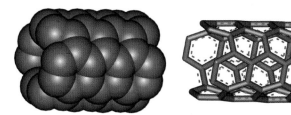

These tubes of carbon atoms that look like molecular versions of coils of hexagonal chicken wire have enormous promise that is just starting to be realized. The tubes are immensely strong, and, if they could be produced in metre lengths, rather than the sub-millimetre lengths currently available, would be stronger that steel cables and permit feats of fabulous engineering and architecture that would please Buckminster Fuller infinitely. There are two versions of the tubes, one with hexagons running parallel to the axis of the tubes and the other with hexagons forming a helix around the axis. The former are metallic conductors and the latter are semiconductors. Such tubes hold great promise as components of microminiaturized circuits, and transistors made from carbon nanotubes have already been produced. There has long been a fear that silicon might replace us carbon-based thinking machines; in fact, the real fear might come from carbon itself, pure carbon, divested of all the messy clutter that renders it 'organic'.

17 ETHANE C_2H_6

18 PROPANE C_3H_8

19 BUTANE C_4H_{10}

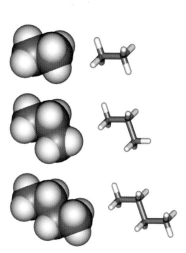

Now we begin to allow the methane molecule to grow. A simple adaptation is to imagine breaking one C–H bond and inserting a $-CH_2-$ group between the carbon and hydrogen atoms. This slightly more complicated hydrocarbon is *ethane*, and with it carbon begins to show its ability to combine with itself – an ability that can result in magically complex molecules.

Ethane, a colourless, odourless gas, accounts for about 30 per cent of the natural gas from Pennsylvania and is also extracted from oil wells. Ethane molecules interact slightly more strongly with each other than do methane molecules (this is a common result of increasing the number of atoms, and hence electrons, in a molecule); although it is a gas at room temperature, it condenses to a liquid at a higher temperature (−89 °C) than does methane (−162 °C). Much more important than ethane is its cousin ethylene (47), to which most of the ethane in natural gas that is not simply burned is converted.

Both propane and butane are used as LPG and are carried around as camping gas. Both are gases at normal temperatures, but butane condenses to a liquid at 0 °C and so cannot be used for camping under cold conditions. Thus, butane tends to be used in the southern parts of the USA, and propane (which condenses at −42 °C under normal pressure) in the north. Butane is also used as a fuel in cigarette lighters, where, under pressure, it is stored as a liquid. For both, complete combustion is like that of methane, with the end products being carbon dioxide and water. More than 20 per cent of the propane obtained from natural-gas sources is converted into ethylene and its relative propylene (49). Butane's destiny is often synthetic rubber (69).

20 OCTANE C_8H_{18}

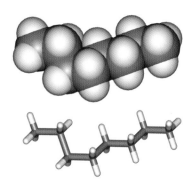

The octane molecule results when we continue the hypothetical process that led from methane (15) to butane (19). Now a sufficient number of −CH$_2$− units has been introduced into the original C–H bond of methane to make the chain eight carbon atoms long (hence the name *octane*).

Hydrocarbon molecules that contain half a dozen or so carbon atoms interact just strongly enough with each other to give a liquid at room temperature, so they are convenient to transport in tanks. However, the liquid is still volatile and not too viscous to form a fine spray in the carburetor of an engine. Octane is representative of the size of the hydrocarbon molecules present in gasoline. Diesel fuel is less volatile: its molecules are typically hydrocarbons with about sixteen carbon atoms in a chain. The sixteen-carbon analogue of octane is called either *cetane* or, more formally, *hexadecane.*

Octane is called a *straight-chain hydrocarbon* because all its carbon atoms lie in a line, with no branches; but do not be misled into thinking that 'straight-chain' means geometrically straight. Every hydrocarbon chain is actually a zigzag of carbon atoms and is flexible as well; moreover, each atom can be twisted around the bond joining it to its neighbour. Gasoline therefore contains some octane molecules that are rolled up into a tight ball, others that are stretched out but still zigzag, and others in the various intermediate conformations (see the Introduction). Octane molecules are constantly writhing and twisting, rolling and unrolling, so that gasoline is more like a mass of molecular maggots than a pile of short sticks.

21 2,2,4-TRIMETHYLPENTANE C$_8$H$_{18}$

The name 2,2,4-trimethylpentane is a descriptive label indicative of the structure of this molecule. Because it is such a mouthful, the molecule is usually (and more affectionately) known as *iso-octane*, the name we shall use from now on. Like octane itself, iso-octane has eight carbon atoms.

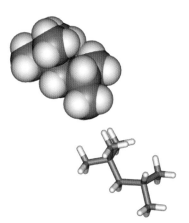

Iso-octane is an example of a *branched-chain hydrocarbon*, for its carbon chain has 'side groups' like the branch lines of a railway. Iso-octane and other branched hydrocarbon molecules are more desirable than straight-chain molecules in commercial gasolines. A straight-chain hydrocarbon ignites explosively in an engine because its chain of carbon atoms is so exposed, causing 'knocking' and power loss. A branched-chain hydrocarbon, however, burns smoothly, heating the gas in the cylinder throughout the power stroke, and so exerts a steady pressure on the receding piston. The ability to burn smoothly is measured as the *octane number* of the fuel. Iso-octane, with excellent non-knocking properties, is ascribed an octane number of 100. Heptane, which is like octane but with only seven carbon atoms, knocks horribly and is ascribed a value of 0. A 95-octane fuel is then equivalent in knocking characteristics to a mixture of 95 per cent iso-octane and 5 per cent heptane by volume.

The corresponding scale for diesel fuel involves the *cetane number* (cetane is hexadecane, as mentioned in 20). A cetane number of 100 corresponds to pure cetane, and 15 (the lowest point on the cetane-number scale) corresponds to pure heptamethylnonane, one of cetane's branched-chain isomers in which seven CH$_3$ groups are strung out like warts along a chain of nine carbon atoms. It is worth noting that, for a diesel engine, a rapidly igniting straight-chain molecule is desired. This reflects the different operating principles of the two types of engine: in a diesel engine, the fuel is sprayed in during the power stroke (not, as in a gasoline engine, all in one shot prior to compression) and needs to ignite as it enters the cylinders.

Gasoline companies strive to increase the proportion of branched-chain molecules in their products through *catalytic reforming*. In this process, straight-chain molecules are heated in the presence of a catalyst; this causes groups of atoms to break off the straight-chain hydrocarbons and then reattach elsewhere, forming branches on the molecules.

22 BENZENE C$_6$H$_6$

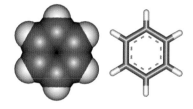

The planar, hexagonal arrangement of carbon and hydrogen atoms characteristic of benzene plays a very special role in chemistry because it is exceptionally resistant to attack. It appears in many of the larger organic molecules, acting as a kind of relatively inert platform to which other groups are attached. The reasons for its stability are complex, but they include the lowering of energy that results when bonding electrons are able to spread around the ring of carbon atoms. As mentioned in the Introduction, the spreading of double-bond character round the ring is denoted by the dotted line in the tube diagram.

Benzene was once called *benzol*, but this name is now reserved for the less pure grades of benzene. It is relevant to the present discussion because the combustion characteristics of hydrocarbon fuels are improved if compounds based on benzene are included in gasoline. These compounds are collectively known as *aromatic compounds*, for many, including benzene, have characteristic (but not always pleasant) odours. They are produced by catalytic reforming of petroleum and, to a much smaller extent currently, by the distillation of coal.

Coal was formed during the Carboniferous period (about 300 million years ago), when land areas were covered by luxuriant vegetation, particularly ferns, and swamps. As the vegetation died, it was submerged under water, where it decomposed anaerobically and gradually formed peat bogs. These bogs were covered by layers of sand and mud and the pressure hardened the deposits into coal. *Peat* has a low carbon content and a high moisture content; *lignite*, *bitumous coal*, and *anthracite* have successively higher carbon contents.

Coal is an extremely complex mixture of compounds with an internal structure like that depicted in the illustration. Note the many benzene-like rings; when coal is heated in the absence of air, these great sheets vibrate and, ultimately, break up. The molecular fragments can be driven off, collected, and separated. The proportions of the different types of molecules in the product depend on the temperature to which the coal has been heated; the smallest fragments,

A schematic representation of a portion of the molecular structure of coal. Note the large number of benzene-like rings. These sheets break up when the coal is heated.

including carbon monoxide and methane, are obtained at the highest temperatures, when the violence of the break-up is greatest. The residue left after this distillation process is called *coke*. It is formed by heating coal to 350–500 °C, at which point it softens and partly decomposes, and then raising the temperature to about 1000 °C. Coke is mainly carbon but includes some mineral and remaining volatile matter. A great deal is used in blast furnaces in the manufacture of iron from its ores; different coals are blended to obtain cokes of sufficient strength to withstand the weight of ore in the furnace.

23 TOLUENE C_7H_8

24 XYLENE C_8H_{10}

The toluene molecule can be viewed as a benzene molecule in which one C–H bond has been broken open to accept a $-CH_2-$ group. The xylene molecule is then a toluene molecule in which that process has been repeated elsewhere in the benzene ring. In fact there are three different xylenes, with the second $-CH_2-$ group inserted next to the

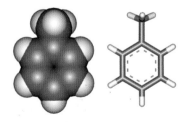

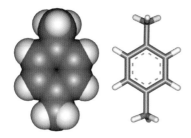

one already present, one carbon atom away from it, or diametrically opposite it across the ring. Only the last of these three isomers is shown.

Both these aromatic hydrocarbons occur in gasoline, and, together with benzene, they are the main contributors to high-performance *BTX* gasoline. Their concentrations are enhanced by catalytic reforming in an oil refinery. Toluene, which is also the basis of TNT (188), is so called because it was originally obtained from *Tolu balsam*, the yellow–brown, pleasant-smelling gum of the South American tree *Toluifera balsamum*. This balsam has been used in cough syrups and perfumes.

All three xylenes are liquids with characteristic odours. They were once obtained by distilling wood in the absence of air (*xulon* is Greek for 'wood'). When it is not being burned in engines, *para*-xylene (the isomer shown) is used for the production of artificial fibre (73). A mixture of the three isomeric xylenes is called *xylol* and is used as a solvent.

ALCOHOLS AND ACIDS

'Alcohol' is a general term denoting an organic compound that contains the group –OH. The origin of the name, from the Arabic *al kohl* for 'the fine powder', is rather curious. The Egyptians stained their eyelids with the black inorganic compound antimony sulfide. This substance was obtained by grinding, but eventually the name was applied to the essence of anything, including the liquid obtained by distilling wine. This section describes a series of compounds related to wine and its oxidized product vinegar, and introduces three more types of compounds that figure frequently in the following pages: *aldehydes*, which are compounds that contain the group –CHO; *carboxylic acids*, which are organic compounds that contain the *carboxyl* group; and *esters*, which are combinations of alcohols and carboxylic acids.

25 METHANOL CH$_4$O

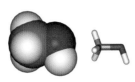

Methanol, or methyl alcohol, is also known as *wood alcohol*, because it was originally obtained by distilling hardwoods. (*Methy* is a Greek word that usually means 'intoxicate' but sometimes means 'wine', and *hule* is the usual Greek word for 'substance' but is sometimes used for 'timber' or 'group of trees'.) Methanol also occurs in wood smoke, and traces are present in new wine, where it contributes to the bouquet. Although methanol does cause an initial inebriation when it is drunk, it is an indirectly poisonous liquid. Its toxicity is largely due to the formation of formic acid (30) and formaldehyde (28) by the enzyme *alcohol dehydrogenase* in the body. These compounds attack the ganglion cells in the retina, cause degeneration of the optic nerve, and can cause permanent blindness. Death usually follows ingestion of about 50 millilitres or more.

Methanol is a colourless, mild-smelling, almost tasteless liquid. It is highly flammable and burns with an almost invisible blue flame because there is insufficient carbon in each molecule to aggregate into incandescent soot particles. *Methylated spirits* consists of ethanol (26) made undrinkable by the addition of about 10 per cent methanol and a warning purple dye. One antidote for methanol poisoning is ethanol, which is metabolized more rapidly than methanol. As a result, the methanol persists longer but its toxic products are produced more slowly and a greater proportion of the methanol is excreted in the urine and breath.

26 ETHANOL C_2H_6O

27 γ-AMINOBUTANOIC ACID $C_4H_9O_2N$

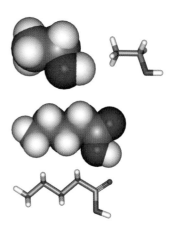

Ethanol, or ethyl alcohol, is what the world calls *alcohol*. Since antiquity (though neither universally nor without remission), it has been the essential component of most socially acceptable intoxicating liquids. Fermentation was probably stumbled onto by accident, when honey, grain, and fruit juices were found to yield mead, beer, and wine and concentrated by distillation, discovered in 800 CE by Jabir ibn Hayyan (known to the West as Gerber). Physiologically, alcohol acts as a depressant, like a general anaesthetic; in the later stages of intoxication, the effects are akin to those induced by a frontal lobotomy. Ethanol seems to its imbiber to be a stimulant, but in fact it acts by freeing parts of the cortex from inhibitory controls.

To understand the action of ethanol, it is necessary to delve into the brain and examine the role of the neurotransmitters, the molecules that communicate between nerve cells at synapses (nerve-cell junctions). Several of these neurotransmitters will be mentioned in later sections, but here we focus on γ-aminobutanoic acid (GABA): the 'gamma' tells us that the amino group, $-NH_2$, is attached to the third carbon atom from the carboxylic carbon atom.

The GABA molecule normally acts by binding to a protein molecule on the surface of the presynaptic (upstream) nerve cell of a synapse, and inhibits the cell's activity. This inhibition results from a distortion the GABA causes in the local structure of the cell membrane, in essence widening the channels through which chloride ions (Cl^-) can pass into the cell. The higher internal concentration of these ions changes the voltage difference across the membrane, and the cell becomes unable to fire. An ethanol molecule doesn't have its own receptor, but appears simply to dissolve in the lipid membrane of the neuron and increase its fluidity. This fluidity in a particularly sensitive region of the membrane forming the synapse enhances the binding of GABA and thereby encourages the enlargement of the chloride-ion channels, or the fluidity results in the enlargement of these channels directly, in either case amplifying GABA's inhibiting effect. Some tranquilizers and sedatives have a similar outcome; for example, the benzodiazepines (172) link to the same protein molecule as GABA, but at a different location. Because the benzodiazepines and ethanol attach to the same protein molecule, there is a strong synergistic interaction between them: Valium can be lethal when it is taken with alcohol. Ethanol might also disrupt nerve membranes more generally, and inhibit the propagation of action potentials, giving rise to another, more generalized inhibition.

Other physiological effects that accompany the ingestion of ethanol include its interference with the production of some antidiuretic hormones, which leads to an enhancement of the secretion

A synapse, showing the presynaptic vessels containing the neurotransmitter and the postsynaptic binding sites to which they migrate when they are released.

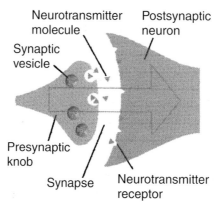

Neurotransmitter molecule

Postsynaptic neuron

Synaptic vesicle

Presynaptic knob

Synapse

Neurotransmitter receptor

of water, urination, and hence a sense of dehydration. Alcohol also causes blood vessels to dilate; the resulting increase in flow of blood through the capillaries beneath the skin gives a rosy hue to the complexion, together with a feeling of warmth. Alcohol is absorbed only slowly in the stomach, but rapidly once it reaches the upper portion of the small intestine. Because blood from the digestive tract flows through the liver before going back to the heart, any alcohol it carries might kill liver cells, which are replaced by fibrous tissue, resulting in the condition known as *cirrhosis* (specifically, Laënnec's cirrhosis) of the liver (the name cirrhosis is derived from the Greek word *kirrhos*, for 'tawny').

Ethanol is made by fermentation of carbohydrates (93) and, industrially, from ethylene (47). It mixes with water in all proportions, since its –OH group can form hydrogen bonds with the water molecules. Wine, beer, and spirits all contain ethanol, and for table wines about one molecule in twenty is ethanol (the bulk of the remainder being water). *Proof* in the USA denotes twice the percentage of ethanol by volume, so that 100 proof spirits are 50 per cent ethanol and each degree of proof corresponds to 0.5 per cent alcohol. In the UK, 100 proof spirit corresponds to 114.12 proof spirit in the USA. Spirits contain numerous other components as well; Scotch whisky (from the Scottish Gaelic *uisge beatha* and the Irish Gaelic *uisce beathadh*, meaning 'water of life'), for example, includes volatile organic molecules from the smoke of *peat*, which is decaying organic matter that is burned in the process of drying the barley malt (partially germinated kernels) before it is fermented. *Vodka* (Russian for 'little water') is virtually only ethanol and water: it can be prepared from the cheapest sources, usually grain, because most of the flavour is removed by passing it through charcoal. Metabolism of ethanol in the liver produces acetaldehyde (29) and, as explained there, hangovers.

Asians are more sensitive to alcoholic drinks than are Europeans: much less alcohol is needed to induce the vasodilation responsible for facial flushing and tachycardia. The cause can be traced to the presence of a form of mitochondrial aldehyde dehydrogenase, an enzyme that converts acetaldehyde to acetate, that allows a higher steady-state level of acetaldehyde in the blood than occurs in Europeans.

A 'hangover' refers to the headache, vertigo, nausea, pallor, sweating, and tachycardia experienced after overindulgence in alcohol. The causes include dehydration and electrolytic imbalance, hypoglycaemia, and the presence of excess acetaldehyde and lactic acid. A dangerous, slippery-slope, remedy is the 'hair of the dog', another alcoholic drink to help overcome these withdrawal responses.

28 FORMALDEHYDE CH₂O

We can think of the formaldehyde molecule as being obtained by removing two hydrogen atoms from a methanol molecule – one from the carbon atom and one from the oxygen atom. That is, in fact, the origin of the name *aldehyde* (from *al*cohol *dehydro*genated) given to compounds like formaldehyde that contain the group –CHO. It also describes the industrial preparation of formaldehyde, in which oxygen is used to pluck hydrogen atoms off the methanol molecule and carry them away as water. Billions of kilograms of formaldehyde are produced in this way each year, for its high reactivity makes it a valuable industrial intermediate.

Formaldehyde (more formally, methanal) is a pungent, colourless gas with a suffocating odour. It is soluble in water and is often encountered as the 40 per cent aqueous solution called *formalin*, which is used for preserving biological specimens. Formalin retains its sterilizing capacity even when it is diluted to 10 per cent, and it is used to kill anthrax spores during the treatment of wool and

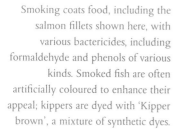

Smoking coats food, including the salmon fillets shown here, with various bactericides, including formaldehyde and phenols of various kinds. Smoked fish are often artificially coloured to enhance their appeal; kippers are dyed with 'Kipper brown', a mixture of synthetic dyes.

hides. Formaldehyde is also present in wood smoke and is one of the agents responsible, by attacking bacteria, for the preservative action of the smoking of foods. The mode of action in all these cases appears to be its ability to react immediately with the –NH and –NH$_2$ groups characteristic of proteins (80) and to link together neighbouring protein chains, hardening the substance and taking the protein molecules out of commission. The same type of reaction is used to produce synthetic resins and adhesives (as with urea, 147), to obtain *urea–formaldehyde resins*. The preserving and sterilizing action of formaldehyde can therefore be thought of as the formation of resins from living starting materials.

Methanol is metabolized to formaldehyde by the enzyme *catalase*. Because that enzyme is also involved in the chemistry of vision (Chapter 5), it occurs in the retina of the eye, and a significant amount of metabolism occurs there. The formaldehyde that is produced cross-links the retinal proteins, removes them from active participation in vision, inhibits oxygenation of the retina, and, for doses in excess of about 2 grams, results in blindness.

29 ACETALDEHYDE C$_2$H$_4$O

Acetaldehyde, a pungent, colourless liquid that boils at about room temperature, is the primary metabolic product of ethanol on its route to becoming acetic acid (31). It is produced by the enzyme *alcohol dehydrogenase*, which occurs primarily in the liver but also to a small extent in the retina. Generally, the bigger the person, the bigger the liver and the quicker the alcohol is metabolized and removed from circulation. Acetaldehyde is one of the chemical agents responsible for a hangover, although there are also numerous complicated and interrelated contributions from the physiological changes that occur as the body responds to unnaturally high levels of ethanol and the mild narcosis, acid imbalance, and dehydration it induces. Evidence is accumulating that acetaldehyde is carcinogenic, which should be a sobering thought for those with hangovers.

Alcohol dehydrogenase is present in our bodies not as evolution's anticipation of partygoing but because we need to metabolize the alcohol produced in small amounts by the normal digestion and breakdown of carbohydrates, and in large amounts by the bacteria in our intestines. Some substances (notably other alcohols) compete with ethanol for alcohol dehydrogenase. These alcohols are present in *fusel oil* (from the German word for 'rotgut' or, more specifically, 'evil spirit'), a byproduct of fermentation, and hence are present in some distilled spirits. The metabolism of wine and spirits is therefore slower than that of vodka, which is largely free of everything except ethanol and water. Acetaldehyde is also a product of the action of *Saccharomyces cerevisiae*, a yeast that is allowed to develop on fino and amontillado sherries and imparts to them their nutty flavour. It also contributes to the odour of ripe fruits.

Acetaldehyde is a volatile component of cotton leaves and blossoms, is found in oak and tobacco leaves and is a natural component of many ripe fruits. It also contributes to odours of rosemary, daffodil, bitter orange, camphor, angelica, fennel, mustard, and peppermint. It is used as a synthetic flavouring ingredient in margarine and has been detected in mothers' milk. Acetaldehyde is a component of cigarette smoke and is present at higher concentration in marijuana smoke: a marijuana cigarette produces about 1.2 milligrams. The combustion of a kilogram of wood produces just under a gram of acetaldehyde.

30 FORMIC ACID CH$_2$O$_2$

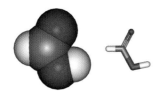

The 'formic' in formic acid's name reflects its origin: formic acid is one component of the venom injected by stinging ants (*formica* is Latin for 'ant'). It is also a component of the fluids injected by stinging caterpillars and bees. The sting of the common nettle, *Urtica*, like that of the hornet, is largely due to a mixture of histamine and acetylcholine: note the convergent evolution of weaponry, with unrelated insects and plants stumbling onto the same solution. Formic acid is also a product of the metabolism of methanol via formaldehyde, analogous to the conversion of ethanol to acetic acid (31) via acetaldehyde. The heightened levels of acidity due to this production of formic acid damage proteins and contribute to the internal devastation that methanol causes.

31 ACETIC ACID C$_2$H$_4$O$_2$

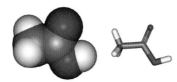

Acetic acid, a colourless liquid with a sharp odour, is the acid component of *vinegar* (a name derived from the French words *vin aigre*, meaning 'sour wine') and is responsible for its characteristic smell. Since acetic acid is an oxidized form of ethanol, it is produced when wine stands exposed to air and the ethanol undergoes aerobic oxidation by the bacteria *Acetobacter*. It is also what gives poor wines their vinegary taste. Acetic acid is also produced in dough leavened

This black ant (*Lasius niger*, from the subfamily *Formicinae*) injects a venom rich in formic acid.

Macrophotograph of stinging hairs on the stem and leaves of a stinging nettle, *Urtica dioica*. The stings are long hairs with sillicified tips which break off when the nettle is brushed against or stroked. The sting penetrates the skin, injecting a mixture of acetylcholine and histamine; the former is a neurotransmitter substance normally produced at nerve terminals, the latter a chemical mediator normally produced by mast cells and responsible for many of the unpleasant symptoms of allergic reactions. Only the long hairs on the leaf are stinging hairs; the short ones are harmless.

with the yeast *Saccharomyces exiguus*, which cannot metabolize the sugar maltose; as a result *Lactobacillus sanfrancisco*, a particular group of bacteria that depends on maltose, can thrive and excrete acetic acid and lactic acid (32). The dough becomes *sourdough* and is baked into the bread for which the San Francisco area is known.

32 LACTIC ACID $C_3H_6O_3$

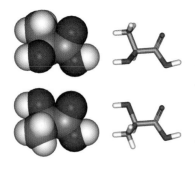

Lactic acid is a solid at room temperature. Of itself, that fact is not very interesting; however, it does serve to illustrate how a second –OH group provides new opportunities for the formation of hydrogen bonds between neighbours and hence a stronger attraction between them. Slightly more interesting is the existence of two kinds of lactic acid molecule, an L form (for *laevo*, from the Latin word for 'left') and a D form (for *dextro*, from the Latin word for 'right'). Each is the mirror image of the other; and just as a left glove cannot be superimposed on its mirror image, a right glove, so a molecule of L-lactic acid cannot be superimposed on a molecule of D-lactic acid. The two forms, which are called *optical isomers*, have virtually identical chemical properties but differ in their effect on plane-polarized light: L-lactic acid rotates its plane of polarization – the plane in which the electric field lies – to the right, whereas D-lactic acid rotates it to the left as the light approaches the observer. This phenomenon is known as *optical activity* and is a general property of *chiral* molecules – which are molecules that cannot be superimposed on their mirror images. As we shall see, chirality is far from being of only academic interest (175).

D-Lactic acid is obtained by the action of bacteria on meat extract, and L-lactic acid is obtained from the fermentation of sucrose with *Bacillus acidi laevolactii*. A mixture of the two is obtained from sour milk. Lactic acid is widespread in nature. A clue to the reason for its ubiquity is obtained by comparing its molecular formula, $C_3H_6O_3$, with that of glucose (87), $C_6H_{12}O_6$: a lactic acid molecule is, in essence, half a glucose molecule. Indeed, a widespread source of lactic acid is the anaerobic fermentation of sugars and the action of enzymes on glucose supplies. Fresh milk rapidly becomes populated with bacteria that act on the milk sugar *lactose*, break it up for its energy, and excrete lactic acid. The acid causes the fatty droplets to coalesce, and the milk curdles. This process is encouraged in a controlled way in the production of yogurt, which depends on production of lactic acid by a mixed culture of the bacteria *Lacto-bacillus bulgaricus* and *Streptococcus thermophilus.* Fermented pickles also owe their tartness to lactic acid. *Sauerkraut* is produced by steeping fresh cabbage in brine, which suppresses the growth of some bacteria and gives others – first *Leuconostoc mesenteroides* and then *Lactobacillus plantarum* – a chance to thrive. As they do, they consume glucose units and excrete lactic acid, sharpening the taste of the cabbage.

Lactic acid is also produced from glucose by enzymes in our sweat glands (which accounts for the acid taste of perspiration) and by the action of bacteria on the lining of the vagina, which is a rich store of glucose. It is also produced, as a last resort, in muscle that has exhausted its immediate oxygen supply and cannot metabolize glucose aerobically. Thus sprinters may unconsciously resort to the anaerobic energy supply of their ancestors, and make do with the energy released by slicing glucose molecules into lactic acid halves. Unfortunately, this process builds up the concentration of acid in the muscles, interfering with their operation so that they feel heavy and weak, and may produce cramp.

Bodily lactic acid becomes involved indirectly with heavy drinking: because the metabolizing capacity of the liver may be saturated by the demands of the ethanol, lactic acid is not removed so efficiently. It may then build up in the bloodstream, raise the acidity of the muscles, and lead to the kind of fatigue experienced – for different reasons – by an athlete. Lactic acid can also influence the deposition of solid compounds (specifically, salts of uric acid, a derivative of the class of molecules known as purines, 199). These compounds are normally excreted in the urine, but their excretion is inhibited by lactic acid and they may, instead, be deposited in the joints, causing the painful condition known as *gout.* The first deposits usually occur in small joints, particularly the metatarsal–pharyngeal joint (of the big toe); their formation is encouraged by alcohol and foods that are rich in purines, including the classic accompaniments of cartoon gout – claret and port.

Sutures used in internal surgery are biodegradable polymers that are decomposed by reaction with water or as a result of the action of enzymes. One such polymer is poly(lactic acid), in which long stringy molecules are made when the acid functional group of one lactic acid molecule reacts with the hydroxyl group of another, producing an ester-like link. In the body, the polymerization is reversed and the lactic acid so released is incorporated into normal metabolic processes. A stronger polymer is obtained by using glycolic acid, which is like lactic acid but with one carbon atom fewer, either alone or in conjunction with lactic acid.

Lactic acid is used as a softener of the cuticle that grows at the base of nails: it works by dissolving the keratin of the *stratum corneum*, which removes any dead skin. Further up the body, lactic acid is used in preparations designed to enhance the shave obtained with electric razors: the alcohol in the preparation removes the film of perspiration on the skin and, in conjunction with lactic acid, acts as an astringent that tightens the skin and causes the hairs to protrude. More generally, lactic acid and its near relatives, the other *α-hydroxy acids*, are used in anti-wrinkle creams. They cause the skin to peel, to reveal the newly growing layers below. Because lactic acid is found in milk, perhaps there was some justification for bathing in ass's milk.

FATS AND OILS

There is no deep, fundamental distinction between a fat and an oil: at room temperature a fat is a solid and an oil is a liquid, and what is a fat in one household may be an oil in another. The alcohol part of most fats and oils is *glycerol* (the commercial product is also known as *glycerine*). The carboxylic acid part generally has a long chain of carbon atoms and is called a *fatty acid*. Different fats and oils correspond to different fatty acids being attached to the glycerol unit, and each combination has its own melting point. The fats and oils from different sources actually consist of different mixtures of these molecules.

33 GLYCEROL $C_3H_8O_3$

A glance at the structure of glycerol suggests that it will be markedly different from its parent, the gas propane. In particular, the three oxygen atoms sensitize their hydrogen atoms for the formation of hydrogen bonds and the glycerol molecules stick together in a viscous, syrupy liquid. Moreover, the molecules not only bond strongly to each other but also form strong hydrogen bonds with any water in the vicinity. What is less obvious is that the hydrogen bonds and the oxygen atoms give the molecule a sweet taste (Chapter 4).

Glycerol's ability to bond water molecules to itself accounts for its widespread use in cosmetics, pharmaceutical preparations, and foods. It is used as an *emollient* (a softener) and a *demulcent* (a soother) in cosmetics, and as a *humectant* (antidrying medium) in toothpaste. It has further advantages for use in toothpaste, for it gives the preparation a shiny surface and has a sweetening function. Glycerol is also added to mouthwashes to increase the viscosity of the liquid, to give a better aftertaste; mouthwashes without humectants are said to have an unpleasant 'chemical' taste. It is added to candies to prevent their crystallization and is sprayed on tobacco to act as a humectant, to keep the leaves moist and prevent them from crumbling before processing (184). Glycerol is also added to glues to prevent them from drying too fast, and it is incorporated into plastics – notably cellophane (93) – as a *plasticizer*, to keep them supple by acting as a lubricant between polymer molecules.

The sweetness and smoothness (essentially, viscosity) of the liquid and its solutions makes glycerol a desirable component of wines. Nature contributes to our pleasure in this respect through the unlikely agency of 'noble rot', caused by the fungus *Botrytis cenerea*, which, under certain conditions, attacks grapes on the vine, injures their skins, and allows some water to evaporate. The botrytized grapes are left richer in glycerol, and the resulting wines (most notably some Sauternes) are sweet and smooth to the palate. Glycerol is also present in red wines, particularly good Burgundies. The formation of 'legs' on the side of a glass after the wine has been swirled, which is a sign of quality, is due to the presence of slow-draining, viscosity-enhancing glycerol.

Apart from its ability to anchor water molecules, each –OH group of glycerol can combine chemically with and anchor one carboxylic acid molecule. The resulting esters are the fats and oils described in this section. If only one –OH group is esterified, the ester is a *monoglyceride*; if two, a *diglyceride*; and, if all three, a *triglyceride.*

34 STEARIC ACID $C_{18}H_{36}O_2$

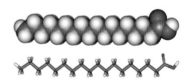

Stearic acid is an example of a fatty acid. In particular, it is a typical *saturated* carboxylic acid. In a saturated compound there are no double bonds between neighbouring carbon atoms. One consequence

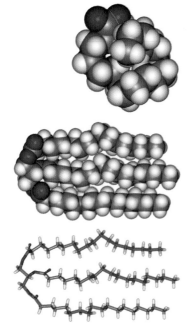

of stearic acid's having only single carbon–carbon bonds is that its hydrocarbon chain is flexible like that of octane (20), because each –CH$_2$–CH$_2$– unit acts as a flexible hinge. Hence, as well as stretching out into a zigzag, stearic acid can roll up into a compact ball.

35 TRISTEARIN C$_{57}$H$_{110}$O$_6$

The tristearin molecule is one of the most complicated we have met so far, but its structural theme is very simple. The molecule is a triglyceride (33); think of it as a three-fold ester in which each of three stearic acid molecules is anchored to a glycerol –OH group. The glycerol chain can be seen on the left of the illustration, with the three hydrocarbon chains of the fatty acid spreading away to the right like three streamers. In a normal sample, the streamers would be coiled and tangled with each other and with the streamers of neighbouring molecules.

Some of the properties of tristearin can be guessed by glancing at its molecular structure. One important feature is the absence of hydrogen atoms attached to oxygen atoms, leading us to expect that the molecules do not form hydrogen bonds. That is the case, and it is the reason why fat and water do not mix. That is an advantage for organisms, for then fat molecules are less likely to be washed out of their bodies and can therefore be stored. Moreover, since their molecules do not stick together tightly, we can also expect fats to be less dense than water, and to float on it. Tristearin molecules can maximize their packing by rolling their flexible fatty-acid side chains up into moderately compact balls; hence tristearin is a fat at body temperature, not an oil.

Compounds that are predominantly hydrocarbon-like often dissolve in hydrocarbon-like solvents such as benzene (22). Naturally occurring substances that dissolve in hydrocarbons and alcohols but not in water are called *lipids*; fats and oils are lipids.

Tristearin is one of the principal components of beef fat and cocoa butter (the major component of chocolate). The role of fat is principally as a reserve fuel supply; when it has to be used it is ultimately oxidized to carbon dioxide and water. It is therefore interesting to note that animals use a fuel very similar to that adopted for automobiles: apart from their slightly longer carbon chain, the side chains of fats differ from diesel fuel only in that they are already very slightly oxidized (as a result of the presence of the six oxygen atoms in

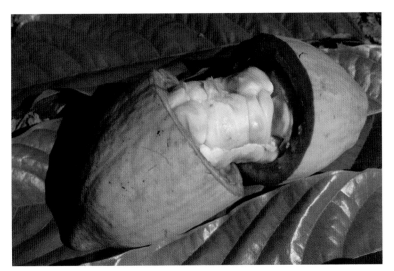

Cacao is the source of chocolate and cocoa. The pods grow along the branches and on the trunk of the tree and contain the seeds known as 'beans'. These 'beans' were once used by the Aztecs as currency.

tristearin). The length and saturation of the stearic acid chains is an advantage to animals, because fat need not be stored in a special tank and can be used as an insulating coat. The camel's hump is an elegant solution to the double problem of storing energy and water, for, as noted above, oxidation of tristearin produces water, so a camel can drink its own exhaust.

Fats have various secondary roles in food. They act as solvents for many flavour components and for some of the molecules that cause colour. Beef fat, for example, is coloured slightly yellow by carotene molecules (158) acquired from grass. Fats also increase the satiety value of meals, since they leave the stomach slowly and hence delay the onset of hunger. This is also why drinking milk before alcohol slows the latter's absorption; it takes longer for the alcohol to reach the intestine, where it is absorbed. Just as the automobile designer uses long-chain hydrocarbon molecules to lubricate moving parts, so animals use fats to lubricate their own meat fibres. A part of the tenderness of beef is due to lubrication by tristearin and its analogues.

36 OLEIC ACID $C_{18}H_{34}O_2$

Oleic acid, with its double bond between two neighbouring carbon atoms, is an *unsaturated* fatty acid. In general, an unsaturated compound is one with at least one carbon–carbon multiple bond; it is so called because it does not have its full complement of hydrogen atoms. The double bond has a marked effect on the shape of the molecule and on the shapes of the triglycerides it forms with

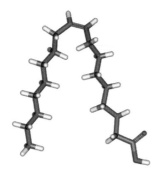

glycerol. Since the molecule cannot be twisted around the double bond, the chain is much less flexible than that of stearic acid and cannot roll up into a ball. The molecules of the esters it forms are much less compact than tristearin, the molecules do not pack together so well, and the compounds are oils rather than fats. Salad oils are often 'winterized' (chilled and filtered) to remove the fats that harden and lead to cloudiness at refrigerator temperatures. Pork, lamb, and poultry fats contain a higher proportion of unsaturated fats than beef does and hence are softer to the touch.

Plant stems and leaves are under less evolutionary disadvantage from weight than animals are and can store their food reserves as carbohydrate. However, their seeds need to support the developing embryo until it is self-sufficient, and for this a compact, efficient food reserve is required. Hence, oils are more abundant in seeds than they are in stems. Corn, cottonseed (*Gossypium*), soybean (*Glycine max*), safflower (*Carthamus tinctoris*, the petals of which also supply the red dye used for some kinds of rouge), and the solar-tracking sunflower (*Helianthus annuus*) are all good sources of unsaturated fatty acids, and oleic acid itself is the principal fatty acid in the olive oil pressed from the ripe fruit of the olive (*Olea europaea*).

The oleic acid present in the cocoa butter of chocolate gives it an interesting property. Most fats and oils are complex mixtures of triglycerides, with the result that they do not melt at a precise temperature but rather soften over a range. Cocoa butter, however, is markedly uniform in composition: each triglyceride molecule has an oleic acid streamer springing from the central carbon atom of the glycerol anchor, and the two other streamers are often either stearic acid or the closely related palmitic acid. This uniformity results in a much sharper melting point than is common for fats, and chocolate is brittle almost up to its sharp melting point of 34 °C (just below body temperature, 37 °C). Moreover, the melting is so sudden and energy-absorbing (like any melting) that, when it occurs in the mouth, it gives a feeling of coolness.

One problem with cocoa fat is that the molecules can stack together in six ways, labelled I to VI, with type I the least stable and type VI the most, giving solids with different melting points. The chocolate industry aims to get cocoa butter to crystallize as type V, for this form has a glossy appearance. The shiny surface of type V chocolate consists of a lot of tiny reflective crystals. If small, spiky type VI crystals grow on the surface (which will happen eventually, as type VI crystals are more stable), it becomes dull and they appear as a greyish 'bloom' on the otherwise hard, shiny surface. Incidentally, chocolates with praline interiors are typically first to develop a bloom because the nut-based filling contains a fat that is liquid at room temperature. This fat slowly migrates to the outer surface of the chocolate casing, dissolves some cocoa butter on the way, and deposits it on the surface, where it recrystallizes to give the bloom. There is also a migration of cocoa butter into the centre, which causes it to harden.

Biodiesel fuels are fuels that have been obtained from vegetable oils. In a typical process, a vegetable oil like triolein is heated in the presence of methanol (25) and sodium hydroxide (caustic soda, NaOH). The long hydrocarbon chains are snipped off the glycerol fragment and attach instead to the methyl groups of methanol. Instead of being pinned together as a triglyceride, each chain is a single ester molecule and forms a liquid that is much less viscous than the original oil. One major advantage of this type of fuel is that it makes no net contribution to global warming, because, although it releases carbon dioxide when it burns, the fuel itself is built (by photosynthesis) from carbon dioxide in the atmosphere. Growth, conversion, and combustion simply recycle the carbon dioxide already present rather than, as in the combustion of fossil fuels, releasing carbon that had been locked away.

37 LINOLENIC ACID $C_{18}H_{32}O_2$

Linolenic acid, which occurs in linseed oil (produced from flaxseed, *Linum usitissimum*), is closely related to oleic acid (36) but, since it has three double bonds, it is an example of a *polyunsaturated* fatty acid. The stiffness of the molecule that results from these multiple bonds leads us to expect the triglycerides built from it to be oils. We should also expect the oil to be chemically reactive, since carbon–carbon double bonds are chemically sensitive regions of molecules and are liable to suffer attack by other substances.

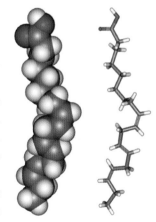

Linseed oil is in fact an example of a *drying oil*, one that produces a film when its double bonds are attacked by oxygen on exposure to air. The mechanism involves first the creation of radicals as a result of attack by oxygen, followed by a chain reaction that results in links being formed between neighbouring molecules. The process, which is called *polymerization*, occurs in polyunsaturated vegetable oils that are left to stand and is accelerated when they are used in frying. Old

The seeds of the rape plant (*Brassica napus*) provide rapeseed oil, a source of linoleic acid.

frying oil should not be topped off with new oil, for surviving radicals in the former will cause polymerization to proceed in the latter.

Air-induced polymerization is desirable for oil-based paints, which consist of a pigment in suspension in a drying oil, traditionally linseed oil plus some catalyst (drier) to accelerate the oxidation. The setting qualities of alkyd paints (62), which include polyunsaturated fatty acids as components, depend on the development of cross-linking between chains of those acids.

The fatty acids that occur in fish oils such as cod liver oil are rich in polyunsaturated fatty acids. This richness is probably the outcome of evolutionary pressure, because the triglycerides they form pack so poorly together that they remain liquid even in the cold environments the fish inhabit.

38 CHOLESTEROL $C_{27}H_{46}O$

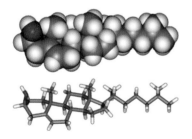

The ingestion of saturated fats provides the body with long, flexible chains of carbon atoms. It can use them to produce cholesterol, and hence raise the concentration of cholesterol in blood plasma to above its usual level. Note that, although this molecule has an elaborate, rigid, hydrocarbon framework, its business end (more formally, its *functional group*) is primarily the –OH group. In other words, cholesterol is chemically an elaborate alcohol (hence the *-ol* in its name).

Cholesterol, which is normally produced in the liver, plays an essential role in metabolism in the body. It is a precursor of various hormones (195), is used in production of cell membrane, and is the starting point for the production of bile compounds (its name comes from the Greek words for 'bile solid'). These compounds act as follows. When lipids enter the duodenum, they are partially broken down into fatty acids by the enzyme *pancreatic lipase.* However, the long-chain fatty acids characteristic of fats and oils are too much like hydrocarbons to be soluble in water; left to themselves, they would precipitate as a greasy solid and clog the intestine like a blocked drain. The bile compounds, however, act as detergents, surrounding the droplets of fatty acid and scattering them as groups of about a million molecules, giving a clear dispersion that can flow through the gut and can be absorbed and stored as fuel.

So much for the good and biochemically necessary side of cholesterol. Most people think of the shady role of cholesterol as the encourager of *atherosclerosis* (from the Greek words *atheroma*, meaning 'porridge', and *skleros*, meaning 'hard'), a hard porridgy deposit on the intima, or smooth walls of arteries. These deposits are an accumulation of lipids, mainly cholesterol, and complex carbohydrates; they are hardened as calcium ions from the blood plasma seeping through accumulate in them. As the porridge hardens, it blocks the flow of oxygenated blood to the myocardium, can act as the site of thrombus formation, and can lead to ischaemic (oxygen-deficiency) heart disease. The formation of gall stones, technically *cholesterol cholithiasis*, by the aggregation of cholesterol molecules occurs when there is an excess of cholesterol in the bile.

There is some evidence that unsaturated fats are less likely to enhance the cholesterol level than are saturated fats; apparently this involves a change in the way cholesterol is distributed between the plasma and various cellular compartments where it is stored, rather than a change in its rate of formation or decay. Manufacturers of edible fats are faced with the problem of maintaining a desirable level of unsaturation without producing a runny, inconvenient oil that could rapidly become rancid.

BUTTER AND MARGARINE

Butter is prepared from milk by churning its cream, which causes the fat globules to coalesce; the liquid that remains is called *buttermilk.* The original margarine (from the Greek word *margaron*, for 'pearl') was invented as a butter substitute in 1869 and was a mixture of milk, chopped cow's udder, and beef fat. Today's margarine has somewhat more obscure origins, being synthesized from substances that include vegetable oils and petrochemicals (chemicals obtained from petroleum). Some of the components of butter and margarine, including the decomposition products that contribute to rancidity, are discussed in this section. The molecule responsible for the colour of butter and margarine is carotene (158).

39 BUTANOIC ACID C$_4$H$_8$O$_2$

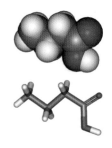

The fats and oils in cow's milk are rich in triglycerides built from short-chain fatty acids, including butanoic acid. Because the various short chains can be pushed past each other reasonably easily, butter is soft rather than waxy like fats, such as tristearin (35), and the long-chain hydrocarbons called *paraffin waxes*. However, milk is a complex mixture that includes stearic acid (34) and oleic acid (36) fats. As with other animal fats, the complexity of the mixture results in a solid that melts over a range of temperatures rather than remaining hard up to a sharp melting point.

Human-milk fat is rich in linoleic acid (41) and contains an unidentified substance that promotes the growth of the bacterium *Lactobacillus bifidus*. The bacteria produce large amounts of lactic acid (32) in the infant's intestine, and the acidity suppresses the growth of harmful bacteria.

In milk, the fats form an emulsion with water, in which they are dispersed in small droplets. Each droplet is surrounded by detergent-like molecules, including cholesterol (38); these molecules have a hydrocarbon part that can mingle with the hydrocarbon chains of the fats and one or more –OH groups that can form hydrogen bonds with the surrounding water. They stabilize the fat droplets by surrounding them with a water-favouring sheath. Coagulation of the droplets occurs because of a protein present in the milk, which links the droplets together and, as they coalesce, they form a head of cream. Heating milk also causes a protein, lactalbumin, to coagulate; a skin is formed as water evaporates. The skin is moderately impervious to steam, and enough pressure can build up beneath it to cause the milk to overflow from the vessel. Some of the protein is destroyed and creaming is reduced by heating the milk briefly to about 100 °C.

Cream on the top of milk is less common now that *homogenized milk* is widely available. Homogenized milk is formed by forcing milk through small openings under pressure. This breaks the fat globules into smaller particles, resulting in a more viscous, whiter, blander tasting, and more stable emulsion.

40 BUTANEDIONE $C_4H_6O_2$

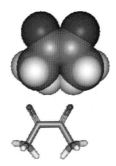

Molecules that contain the *carbonyl group* $>C{=}O$ are called *ketones* and are responsible for many natural flavours and odours (Chapter 4). Butanedione (also known as *diacetyl*) is a volatile yellow liquid ketone with a cheese-like smell. It is, in fact, the molecule that gives butter its characteristic flavour and the molecule you should have in mind when you smell it; for, when cream is incubated with bacteria, they produce some butanedione. After incubation, the cream is churned. This breaks down the sheaths around the fat droplets, and they coalesce into a soft, solid mass. Sheep's milk and goat's milk are richer in short-chain triglycerides than is cow's milk, and cheese made from them (such as Roquefort) is richer in pungent molecules.

You may be able to smell butanedione by sniffing your armpits or someone's unwashed feet, because it is a contributor to the odour of fermenting perspiration. Fresh sweat is almost odourless, but the action of the bacterium *Streptococcus albus*, which is present on the skin, increases its acidity and makes it an inviting feast for other bacteria; they, in turn, excrete pungent compounds including butanedione. Deodorants, some of which use acidic aluminium salts as the active agent, act by killing the bacteria. Antiperspirants, which are also often aluminium compounds, appear to act by coagulating and blocking the sweat ducts.

41 LINOLEIC ACID $C_{18}H_{32}O_2$

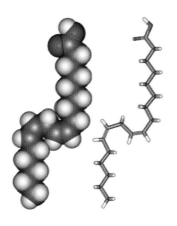

Linoleic acid is the principal fatty acid in many vegetable oils, including cottonseed oil, soybean oil, and corn oil. It is also abundant in rapeseed oil (from members of the mustard family, *Brassicaceae*) and is used in the manufacture of margarine, shortening, and salad and cooking oils.

Triglycerides (the analogues of tristearin, 35) built from linoleic acid are oils because of the double bonds in their chains of carbon atoms. Since manufacturers of margarine and shortening want a soft solid, they bubble hydrogen through the oil in the presence of a nickel catalyst. This *hydrogenation* process brings about several changes, including the partial saturation of the carbon chains (that is, the elimination of double bonds) as hydrogen atoms attach to carbon atoms that were originally joined by double bonds. The replacement of carbon–carbon double bonds with single bonds allows the carbon chains to become flexible. As a result, the molecules can pack together closely and the oil is converted into a fat. The hydrogenation process is stopped sooner if the oils are destined to become softer (tub) margarines.

The elimination of the double bonds during hydrogenation also reduces the likelihood of attack by oxygen, so the fat remains fresh longer. Nasty-smelling molecules are removed by passing super-heated steam through the molten fat. This procedure also removes the molecules responsible for colour, so carotenes (158) of various kinds are added to restore a butter-like appearance. The odour of butter is simulated by adding butanedione (40). The flavour is enhanced and sharpened by emulsifying the fats with skimmed milk that has been cultured with bacteria that produce lactic acid (32). The nutritional value is improved by the addition of vitamins A and D. Finally, natural

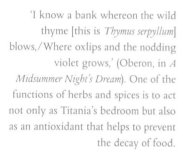

'I know a bank whereon the wild thyme [this is *Thymus serpyllum*] blows,/Where oxlips and the nodding violet grows,' (Oberon, in *A Midsummer Night's Dream*). One of the functions of herbs and spices is to act not only as Titania's bedroom but also as an antioxidant that helps to prevent the decay of food.

surfactant molecules (*lecithins*, which are triglyceride-like substances with one side chain containing a phosphate-like group) are added to ensure that the entire concoction hangs together.

42 2-*tert*-BUTYL-4-METHOXYPHENOL $C_{11}H_{16}O_2$

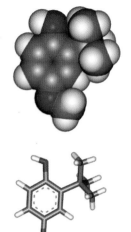

The shelf life of margarine and other fats and oils is improved if attack by oxygen on the double bonds in the carbon chains can be repelled or at least diverted, for then they do not turn rancid. The compound 2-*tert*-butyl-4-methoxyphenol or, more succinctly, BHA is an *antioxidant*, a substance that inhibits oxidation; it acts by interrupting the chain reaction in which oxygen combines with double bonds and slices molecules in two. It does so by combining with peroxide radicals (radicals of the form X–O–O, where X is the rest of the molecule, 11) before they have time to attack other molecules and continue the chain.

Antioxidants have been used in foods for thousands of years, but their role has only recently been appreciated. Among the more familiar ones are spices, which not only mask unpleasant odours but also help to prevent their formation. Sage, cloves, rosemary, and thyme all contain phenolic compounds resembling BHA, which interrupt the chain reactions and stabilize fats against oxidation. Thyme oil is also effective against bacteria and has been used in gargles, mouthwashes, and disinfectants. Animals contain antioxidant-like materials, including vitamin E, that serve a similar purpose – to prevent them from going rancid while they are still alive.

SOAPS AND DETERGENTS

Soaps and detergents are used to remove grease and dirt from fabrics in water. Their molecules have in common a long hydrocarbon chain with, at one end, a water-soluble group called the *head group*. The hydrocarbon end of these molecules can mingle with grease, which is typically a mixture of fats and oils. As a result, the surface of a grease droplet becomes surrounded by a sheath of head groups, which do not mix with the grease. The head groups form hydrogen bonds with water, which lifts them and the droplet and washes them away from the fabric.

Soap bubbles. Pure water, like any pure liquid, does not form a foam. Foaming is a sign of the presence of molecules that accumulate at the surface of the liquid and help to stabilize thin films. Among these molecules are the compounds we use as soap.

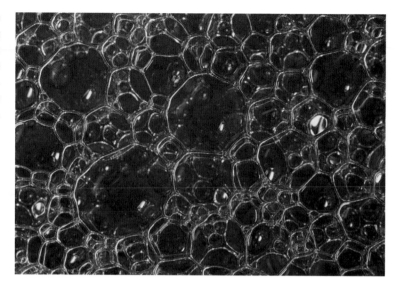

Molecules that have the dual character of being both *hydrophobic* (water repelling) and *hydrophilic* (water attracting) are called *surface-active agents*, since they accumulate and act at the surfaces where substances meet. This name is normally shortened to *surfactant*. A soap is a surfactant. A *detergent* is usually a synthetic surfactant with various additives, some of which are described in this section. Others present in a typical formulation include phosphates and brighteners.

Phosphates are added to detergents to provide the optimal acidity for the functioning of the surfactant molecules, to remove calcium and magnesium ions by wrapping around them and hiding them away from other ions with which they might precipitate and form a scum, and to attach to dirt particles that have been washed off the fabric to prevent them from redepositing. Unfortunately, since phosphates are fertilizers, the waste water from a load of wash is highly nutritious and can promote the growth of microorganisms in rivers and lakes. This can lead to *eutrophication*, or over-nourishment, which leads to clogging by organic growth, perhaps to the point of transforming a lake into a swamp. *Brighteners* are fluorescent dyes that absorb some ultraviolet light and re-emit it as visible light, hence making fabrics look brighter.

Proteolytic (protein-molecule-destroying) enzymes excreted by *Bacillus subtilis* and *B. licheniformis*, which can survive the conditions encountered during laundering, are also included in some detergents. Their action is sufficiently specific that they will modify the composition of the proteins in dirt particles without affecting the fibres of the fabric.

43 SODIUM STEARATE $C_{18}H_{35}O_2Na$

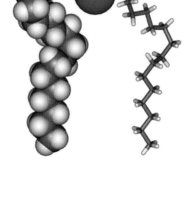

Tallow, the mixture of fat obtained by treating cattle fat with steam and collecting the material that floats on the water, contains tristearin (35). When tristearin is heated with sodium hydroxide (caustic soda), the stearic acid chains are broken off the glycerol (33), and the resulting mixture contains sodium stearate. In a detergent or soap, the long hydrocarbon chain is the part of the stearate ion that mixes with grease, and the ionic $-CO_2^-$ group is the head group that bonds with water.

Other fats and oils are also used in soaps – especially coconut oil, which contains a high proportion of lauric acid, a fatty acid with eleven carbon atoms in its tail. A typical toilet soap consists largely of fatty-acid salts made from tallow and coconut oil and so contains a high proportion of sodium stearate and sodium laurate. The combination of a potassium soap with excess stearic acid gives a slow-drying lather and is used in shaving soap. Post-foaming gels, which foam only when they are applied to the face, are soap solutions that are stabilized by a gelling agent and contain a small percentage of foaming agent, such as a low-boiling-point hydrocarbon that vaporizes on contact with the warm skin. Little by little, the shaver can remove the nearly ten metres of softened and lubricated hair that he grows in a lifetime, at less than half a millimetre a day.

44 SODIUM ALKYLBENZENESULFONATE $C_{18}H_{29}SO_3Na$

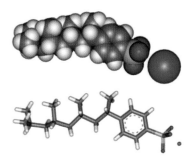

Synthetic detergent molecules like this one are an improvement on soaps for the following reason. When soap (43), a carboxylic-acid derivative, is used in *hard water* (water containing calcium and magnesium ions picked up as it trickled through limestone hills), insoluble calcium and magnesium carboxylate salts precipitate out and form an unpleasant *scum*. This does not happen with sulfonate detergents because calcium and magnesium sulfonates are more soluble in water than the carboxylate salts.

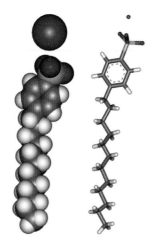

45 SODIUM *para*-DODECYLBENZENE SULFONATE $C_{18}H_{29}SO_3Na$

The dodecylbenzene sulfonate ion shown here is very similar to the ion shown in (44), the only difference being that the twelve carbon atoms of the side chain are strung out linearly rather than being branched. However, that difference has important ecological consequences: one problem with early detergents like the one in (44) is that when they run as effluent into rivers, bacteria are unable to destroy their branched side chains and the rivers foam. This has been overcome by synthesizing detergents with unbranched hydrocarbon chains, for such molecules can be digested by bacteria under aerobic conditions. That is, they are *biodegradable*. They also have detergent powers just as good as those of their branched isomers.

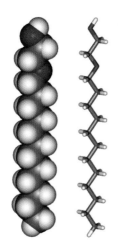

46 POLYOXYETHYLENE $C_{14}H_{30}O_2$

The head group in this molecule consists of the $-O-CH_2-CH_2-OH$ group, which links to water by means of the hydrogen bonds it can form: at least two oxygen atoms are needed in order to ensure that, despite its long hydrocarbon chain, the molecule interacts strongly enough with water. The advantage of this *non-ionic detergent* is that it is less efficient at stabilizing (or maintaining) foam than are the ionic varieties (44, 45) and so its use results in less foam. It is also more effective at removing grease and soil at low temperatures. The former property makes it suitable for use in washing machines, and the latter leads to the economical use of energy, because cooler water can be used. The $-O-CH_2-CH_2-$ group is sometimes repeated several times in the same molecule to enhance the water-attracting power of the molecule, and the hydrocarbon chain can take various forms.

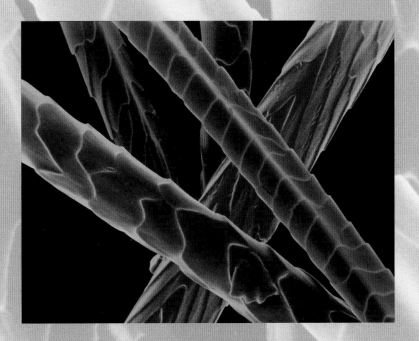

3

Synthetic and natural polymers

Much of nature's artistry depends on its ability to spin complexity from simplicity by linking small, mobile, and easily transportable molecules into chains and webs. The resulting linked molecules, or *polymers*, are fibres, sheets, and blocks that we know as rubber, silk, hair, and wood. Chemists have sought to understand and emulate nature in this achievement as in so many other ways and have made passable imitations and improvements. Now, not only do polymers grow from forest floors as trees, sprout from skin as hair and wool, and exude from insects as silk, but they are also carried in lorry-loads from factories as plastics, textiles, and coatings.

POLYMERS AND PLASTICS

The synthetic substances that transformed the world during the twentieth century are the synthetic plastics. These substances have the common feature that they are built by stringing together many small molecules. The individual small molecules, most of which are obtained from petroleum, are called *monomers* (from the Greek words for 'one part'), and the chains and nets they form are called *polymers* (for 'many parts'). In some cases, two or more different types of monomer are linked together; the resulting material is then called a *copolymer.*

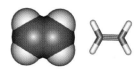

47 ETHYLENE C$_2$H$_4$

The ethylene molecule (more formally, ethene) is the parent of an exceptionally important class of compounds, and we shall use it as a seed just as we used methane (15) to develop the hydrocarbons described in Chapter 2. We can think of an ethylene molecule as an ethane molecule (17) in which one hydrogen atom has been removed from each carbon atom; the carbon atoms then form another bond to each other, so producing a double bond between them.

The presence of the double bond in ethylene makes it much more reactive than ethane, for the carbon–carbon double bond can open when it is attacked by other substances and form bonds with them, leaving the carbon atoms joined by a single bond. Because of its chemically tender double bond, ethylene is too reactive to be found in appreciable amounts in natural gas. It is formed during the refining of crude oil, especially when the bigger hydrocarbon molecules are torn apart during the process known as *cracking.*

One important use for ethylene itself is in the ripening of fruit, for plants generate the gas when they are ready to ripen. It appears to stimulate the metabolic processes involved, perhaps by dissolving in and increasing the permeability of cell membranes or by destroying chlorophyll (157) and unmasking the molecules responsible for the red and yellow colours (158, 162). Fruit shippers often transport their produce unripe and then expose it to ethylene gas at its destination. This practice goes back to the ancient Egyptians, who exposed figs to gas to stimulate their ripening, and to the Chinese, who burned incense to induce the ripening of pears. The active principle has been identified as ethylene in the gases. In the later nineteenth century it was observed that trees located near gas street lamps exhibited the 'triple response' of stunted growth, twisting, and thickening of stems: the cause of this response was traced to ethylene present in escaping town gas. Ethylene also stimulates the formation of flowers in Bromeliads, stimulates flower opening and the senescence of flowers and leaves, and induces femaleness in dioecious flowers. (Dioecious plants, from 'two houses', are plants that have their male and female flowers – their staminate and carpellate flowers – on separate plants, and include the willows, junipers, *Ginkgo*, and cycads.) One reason why ozone (4) is so damaging to plants is that it attacks the ethylene they produce, leading to the formation of dangerously reactive free radicals.

48 POLYETHYLENE $(-CH_2CH_2-)_n$

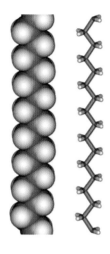

One of the substances that can add to ethylene is ethylene itself. *Polyethylene* is formed when ethylene molecules join together, and the process can continue until the string of linked $-CH_2-CH_2-$ units has grown to an enormous length – perhaps thousands of units long. When you touch a polyethylene article, you can feel the characteristic waxy texture of a hydrocarbon.

In a typical sample of polyethylene there are molecules of many different lengths; each chain has many side chains – some containing a thousand carbon atoms – where the process of polymerization has led to attack on an existing polymer chain. All the molecules are tangled together into a microscopic version of a plate of hairy spaghetti. Pure polyethylene is translucent for the same reason as a slurry of ice is translucent. In the latter, the numerous small ice crystals lie at random orientations to each other, and light passing through is scattered in many different directions. This effect can be so powerful that some

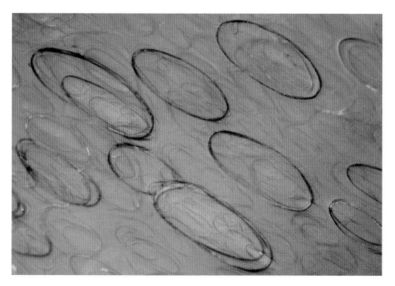

Polyethylene film, which is widely used for packaging, is formed by extruding the molten plastic through a ringlike gap and inflating it like a balloon. This picture is a micrograph of a textured film of plastic used for coating printed materials.

substances look brilliantly white even though they are composed of a colourless, transparent substance: milk is one example, and white house paint, which contains colourless, transparent titanium dioxide (TiO_2), is another. In polyethylene, domains in which the molecules lie next to each other in an orderly fashion to form crystalline regions alternate with domains in which the chains are jumbled together in disorderly amorphous regions. The crystalline regions lie at random orientations to each other and scatter the light like the cracks and bubbles in ice (5).

For *high-density polyethylenes*, the reaction conditions are chosen so that the carbon chains are 10 000 to 100 000 carbon atoms long, are of reasonably similar lengths, and have fewer than one side chain per hundred carbon atoms. The molecules then pack together more effectively, so that the solid is denser, more crystalline, and stiffer than ordinary polyethylene. This procedure can be taken further, to produce ultrahigh-density polyethylenes, so-called *ultrahigh-molecular-weight polyethylenes* (UHMWPEs) with unbranched chains up to 400 000 carbon atoms long. The resulting crystalline materials are so tough that they can be used for bullet-proof vests, and large sheets have been used in place of ice by skaters on ice-rinks.

Polyethylene has excellent electrically insulating properties. These partly stem from the tightness with which the electrons are trapped in their C–C and C–H bonds, so a current cannot flow through the solid. They also reflect the inability of water and ions to penetrate into the oil-like hydrocarbon interior of the solid. Moreover, the molecules are uniformly electrically neutral, with no regions of enhanced positive and negative charge, unlike nylon (79), for example. As a result, they barely respond to electric fields. In particular, they do not begin to oscillate when they are exposed to an alternating electric field and so do not absorb and dissipate its energy. That is one reason polyethylene was so important to the development of radar in the 1940s, for an insulator was needed for cables carrying high-frequency alternating current; it is also why polyethylene is still widely used as an insulator.

The hydrocarbon character of the interior of a lump of polyethylene makes it a congenial home for other hydrocarbon-like molecules. Thus, polyethylene is a solvent for fats, oils, and grease; but, since the polymer molecules are not very mobile, dissolving occurs very slowly, especially in the high-density polymers. Nevertheless, polyethylene articles do slowly absorb grease, become stained by it, and lose some of their electrically insulating qualities.

49 PROPYLENE C_3H_6

50 POLYPROPYLENE $(-CH(CH_3)CH_2-)_n$

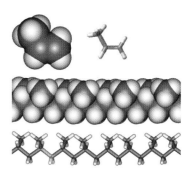

Many polymers are variations on the theme of polyethylene, differing only by having one or more of the hydrogen atoms in the monomer molecule replaced by other atoms or groups. The propylene molecule is an ethylene molecule in which a $-CH_3$ group has replaced one hydrogen atom. It is obtained together with ethylene when hydrocarbons are cracked, and it can be polymerized to form long molecules with a polyethylene-like backbone and $-CH_3$ groups on alternate carbon atoms. Special catalysts are used in order to ensure that there is little chain branching and that the $-CH_3$ groups all point in the same direction. Such an orderly polymer is said to be *isotactic*, and most of the polypropylene available commercially is of this kind.

The molecules in an isotactic polymer can lie close together, giving an extremely orderly solid. Because of its orderliness (its *crystallinity*), polypropylene is stiff, hard, and resistant to abrasion and has a high enough melting point for objects made from it to be sterilized. However, because the $-CH_3$ groups are liable to undergo oxidation, polypropylene articles usually have antioxidants (42) incorporated into them to divert attack by oxygen. Because polypropylene is a kind of frozen oil, unlike nylon (79) it does not absorb water and is resistant to discolouring; these properties make it suitable for outdoor carpeting.

51 VINYL CHLORIDE C_2H_3Cl

52 POLY(VINYL CHLORIDE) $(-CHClCH_2-)_n$

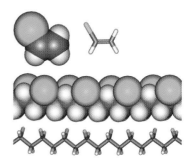

Now we let the ethylene molecule acquire a bit more personality by replacing a hydrogen atom with a chlorine atom (the green sphere in the illustration). At first sight, the new personality is unattractive, for vinyl chloride (more formally, chloroethene) is a carcinogenic gas. Nevertheless, it is manufactured (from ethylene) in huge amounts each year, since it can be polymerized to form poly(vinyl chloride), or PVC, one of the most useful and adaptable of all plastics. Because of the big chlorine atoms present on the chains, PVC molecules do not pack together so well that they form a rigid solid. However, to increase the flexibility of the solid, commercially produced PVC also contains large organic molecules (including esters of alcohols containing about ten carbon atoms) that act as *plasticizers* that lubricate the PVC molecules and allow them to move easily past each other when the solid is bent. These lubricants are not bound chemically to the polymer chains and slowly migrate to the surface. There they are lost and, as a result, the plastic becomes brittle and stiff. Since some bacteria enjoy a good meal of hydrocarbon chains and will eat parts of the plasticizer molecules, biocides are sometimes incorporated into PVC.

PVC is produced in such huge amounts because it is so versatile. It can be mixed with a very wide range of additives chosen to tailor its properties to many different applications. When it is properly protected by its additives, it is also chemically resistant to attack and degradation.

53 VINYLIDENE CHLORIDE $C_2H_2Cl_2$

54 POLY(VINYLIDENE CHLORIDE) $(-CCl_2CH_2-)_n$

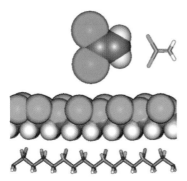

Now we replace a second hydrogen atom with a chlorine atom to give vinylidene chloride, a substance that may be polymerized on its own or copolymerized with vinyl chloride to give the polymers known collectively as *saran*. Saran A is poly(vinylidene chloride), the polymer obtained from vinylidene chloride alone; it consists of long chains in which $-CCl_2-$ groups alternate with $-CH_2-$ groups, as shown here. There are also $-CCl_2-CCl_2-$ and $-CH_2-CH_2-$ groups along the chain. Saran B, the more common variety, is the copolymer with vinyl chloride (51).

Poly(vinylidene chloride) molecules have a very regular structure, which is preserved to some extent in saran B. As a result, the molecules can pack together closely, giving a dense, high-melting-point substance. One consequence of this dense packing of the molecules is the high impermeability of saran films to gases, which results in their use for cling wrapping ('Cling film' and 'Saran wrap'). A second consequence is their low solubility in and impermeability to organic liquids – one reason why saran is so widely used for automobile-seat covers.

55 TETRAFLUOROETHYLENE C_2F_2

56 POLY(TETRAFLUOROETHYLENE) $(-CF_2CF_2-)_n$

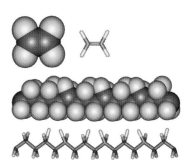

Tetrafluoroethylene is an example of a *fluorocarbon*, a compound of fluorine and carbon. In the molecule, all four hydrogen atoms of ethylene (47) have been replaced by fluorine atoms.

Fluorocarbon compounds constitute almost another world from organic and inorganic chemistry, in part because the C–F bond is so strong that it almost totally ignores attack by other molecules. In the small fluorine atoms, the highly charged fluorine nucleus exerts such tight control over the electrons in its vicinity that they take much less part in weak intermolecular bonding than do the electrons in hydrocarbons. As a result, fluorocarbons are generally more volatile than the corresponding hydrocarbons.

Fluorocarbons came into prominence after World War II, when the nuclear industries grew and supplies of fluorine became available (fluorine is used in the manufacture of uranium hexafluoride, UF_6, a volatile solid used in the separation of uranium isotopes). Tetrafluoroethylene is a colourless, odourless, tasteless gas; its principal destination is polymerization to give the fluorocarbon analogue of polyethylene, polytetrafluoroethylene, or PTFE.

PTFE consists of very long chains, composed of about 50 000 $-CF_2-$ groups each, with very little cross-linking between them. As a result, the molecules pack together to give a dense, compact solid with a high melting point. Even when the material is molten, the chains are so closely packed that they flow past each other only very slowly. Molten PTFE is so viscous that most PTFE articles are made by heating and compressing the powder to obtain a dense, strong, homogeneous lump.

The chemical and thermal stability of PTFE can be traced to two features. One is the considerable strength of the C–C and C–F bonds, which keeps the molecules from decomposing even when they are moderately heated. The second feature is the match between the sizes of the fluorine and carbon atoms, which results in the fluorine atoms forming an almost continuous sheath around the chain of carbon atoms, protecting it from chemical attack. In effect, the fluorine atoms act as chemical insulation around the carbon-atom 'wire'. Grease and oil do not form bonds with PTFE, so surfaces coated in it are 'non-stick'; that is, PTFE is an *abherent* coating, the opposite of an adherent substance like glue. Because fats and oils do not form bonds with the alien PTFE molecules, PTFE feels slippery to the touch. Its molecules pack together so densely that the solid does not absorb water, making it an excellent electrical insulator.

Tetrafluorethylene is often copolymerized with other fluorocarbons to produce the range of plastics widely known as *Teflons* (a Du Pont trade name). One Teflon is PTFE itself. Another, Teflon-FEP (the initials stand for fluorinated ethylene–propylene), is a copolymer of tetrafluoroethylene and the fully fluorinated version of propylene (CF_3–CF=CF_2). The $-CF_3$ groups of the fluoropropylene molecule are bumps on the $-CF_2-$ backbone of the copolymer and result in less close packing. As may be suspected, the solid melts at a lower temperature than does PTFE and forms a liquid with a lower viscosity. It can therefore be moulded by conventional (injection) techniques but retains the desirable thermal and chemical stability of PTFE.

57 STYRENE C_8H_8

58 POLYSTYRENE $(-CH(C_6H_5)CH_2-)_n$

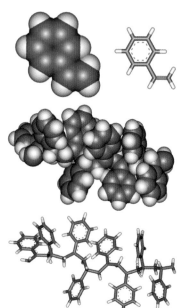

We now return to less extreme modifications of ethylene, for many changes can be made by attaching various groups to the CH_2=CH_2 unit. A styrene molecule, for example, is obtained from an ethylene molecule by replacing one of the latter's hydrogen atoms with a benzene ring (more precisely, with the *phenyl* group ($-C_6H_5$), a benzene ring with one hydrogen atom removed). The benzene ring brings enough electrons to the molecule – and with them stronger intermolecular interactions – to convert ethylene, a gas, into styrene, a colourless liquid. Styrene's principal fate is polymerization to give polystyrene.

Styrene polymerizes to give long chains of $-CH(C_6H_5)-CH_2-$ units with very little cross-linking between them. Although chains in which the benzene rings are all on one side can be prepared, the material that results is too brittle for general use. Most polystyrene consists of *atactic* molecules, in which the benzene rings point in random directions.

Because of the strength of the interactions between benzene groups on the chains, and because these groups obstruct the movement of chains past each other, polystyrene is less flexible than polyethylene. Flexible polystyrene can, however, be obtained by incorporating molecules that act as lubricants, as is done for PVC (52). The highly knobby character of its polymer chains and the random, disorderly way in which they pack together account for the great transparency of pure polystyrene. Lucite and Perspex (61) have a similar transparency for a similar reason.

The benzene rings make polystyrene susceptible to damage by ultraviolet radiation and other high-energy forms of radiation, so antioxidants are generally mixed in with polystyrene; they are especially needed in polystyrene fluorescent-light fixtures, for fluorescent lamps generate some ultraviolet light. There is enough energy in sunlight to induce radiation damage, and, unless polystyrene is protected by antioxidants, it rapidly yellows and degrades. The colour may be due to absorption of light by an oxygen molecule lying close to and loosely bonded to one of the benzene rings.

High-impact polystyrene is a copolymer of styrene and butadiene in which long rubber-like chains of polybutadiene hang from long polystyrene chains. The polybutadiene branches do not mix well with the polystyrene main chain and form little coiled regions among the polystyrene chains to which they are anchored. These little coils absorb energy when the polymer is struck. As a result, the polymer is less brittle than polystyrene alone and can survive hard impacts.

59 AZODICARBONAMIDE $C_2H_4O_2N_4$

This small but action-packed molecule can be thought of as being derived from a nitrogen (N_2) molecule by opening one of its three bonds, attaching a carbon monoxide (CO) molecule to a fragment with a C–N bond, and completing each carbon's bonding tendency with an –NH_2 group, which is a fragment of ammonia (6).

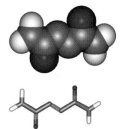

Foamed polystyrene is prepared by adding a foaming agent, of which azodicarbonamide is an example, to the molten plastic. When it is heated, the azodicarbonamide molecule falls apart into the gases carbon monoxide, nitrogen, and ammonia and is captured as bubbles in the molten polymer, like a frozen head of beer.

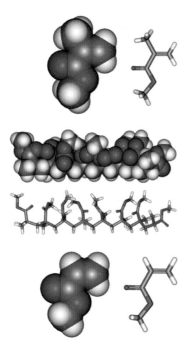

60 METHYL METHACRYLATE $C_5H_8O_2$

61 POLY(METHYL METHACRYLATE)
$(-C(CH_3)(COOCH_3)CH_2-)_n$

62 METHYL ACRYLATE $C_4H_6O_2$

Methyl methacrylate is the monomer from which poly(methyl methacrylate) is made, and poly(methyl methacrylate) is better known as *Lucite*, *Plexiglass*, or *Perspex*. Bulky, irregular side groups attached to the basic ethylene fragment cause the polymer chains to lie together in a very irregular way, so the solid is internally very chaotic. This is just like the lack of arrangement of water molecules in a glass of water. Because the solid is amorphous on a molecular scale, it does not scatter light that passes through. Consequently, blocks of the polymer are brilliantly transparent, like clean water. In fact, poly(methyl methacrylate) is more transparent than glass itself, so it can be used to make thick windows, which may be necessary when the windows have to withstand high pressures (as in aquaria).

One of the side groups attached to the ethylene fragment is a methyl group, $-CH_3$. Its importance in contributing to the knobbiness of the polymer, and hence to its rigidity, can be seen by removing it from the monomer, which turns methyl *meth*acrylate into methyl acrylate (62). This compound can be polymerized in the presence of a surfactant; the resulting *acrylic ester polymer* is in the form of minute droplets that remain suspended as a milky emulsion. This substance is the base for acrylic paints, which are formed simply by adding pigment. When the water evaporates from the emulsion after it has been applied to a surface, the acrylic ester polymer remains as a rubbery, flexible film that retains the pigment. The polymer molecule has the additional advantage of not absorbing the ultraviolet radiation present in the sunlight that reaches the surface of the Earth, so the paint does not degrade, break up, or, provided that the pigment survives, lose its colour.

63 LAURYL METHACRYLATE $C_{16}H_{30}O_2$

64 POLY(LAURYL METHACRYLATE)
$(-C(CH_3)(C_{12}H_{25})CH_2-)_n$

The lauryl methacrylate molecule is a modification of the methyl methacrylate molecule (60) in which the hydrocarbon tail has become a chain of twelve carbon atoms. Poly(lauryl methacrylate), the

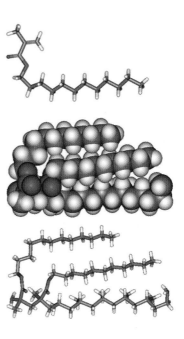

polymer formed from this monomer, and its close relatives are used as additives in viscostatic engine oils. The technological problem to overcome is the loss of viscosity (and the attendant draining from surfaces) that occurs when an oil is heated. An oil that is viscous when the engine is cold is not the answer, because such an oil cannot be pumped over engine surfaces adequately when the engine is first started.

The long hydrocarbon tail of poly(lauryl methacrylate) molecules makes them soluble in the similar environment that exists inside lubricating oils, which consist of hydrocarbons containing more than about twelve carbon atoms, and the thicker the oil, the longer the hydrocarbon chains. At low temperatures, the polymer molecules are coiled into balls, so they do not hinder the flow of the hydrocarbon oil molecules very much. As the temperature is raised, the coils unwind because the atoms in their hydrocarbon chains move more vigorously and flay around through their surroundings. The unwound polymer chains stretch through a greater region of the oil and hinder motion of the hydrocarbon oil molecules much more; thus the oil flows more slowly (is more viscous) than it would if the additive were absent.

Versions of these polymers are also the glues used in repositionable self-adhesive paper notepad sheets (that is, 'Post-It Notes'). The adhesive is applied in tiny microspheres, which disperse the adhesive character over the surface rather than forming a continuous film.

65 METHYL CYANOACRYLATE C$_5$H$_5$O$_2$N

66 POLY(METHYL CYANOACRYLATE)
(−C(CN)(COOCH$_3$)CH$_2$−)$_n$

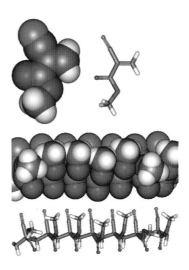

Methyl cyanoacrylate is the substance contained in tubes of *superglue*. When it is spread on surfaces that are to be joined, it begins to polymerize because the surfaces contain traces of water and alcohols. This readiness to polymerize is due to the presence of two very strongly electron-attracting groups that are attached to the same carbon atom. The −C≡N group withdraws electrons from the carbon atom to which it is attached, as does the oxygen-rich −COOH group on the same atom. This electron-pulling effect greatly distorts the electron distribution in the carbon–carbon double bond and makes it very susceptible to attack.

The formation of poly(methyl cyanoacrylate) proceeds quickly once it has been initiated; the surfaces are joined partly because cavities and tiny crevices in them are filled, which locks them together, and partly because the polymer forms chemical bonds with them. The glue sticks hard to skin. This may be a disadvantage for healthy individuals, but it is turned to advantage as illness advances: surgeons can use superglue in place of sutures, and morticians use it to seal, once and for all, the eyes and lips of their clients.

Superglue does have some drawbacks for beauticians, morticians, and their intermediaries, surgeons and veterinarians. First, methyl cyanoacrylate has an irritating odour; and enzymes attack and destroy the bond. Moreover, the breakdown products, cyanoacetate and formaldehyde (28), are toxic and can cause inflammation. Finally, the joints are not very flexible, so they can break. Improved versions of methyl cyanoacrylate, with built-in lubricants, have been developed, in which the tiny little methyl group is replaced by a longer chain of carbon atoms, such as butyl cyanoacrylate (with four carbon atoms in place of the one of the methyl group). More recently, the chain has been lengthened to gasoline proportions, with octyl cyanoacrylate (eight carbon atoms). This superglue also has an antimicrobial effect that can decrease rates of infection in contaminated wounds, perhaps because the microbes start eating along the hydrocarbon chain and end up in the glue.

Superglue is also used to catch felons, not as a flycatcher but as a way of revealing latent fingerprints. The object imprinted with the latent fingerprint is exposed to methyl cyanoacrylate vapour and the molecules adhere to the friction ridges. The print is dusted before the superglue dries, and then lifted and examined in the usual way.

RUBBER

Rubber, so called (by Joseph Priestley) because it can be used to rub out pencil marks, is an example of an *elastomer*, an elastic polymer. Natural rubber is obtained commercially from the coagulated latex of *Hevea brasiliensis*, but it is also present in the hollow stem of the dandelion (*Taraxacum officinale*). It is hard and brittle when cold, and sticky when warm. Its elastic properties are much improved if it is first shaped into the desired form and then heated with sulfur. This is the process of *vulcanization*, which was invented by Charles Goodyear in 1839.

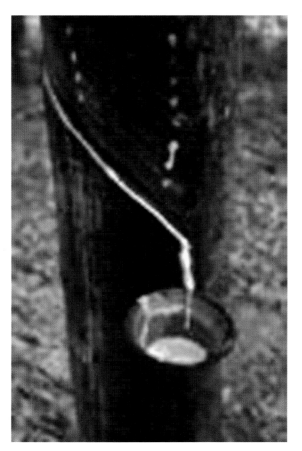

The commercial source of natural rubber is the rubber tree, *Hevea brasiliensis.* Here we see the collection of latex in a pot during the process of rubber tapping in Malaysia.

67 ISOPRENE C_5H_8

68 POLYISOPRENE $(-CH_2C(CH_3)=CHCH_2)_n$

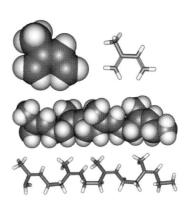

Isoprene, a volatile, fragrant liquid, is the monomer that polymerizes to form natural rubber. Rubber molecules consist of chains of several thousand $-CH_2-C(CH_3)=CHCH_2-$ units. The presence of a double bond in each unit is an important feature, as we will see shortly. Hot rubber smells of isoprene, for some isoprene molecules break loose when it is heated. This smell was the first clue that isoprene was a building block of rubber, yet early attempts to synthesize rubber from isoprene failed. People also smell slightly of isoprene, for a different reason (Chapter 4).

The failure to synthesize rubber was eventually traced to the fact that, when isoprene molecules link together, they can do so in either of two different ways. In natural rubber, all the linking between the units is done by enzymes working under the dictates of genetic control, and *H. brasiliensis* produces a polymer with all links in the *cis* arrangement. Early attempts to synthesize the polymer had produced chains with a random mixture of *cis* and *trans* units, and the material was sticky and useless. This problem has since been overcome by using a suitable catalyst (in this case, a compound that includes aluminium and titanium atoms), and nearly pure *cis*-polyisoprene, with properties very similar to those of natural rubber, can be produced.

Gutta-percha (from the Malay *getah*, gum, and *percha*, a type of tree), the material once used to cover golf balls, to insulate underwater cables, and for dental root canals, is the version of polyisoprene in which all the units are *trans*. It is obtained by boiling the sap of the *Isonandra gutta* tree, a species of tree of the order Sapotaceae, found in South East Asia. It is a much harder material than rubber and more resistant to water.

In unstretched rubber, the long molecules lie in tangled coils. Stretching the rubber unwinds the coils. When the stretched rubber is released, it springs back to its original shape if the polymer molecules have not slipped past each other. In natural rubber, the stretched coils do slide past each other under stress, so natural rubber does not resume its original shape exactly. The problem of molecule slippage is overcome by vulcanization. When natural rubber is heated with sulfur, the sulfur attacks the double bonds that remain in the chains and forms –S–S– bridges between neighbouring molecules. This bridging improves the elastic properties of the solid, because now the molecules are linked into a huge three-dimensional network. Neighbouring molecules can no longer be moved apart, but they are still moderately free to unwind under stress as long as neighbours cooperate. As a result, the solid is more resilient and recovers its shape when the stress is removed. Human hair (85) is naturally slightly vulcanized and so is springy for similar reasons. Heavily vulcanized rubber is the hard, tough, and chemically resistant material called *ebonite* or *vulcanite*.

69 ISOBUTYLENE C$_4$H$_8$

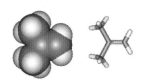

Chemists have tinkered with the composition of rubber and have produced a variety of elastomers. *Butyl rubber* is obtained through the copolymerization of isobutylene with a little isoprene. Without the isoprene, polyisobutylene itself would be just like polyethylene, with methyl groups on every other carbon atom, and no remaining double bonds. However, because the material is a copolymer, the polymer chains have about one isoprene unit per hundred isobutylene units, and there is one remaining double bond (from the isoprene) for every hundred or so monomer units in the chain. As a result, cross-links between the chains can be formed by vulcanization. The resulting elastomer withstands weathering better than other elastomers do because its molecules have fewer places where atmospheric oxygen can attack.

The molecules of butyl rubber pack more closely than do those in natural rubber, so it is much less permeable to gases. It is used for inner tubes of bicycle tyres and for the interiors of tubeless tyres.

Another starting point for new types of rubber is an isoprene molecule in which a –CH$_3$ group has been replaced by a chlorine atom. When this *chlorobutadiene* is polymerized, it gives a structure like that of natural rubber, but with each –CH$_3$ group replaced by a chlorine atom. This 'chloroprene' elastomer is called *neoprene.* It is unusual in that it can be vulcanized by heat alone; with the molecules forming carbon–carbon links directly to each other without the need for bridges built from sulfur. It is resistant to oxidation, oil, and heat and is widely used in automobile parts.

Numerous other elastomers can be tailored to suit a particular task through judicious choices of monomers and of the type of copolymerization. One very widely used composition is *styrene–butadiene rubber*, or SBR, the copolymer obtained from a mixture of styrene (57) and butadiene. Butadiene is like isoprene, but lacks the –CH$_3$ group. Most of this product is vulcanized and used for automobile tyres. A minor use for the raw (unvulcanized) material is as chewing gum.

70, 71 POLYCARBONATES

A polycarbonate is a polymer that includes the carbonate group, –O–CO–O–, in the repeating unit linked together by chains or rings of carbon atoms. Here we show two polycarbonates.

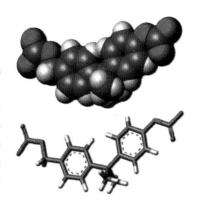

In the first polycarbonate (70), the carbonate groups are joined together by links that include benzene rings. The lumpiness of the polymer chains results in a highly disorganized arrangement of the molecules, and, as for polystyrene (58), the polymer has a glass-like transparency. However, the strong attraction between the benzene rings means that the material is very tough, and so it is used as a kind of shatterproof glass. Ordinary glass consists of silicon and oxygen

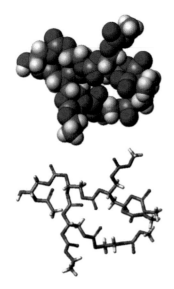

atoms, with some sodium and potassium ions, and all these atoms are heavier than the carbon and hydrogen atoms in polycarbonate glass, so the latter is much lighter than ordinary glass.

The second polycarbonate (71) is radically different from anything we have seen before. The monomer (which we don't show) has a double bond at each end of a carbon chain, so it can take part in polymerization reactions at both ends. As a result, instead of forming chains, it polymerizes to a three-dimensional network. This network doesn't melt once it has been moulded into shape, because the entire object is a single huge molecule – all the parts are joined by chemical bonds and there are no chains to shake apart by raising the temperature. Apart from its transparency and low density, this material has a high refractive index, so it is ideal for making light-weight spectacle lenses.

POLYESTERS AND ACRYLICS

Artificial (as well as natural) fibres should consist of long molecules that can be made to lie parallel to each other as they are drawn out into a thread. One of the ways of achieving this type of polymer is to link together an acid and an alcohol molecule as in the process of ester formation, but in such a way that the ester molecule can go on growing at each end. This results in indefinitely long molecules of repeating units called *polyesters*. To understand this process, we need to consider two typical reactant molecules, one an alcohol and the other an acid, and then see how they react.

72 ETHYLENE GLYCOL $C_2H_6O_2$

The presence of an –OH group at each end of the ethylene glycol molecule has important physical and chemical consequences. Since ethylene glycol can form hydrogen bonds at each end, whereas ethanol (26) can do so only at one end, we can expect it to be a more viscous and less volatile liquid. This is the case. It is also fully miscible with water, with which it can form hydrogen bonds. Less easy to predict is its toxicity, for a dose of little more than 50 millilitres can kill. The molecule is not directly poisonous: its metabolic products are the species that kill. Like hangovers from ethanol, the more permanent hangover from ingestion of ethylene glycol stems from its conversion to glycolic acid, which raises the acidity of the blood.

That acid is then converted into the even more toxic oxalic acid (which is the diuretic found in rhubarb, 100). Once oxalic acid has formed, calcium ions precipitate as insoluble calcium oxalate in various tissues and, with the elimination of calcium, hypocalcaemia and death may occur.

Ethylene glycol is the alcohol used as *antifreeze* for automobile engines. When it mixes with water, it attracts water molecules strongly and interferes with the interactions between them. As a result, they do not crystallize into a solid until the temperature has been lowered to well below the normal freezing point. In a sense, the ethylene glycol molecule acts as a hydrocarbon lubricant for the water molecules, which keeps the fluid supple. The glycol's low volatility is also an advantage, for it does not evaporate from the hot coolant. Ethylene glycol, like glycerol (33), is a humectant, and is used to increase the viscosity of ballpoint-pen inks.

The principal chemical importance of the ethylene glycol molecule lies in its Janus-like double-alcohol character. Since it has an –OH group at each end, it can link with two carboxylic acids. That is, it can be converted into an ester at each end. Why this is important we shall see shortly.

73 TEREPHTHALIC ACID $C_8H_6O_4$

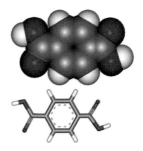

The terephthalic acid molecule is derived, both structurally and in practice, from a *para*-xylene molecule (24) by oxidation of the latter's two –CH_3 groups. Terephthalic acid is a white, crystalline solid. Benzene (22) itself is a liquid, but the carboxyl groups in terephthalic acid, with their ability to form hydrogen bonds, can bind neighbouring molecules together into a solid. Its chemically important property is its two-faced acid character, for, echoing ethylene glycol, it can form an ester at each end.

What, then, will result when this Janus-like acid reacts with that Janus-like alcohol?

74 POLY(ETHYLENE TEREPHTHALATE) $(-O_2C-C_6H_4-CO_2-C_2H_4-)_n$

Long chains of polyester grow when terephthalic acid reacts with ethylene glycol. First, one glycol –OH group combines with one of the acid –COOH groups, linking the two molecules together. However, this still leaves the glycol with one free –OH group that can react with another –COOH group and the acid with a free –COOH group that can react with the –OH group of another glycol molecule. Even after the next round of reaction, there are still a free –OH group and a free –COOH group at opposite ends of the now bigger molecule, and it remains ripe for further reaction. This process stops only when all the reagents have been consumed, a chain combines with its other end, or an impurity with only one –OH or one –COOH group ends the chain. The product, poly(ethylene terephthalate), is known as *Dacron* in the USA, *Terylene* and *Crimplene* in the UK, and *Trevira* in Germany.

The molten polymer is extruded through a spinneret, and the resulting fibre is then stretched to several times its original length. This stretching uncoils the PET molecules and as a result they lie together in an orderly arrangement. Links between the chains are due largely to the interactions between the partial electric charges of the –CO– groups, because the O atom accumulates electrons slightly and becomes slightly negatively charged (red), whereas the C atom of the group is partially stripped of its electrons and becomes slightly positively charged (blue):

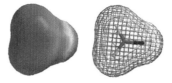

Moreover, the alternating benzene rings stiffen the chains and raise the melting point above that of polymers made up of simple flexible hydrocarbon chains. The stiffness is responsible for the crease resistance of non-iron fabrics made from the fibres. Thin film made from the polymer is sold under the trade name *Mylar* and used for cassette tape; the orderly orientation of the molecules in the film – which leads to a strong, almost unstretchable product – is achieved by stretching it in one direction.

Poly(ethylene terephthalate) has a low softening temperature, and therefore cannot be used to fabricate objects that have to be sterilized, such as recyclable bottles. The interaction between neighbouring molecules can be enhanced by increasing the number of benzene rings, for they can lie face to face and interact by van der Waals interactions. One example is poly(ethylene naphthalate), in which the single benzene ring of terephthalic acid is replaced with two benzene rings fused together along one edge. This substance can be sterilized, and is used to make recyclable bottles.

Other varieties of polyesters are obtained by polymerizing other carboxylic acids with other alcohols. The straightness of the polyester molecules, and hence their fibre-forming character, can be eliminated by choosing an acid that introduces kinks into the molecules. One such is *phthalic acid*, which is like terephthalic acid but has the carboxyl groups as neighbours on the benzene ring. Cross-linking among the chains is achieved if, instead of ethylene glycol, glycerol (33) is used (or included), for its third –OH group can begin a branch line. The polymers that result from such monomers are called *alkyd resins*, and form flexible sheets rather than fibres. They are available as emulsions in water and are widely used as surface coatings, including paints.

75 ACRYLONITRILE C_3H_3N

76 POLYACRYLONITRILE $(-CH(CN)CH_2-)_n$

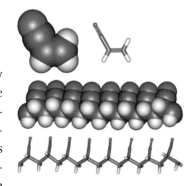

Many people unknowingly spread cyanide over their bodies, for they dress themselves in *acrylics*, polymers obtained by using acrylonitrile as a monomer. The polymerization process is like that for polyethylene (48), and the molecules link together to produce a polyethylene-like chain called *polyacrylonitrile*, in which alternate carbon atoms carry a cyanide group. The polymer, which is sold as *Orlon*, is dissolved in a solvent, and then fibres are spun by squirting the solution through spinnerets into air so that the solvent evaporates. The fibres are then stretched to several times their original length to orient the molecules so that they lie parallel to each other.

Pure polyacrylonitrile resists dyeing, so it is common to introduce small amounts of other monomers, including styrene (57), vinyl chloride (51), and vinylidene chloride (53), to give a copolymer with points of attachment for dyes. These copolymers are called *modacrylics*. Copolymerization with vinyl acetate, a molecule in which a group takes the place of the cyanide group, gives *Acrilan*.

The presence of the chlorine atoms in the vinyl chloride and vinylidine modacrylics results in an improvement in flame resistance; most acrylics used for carpets (and for the personal analogue of carpets, wigs) are actually modacrylics with a substantial chlorine (or bromine) content. Flame resistance is a very complex property to which many features contribute. These include achieving a higher ignition temperature of parts of the polymer chain and introducing weakness into the polymer, so that the fabric drips away before a flame can spread. The combustion products may also smother the fabric and thus help to exclude air. They may also inhibit the spread of a flame by introducing radicals that interfere with the chain reaction that is fire.

One very important derivative of the acrylics is the material known as ABS and widely used in vehicles where light weight and impact resistance is needed. This material is a rather complicated copolymer of acrylonitrile (75), butadiene (like isoprene, 67, with H in place of –CH₃), and styrene (57), hence the acronym ABS. To think of its structure, imagine a polybutadiene backbone, from which hang side chains that are copolymers of styrene and acrylonitrile. The presence of acrylonitrile in the copolymer strengthens the material more than styrene alone. Polystyrene gets its strength, as we saw, from the van der Waals interactions between the benzene rings. However, the –CN groups of polyacrylonitrile are strongly polar with electrons piled up on the nitrogen atom of each –CN group to give it a negative charge (red), leaving the nucleus of the carbon atom slightly exposed, and therefore giving the atom a slight positive charge (blue):

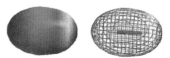

The partial positive and negative charges of –CN groups on neighbouring chains are attracted to each other very strongly, and as a result the material has great strength. It often takes a traffic accident to smash up these interactions. The presence of the polybutadiene backbone, however, gives the material more flexibility than that of polystyrene alone, so the material survives small impacts.

Acrylics are resistant to digestion by microorganisms in the soil and to degradation by sunlight. Hence they are suitable for use out of doors, and artificial grass is sometimes an acrylic.

When polyacrylonitrile itself is heated, an atomic rearrangement occurs, and the –CN groups bite the chain they hang from and form a series of rings:

When this product is heated further, the rings expel atoms and form benzene-like 'aromatic' rings with double-bond character shared round the ring:

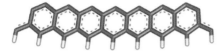

Next, at still higher temperature, two neighbouring chains of rings fuse together:

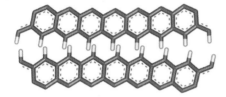

This process continues, until a ribbon-like band of chicken-wire formed mainly of carbon atoms is formed: this is *carbon fibre*:

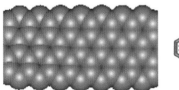 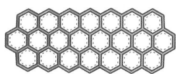

As the ribbons get wider, the 'impurity' nitrogen atoms become less and less significant. Carbon fibre is like a ribbon of pure carbon, a kind of two-dimensional version of diamond, and is immensely strong. It is used to reinforce substances, such as epoxy resins, where lightness but great strength is required, as in the turbine blades of jet engines.

NYLON

Nylon, the first completely synthetic fibre, was developed in the mid-1930s at the laboratories of Du Pont. It still retains a very large share of its market. Nylon is an example of a *polyamide*, a polymer that resembles a polyester (74) but results from the reaction between $-NH_2$ (rather than $-OH$) and a carboxyl group. This difference has important consequences. For example, the extra hydrogen atom in $-NH_2$,

A coloured scanning electron micrograph of the weave of a nylon stocking, showing the fibres slightly stretched. The magnification is about 150 times.

which survives after the amide link has been formed (in contrast to the single hydrogen atom of –OH, which is removed), provides the opportunity for hydrogen bonding. With hydrogen bonding, as we have seen, comes strength.

There are numerous polyamides, and they are named according to the number of carbon atoms in the monomers. The two most important, in the sense of taking the lion's share of the market, are nylon-6,6 and nylon-6. For historical reasons that partly reflect the availability of raw materials, nylon-6,6 originally predominated in the USA, and nylon-6 in Europe. Polyamides in which the monomers are derived from benzene (22), so that they have chains of hexagonal benzene rings joined by amide linkages, are generically known as *aramids* (a word obtained from the fusion of 'aromatic,' a general term for derivatives of benzene, and 'amide').

77 ADIPIC ACID $C_6H_{10}O_4$

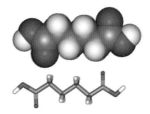

Adipic acid is a white crystalline solid. It resembles terephthalic acid (73) in that it has an acid group at each end; like terephthalic acid, it can also grow at each end. If it is esterified with a double alcohol such as ethylene glycol (72), it will grow into a polyester. We shall soon see how it can grow into a polyamide, and how its six carbon atoms contribute one of the sixes to nylon-6,6.

The two carboxyl groups of adipic acid have just enough hydrogen-bonding capacity to overcome the hydrocarbon character of the $-CH_2-$ chain and to drag it into solution in water, in which it is slightly soluble. It is an approved food additive, and it has been used to acidify soft drinks and control the acidity of cosmetics. Adipic acid also contributes to the sharp taste of beets (*Beta vulgaris*).

78 HEXAMETHYLENEDIAMINE $C_6H_{16}N_2$

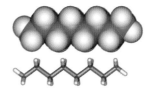

Each $-NH_2$ group in this molecule is called an *amino group*. The amino group is the nitrogen analogue of the $-OH$ group, and amines are to some extent the nitrogen analogues of alcohols. However, as always in chemistry, the replacement of one atom with another often brings in its wake profound consequences, even though the two atoms may be quite similar. The step from the world of oxygen chemistry to that of nitrogen is accompanied by two important changes. One is the presence of an additional hydrogen atom in $-NH_2$, compared with $-OH$. The second is a more subtle change: nitrogen has a weaker nuclear charge than does oxygen and hence has slightly less control over its electrons. This apparently innocuous difference effloresces into the world with great consequences for nylon and for life, as we shall see next.

Hexamethylenediamine is closely related to some evil-smelling amines. To discover how narrow is the line between your clothing and your putrefaction, see putrescine (150).

79 POLY(HEXAMETHYLENEDIAMINE ADIPAMIDE) $(-CO-C_4H_8-CONH-C_6H_{12}-NH-)_n$

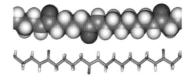

A polymer molecule is formed when a diamine such as hexamethylenediamine is allowed to react with a dicarboxylic acid like adipic acid. The two react to produce an *amide*, a compound characterized by the group of atoms.

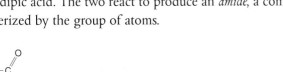

A coloured scanning electron micrograph of Velcro. One half of the fastener is a surface with nylon loops (orange), which are loosely woven strands in an otherwise tight weave; the other is a surface with hooks (red) that are loops woven into the fabric and then cut.

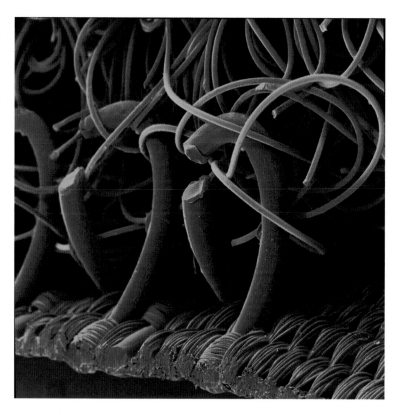

This group is often written –CONH–. As in polyester formation (74), their ability to react is not quenched, because an amino group and a carboxyl group remain at opposite ends of the new molecule; growth can continue there by reaction with other dicarboxylic-acid and diamine molecules. Hence, the chain can grow indefinitely, to produce a very long molecule (only a fragment of which is shown in 79).

One manufacturing problem, that of mixing the diamine and the acid in equal proportions, can be solved elegantly by making use of nitrogen's lesser control over its electrons: its lone pair of electrons can accept a hydrogen nucleus from the acidic carboxyl group. Thus, when adipic acid and hexamethylenediamine are mixed in nearly equal proportions, the double acid and the double amine form a double salt, commonly known as *nylon salt*, with $-CO_2^-$ groups as anions and $-NH_3^+$ groups as cations. Any excess acid or diamine is then removed, which ensures that there is one diamine molecule for every acid molecule. When the salt is heated, the $-CO_2^-$ groups of the acid react with the $-NH_3^+$ groups of a neighbouring diamine, and long chains of polyamide are formed. These chains can be thought of as polyethylene (48) molecules that are interrupted after every few $-CH_2-$ groups by an amide group. As we shall see, the differences between polyethylene and polyamides stem from the presence of the electron-hungry oxygen and (to a lesser extent) nitrogen atoms that are strung along the chain.

The result of polymerizing the six-carbon-atom hexamethylene-diamine with the six-carbon-atom adipic acid is *nylon-6,6*, one of the most important of the polyamides. It is extruded from the reaction chamber and cut into cubes for moulding or else spun from the melt for fibre. The spinning orients the polyamide molecules parallel to each other. In that position the chains can stick together by forming $N–H\cdots O$ hydrogen bonds between the $-NH-$ groups of one chain and the $-CO-$ groups of another. This hydrogen bonding gives the fibres their great strength. It also accounts for the good elastic recovery of nylon, for the hydrogen bonds act like the sulfur bridges in vulcanized rubber (68): they pull the fibres back to their original arrangement when stress is removed. Nylon stockings hug moving legs by virtue of the hydrogen bonds between their molecules; polyethylene stockings would just sag.

Nylon-6,6 is a strong, tough, abrasion-resistant material with moderate resistance to water. It is less water-resistant than a pure hydrocarbon polymer such as polyethylene (48), because water molecules can worm their way in by latching onto the amide groups by means of hydrogen bonding. Although nylon-6,6 is satisfactory as an insulation for wires carrying low-frequency electric current, its electrical properties are not as good as polyethylene's at and above radio frequencies. High-frequency electric fields make the $>C=O$ and $>N–H$ groups (which are absent from polyethylene) waggle; hence the polymer molecules start to vibrate. This motion absorbs energy, and the electric signal is attenuated more rapidly than it would be if polyethylene were used. The presence of the electron-attracting oxygen and nitrogen atoms in the polymer molecules allows a fabric made of nylon to pick up an electric charge when surfaces rub together. This is the origin of the mild electric shock that we sometimes experience after walking across a nylon carpet in a dry atmosphere.

Other polyamides are made with monomers containing different numbers of carbon atoms. As a rule, the longer the carbon chains, the more water-resistant the polymer: a longer chain makes the polyamide more closely resemble a pool of oil, and water is more strongly repelled.

As noted in the introduction to this section, aramids are polyamides obtained by use of monomers derived from benzene, such as terephthalic acid (73) in place of adipic acid. An example is the strong, rigid, and less water-absorbing substance marketed as *Kevlar*. An alternative, made using a version of terephthalic acid in which the –COOH groups are on next-but-one atoms on the ring rather than on diametrically opposite carbon atoms, is sold as *Nomex*. The aramid chains can stack together closely as a result both of the hydrogen bonds between their amide groups and of interactions between benzene rings, which can lie face to face. This gives a highly crystalline and very strong material that is used for tyre cords and bullet-proof vests. One advantage of aramids over non-aromatic polyamides is that they form better fibres. The reason can be traced to the fact that the –CO–NH– bond can adopt one of two arrangements:

The former '*trans*' arrangement results in chains that can be fully stretched out, which is ideal for the formation of fibres; the latter '*cis*' arrangement introduces some kinks into the polymer chain, and they cannot be fully extended. Because the benzene rings of aramids are so bulky, the '*cis*' arrangement is unlikely to be found because that would squash the benzene groups together, so the material is fully extensible.

PROTEINS

Proteins are *polypeptides*, compounds in which the characteristic feature is the repeating unit –CO–NH–C–, with various groups dangling from the second carbon atom. They are nature's version of nylon (79), the principal difference being that there is only one carbon atom between each pair of –CONH– units, rather than the half dozen or so typical of polyamides. Nevertheless, a wide variety of groups can be attached to that carbon, and polypeptides are much more varied than the nylons that have been synthesized so far. Proteins (from the Greek word for 'primary') are the building blocks of living things, and they include constructional substances, such as claws and collagen, and the enzymes, the worker-molecules of the hive of cells in our bodies. The building blocks of the proteins, in turn, are the twenty of so *amino acids* that occur naturally. We examine a few of them in this section.

Humans can synthesize about a dozen of the twenty naturally occurring amino acids, but about eight must be ingested in the diet. The latter are the *essential amino acids*. Because we are so similar to animals, eating their flesh gives a ready supply of all the amino acids we need. Plants have trodden a different evolutionary path from our common ancestor, and we cannot be confident that all the amino acids we need will be present in a single vegetable source. Cereals, for instance, are generally deficient in one amino acid, lysine. Cultures in which meat is either too great a luxury or a moral abhorrence can circumvent such deficiencies by balancing a vegetable thin in a particular amino acid with one rich in it. In some cases there is a further complementarity, for the first vegetable may be rich in an amino acid that is at least partially absent from the second. This is possibly the origin of characteristic ethnic culinary preparations, such as the soybean and rice combination typical of the orient, the bean and corn combinations of Central America, and the macaroni and cheese combination of Italy.

Some amino acids contain sulfur. Indeed, almost all of the 150 grams of the element present in a human body is in the amino acids that are found predominantly in the proteins of the hair, skin, and nails. Brazil nuts (the seeds of the Amazon tree *Bertholletia excelsa*) are a rich source of these proteins, but I am not aware that they have been tested as a cure for baldness or that their oil is used (as a gimmick) in shampoos.

80 GLYCINE $C_2H_5O_2N$

A glycine molecule can be considered as a molecule of acetic acid (31) in which one of the hydrogen atoms of the $-CH_3$ group has been replaced by $-NH_2$, the *amino group*. All the amino groups are built on glycine's pattern, with the amino group attached to the carbon atom immediately adjacent to the carboxyl group. That is, they are all *α-amino acids* (as distinct from β-, γ-, and so on, in which the amino group is progressively further from the carboxyl group).

A glycine molecule thus possesses a carboxyl group (which makes it an acid) next to an amino group. It is this juxtaposition of these two reactive groups that makes α-amino acids, of which glycine is the simplest, so adaptable and important. Glycine units can be strung together in a polyamide chain; the result is *polyglycine*, $(-NH-CH_2-CO-)_n$, the world's most boring protein. However, although it is itself not particularly interesting, polyglycine is the backbone that supports life, as we shall now see.

81 ALANINE $C_3H_7O_2N$

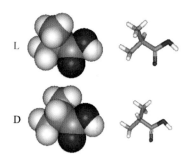

The possibilities for variety in proteins become richer as soon as amino acids in which different groups are attached to the spare carbon atom are obtained. The simplest possibility is to introduce a $-CH_2-$ group between a carbon atom and a hydrogen atom of glycine, obtaining alanine. Alanine is the nitrogen analogue of lactic acid (32), which has an $-OH$ group in place of alanine's $-NH_2$.

If the process of inserting $-CH_2-$ is allowed to continue, various other amino acid molecules can be created, especially if the hydrocarbon tail is allowed to branch. One of them is valine, which has a bulkier and oilier side chain than that of alanine (a point to which we shall return). Only about twenty such substitutions are found in nature, and all the proteins in our bodies are built from permutations of them in a polypeptide string.

An interesting detail of nature is that, although two versions of each amino acid (other than glycine) are possible, only one is found naturally. The two versions of alanine, L-alanine and D-alanine, are shown as (81L) and (81D), respectively. They differ like a left hand (L) and a right hand (D), in that one is the mirror image of the other. The same is true of lactic acid (32); however, unlike lactic acid, all naturally occurring amino acids on Earth, and hence all proteins, are built exclusively from *left-handed* amino acids. Thus life, in a sense, is fundamentally left-handed. The reason for this left-handedness is not known, but some people speculate that it may be connected with a similar asymmetry in the properties of elementary particles.

82 INSULIN

The linking of amino acids into a polypeptide chain occurs when an amino group of one molecule reacts with a carboxyl group of a neighbour, exactly as in the production of a polyamide (79):

where the yellow objects represent attached groups of various kinds. However, with twenty different amino acids available, polypeptides can be formed with almost infinite variety. The order in which amino acids are joined to give a peptide is determined ultimately by the genetic material DNA (203) in the nuclei of cells.

Insulin (shown above: the red cylinders represent helices and the red dots are water molecules) is one of the smaller polypeptides, and consists of two relatively short chains, one of 21 amino acids and the other of 30, linked together by two 'disulfide bridges', which are –S–S– links like those formed in vulcanized rubber (68). The molecules are generated in cells in the 'islands of Langerhans' in the pancreas in response to the concentration of glucose in the blood and the rate at which that concentration changes. Their principal function is to control the uptake, utilization, and storage of glucose, amino acids, and fats. Most glucose is metabolized in the liver, and that is the site where insulin regulates carbohydrate metabolism most extensively. Its function is to retard glycogenolysis (the breakdown of glycogen, see 87) and to promote the synthesis of glucose from sources other than carbohydrate, the net effect being to increase the storage of glycogen.

The blood glucose level is controlled by a variety of mechanisms, some of which involve insulin. After fasting, the glucose level in the blood falls, there is a consequent decrease in release of insulin from the pancreas, and the liver produces more glucose to restore the concentration. After a meal, the abundant supply of glucose stimulates the output of insulin and the excess glucose is stored as glycogen in the liver.

Diabetes mellitus is a disease characterized by hyperglycaemia, a high concentration of glucose in the blood, and arises from insulin deficiency or resistance to its action. In *juvenile-onset diabetes* (more properly *insulin-dependent diabetes mellitus*, or *type-I diabetes*) there is a deficiency in the concentration of insulin-generating cells in the pancreas as a result of the immune system starting to regard them as foreign and setting about destroying them, so the insulin concentration is intrinsically low. The condition can be controlled to some extent by injections of insulin. Oral delivery is not possible because the acid conditions of the stomach destroy the coiled shape of the polypeptide, and with destruction of shape goes destruction of function. In the more common *maturity-onset diabetes* (*insulin-independent diabetes mellitus*, or *type-II diabetes*), which typically occurs in middle-aged, obese people, there is no deficiency of insulin; instead, the disease is related to the availability of insulin receptors. The effect of obesity is seen strikingly in the Pima Indians of Arizona, among whom 40 per cent of the population is diabetic, probably because of a sudden increase in the abundance of food. The treatment includes losing weight, whereupon the number of insulin receptors rises again; insulin is also sometimes required, especially if dieting fails, so as to enhance the effect of the depleted number of receptors. The challenge, when supplying insulin, is to avoid peaks and troughs in blood glucose level.

Insulin is obtained from a number of sources. Bovine (from cows) and porcine (from pigs) pancreas are the classic sources. These insulins are not exactly the same as human insulin: bovine insulin differs in three amino-acid residues and porcine insulin differs in one. Recombinant DNA techniques have now succeeded in producing human insulin from bacteria.

83 HAEMOGLOBIN

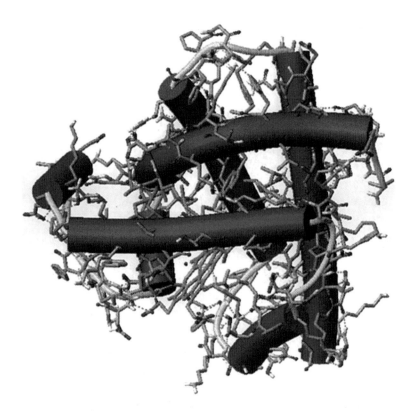

Haemoglobin is the oxygen-transport protein in red blood cells. It consists of four subunits of which one is shown here (the red cylinders represent helices), which are stacked together. Each of the units is very similar to the oxygen-storage protein myoglobin. At the centre of each myoglobin-like unit is a 'prosthetic group', a small, non-peptide molecule bound to the polypeptide chain; the prosthetic group in haemoglobin is called *haem*: it is a flat ring of atoms like the ring in chlorophyll (157) and contains a central iron atom. It is to this iron atom that oxygen attaches or, in the case of carbon monoxide poisoning, a carbon monoxide molecule.

The oxygenated form of haemoglobin is called the *relaxed state* (R state) and the deoxygenated form is called the *tense state* (T state). In the T state the iron atom is linked on one side of the haem group to the nitrogen atom of a histidine amino-acid group; in the R state, that link remains but the O_2 molecule links to the iron atom on the other side of the ring. As it forms a bond, the iron atom shrinks slightly, and instead of lying slightly above the plane of the haem ring, it falls back into the ring. As it does so, the atom pulls the histidine residue with it. As a result, one pair of myoglobin-like groups rotates relative to the other pair. This realignment of two of the units disrupts an interaction between amino acids that helps to stabilize the deoxygenated form, and as a result another of the myoglobin-like groups is more capable of taking up the next O_2 molecule than the originally fully deoxygenated form was. In other words, in haemoglobin there is a *cooperative* uptake of O_2 molecules, and a similar cooperative unloading of O_2 when conditions demand it. The result of this cooperativity is that haemoglobin is very effective at picking up oxygen in the lungs and very good at releasing it at the point of use.

84 LEUCINE $C_6H_{13}O_2N$

85 α-KERATIN

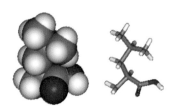

Hair and *wool* consist of α-keratin. This protein is made up of chains of amino acids in which glycine and leucine figure prominently, with about half a dozen others playing an equally important role. Many of the acids have bulky side groups, as does leucine, and some carry sulfur atoms.

The polypeptide chain forms a right-handed helix, called an *α-helix*, with a shape maintained by hydrogen bonds between different amino acids. Three of these right-handed helices wrap around each other in a left-handed coil, where they are held together by more hydrogen bonds and some sulfur bridges. The sulfur bridges reach between amino acids that contain sulfur atoms and resemble the bridges in vulcanized rubber (68). Nine of these coils cluster around two more, giving a *microfibril* of eleven coils, each consisting of

A coloured scanning electron micrograph of fibres of angora wool, the fibre known as mohair (from the Arabic *mukhayyar*, 'goat's hair fabric'). The scales of mohair, which are typical of animal hairs, are fewer in number and lie flatter than in sheep wools and give the fabric a smoother feel. The more numerous scales of wool are responsible for the formation of felt, for they lock the fibres into a dense tangle when they are rubbed together. Angora wool was originally obtained from the White Angora goat, which originated from the district of Angora in Asia Minor, but crosses with other breeds of domestic goat have improved the yield.

three α-helices. Hundreds of these microfibrils are embedded in an amorphous protein matrix to give a *macrofibril*, and these macrofibrils stack together to give a hair cell. A *hair fibre*, in turn, consists of a stack of these cells (see overleaf).

The extensibility of wool and hair is due to the ability of the highly wound structure to unwind, even as far down as to unwind the α-helices, when the hydrogen bonds that support them are broken. The shape is restored when the tension is released because the sulfur bridges survive the stretching (as they do in vulcanized rubber) and snap the polypeptide back into its helical arrangement. In the *permanent waving* of hair, the sulfur bridges are broken and the hair is stretched, after which the bridges reform in a more fashionable arrangement.

Claws, nails, and hooves are also keratin, but they are more highly cross-linked with sulfur bridges (in a sense, more heavily vulcanized, like ebonite, 68) and are more rigid.

The colour of black, brown, and fair hair is due to various concentrations of *melanin* (165). Red hair is coloured by a pigment (trichosiderin) based on iron, like blood and rust. The bleaching of hair is usually an attack on the compounds responsible for its

A strand of hair can be successively taken apart to yield smaller fibres, macrofibrils, microfibrils, protofibrils, and the keratin molecule itself.

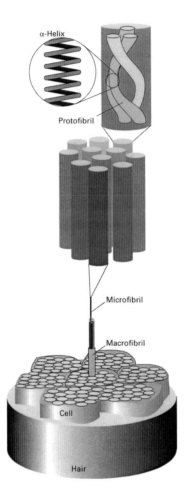

colour; it is almost always accomplished with dilute solutions of hydrogen peroxide (11), which oxidizes the molecules. A side effect of hydrogen peroxide bleaching is that it leads to the formation of more sulfur bridges (by removing hydrogen atoms from some –SH groups, allowing the remaining S atoms to form bonds). This increase in 'vulcanization' makes bleached hair more brittle. For those wishing to reclaim the appearance of youth, grey hair can be blackened with preparations that contain lead acetate. Here, once again, sulfur is the pivot of a cosmetic affectation, for the lead ions combine with the sulfur of the amino acids, in effect forming black lead sulfide. The same blackening was less welcome when it affected white paint, which once contained lead, in industrial areas where sulfur compounds were present in the air (7).

The lustre of hair is its ability to reflect light. Some alkaline hair preparations and shampoos remove hydrogen ions from the keratin molecules, thus altering their distribution of electric charge. As a result, they and the microfibrils coil more tightly and become more reflective, enhancing the lustre. *Hair conditioners* include ionic substances (organic derivatives of nitrogen) that attach to the fibres and modify their electric charge. This increases the electrical repulsion between hairs that happen to approach each other; since they cannot stick together, the hair is given a sense of having 'body'.

86 β-KERATIN

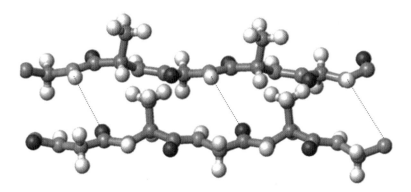

Silk is the solidified fluid excreted by a number of insects and spiders, the most valuable being the exudent of the silkworm, the caterpillar of the silk moth (*Bombyx mori*). *Wild silk* is obtained from the night peacock moth (*Antheraea pernyi*), which does not feed on the mulberry and is not domesticated. Silk, the common name for β-keratin, is a polypeptide made largely from glycine and alanine (and smaller amounts of other amino acids, principally serine and tyrosine). Most of the amino acids do not have the bulky side groups characteristic of those contained in the wool polypeptide. Partly as a result of their smaller side groups, β-keratin molecules do not form a helix; instead they lie on top of each other to give pleated sheets of linked amino acids, with the glycine appearing on only one side of the sheets. The sheets then stack one on top of the other.

This planar structure is felt when you touch the smooth surface of silk. Silk is less extensible than wool because its polypeptide chains are all nearly fully extended (as in pulled nylon fibres). However, it is flexible, because the sheets are only loosely bonded to each

A polarized-light micrograph of processed silk fibres. Silk is a protein extruded by the caterpillar of the Silkworm moth (*Bombyx mori*). In the production of silk (the process known as sericulture) the caterpillar uses the silk to build its cocoon; once that process has been completed, the larva is killed and the single silk fibre, which is several hundred metres in length, is unwound. The thickness of silk yarn is expressed in terms of *denier*, the mass in grams per 9 km of length. Silk is also the substance extruded by spiders: here the web has been spun from the silk extruded from the spider's spinnerets.

other (by hydrogen bonds) and can slide over each other reasonably freely.

SUGAR, STARCH, AND CELLULOSE

Sugars, starch, and cellulose may seem strange bedfellows for nylon and wool. However, they continue the theme of this chapter, in which we have seen complexity being spun by the repetition of a single unit. Like proteins, starch and cellulose are natural polymers; the simple repeated unit is *glucose* or a similar molecule. Edible *starch* and inedible, structural *cellulose* (the most abundant organic chemical on Earth) are examples of *carbohydrates*, substances with formulas that are often multiples of CH_2O, which is (falsely) suggestive of 'hydrates of carbon'. So surely and thoroughly does nature take advantage of the opportunities open to its materials that in *wood* it has already stumbled upon the cellulose analogue of foamed polystyrene.

87 GLUCOSE $C_6H_{12}O_6$

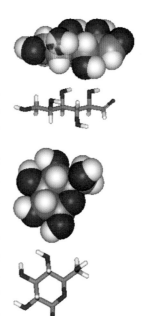

The glucose molecule exists in two main forms. In one, (87 upper), it is a string of six carbon atoms; five of them carry –OH groups and the sixth has a doubly bonded oxygen atom, so that the line ends in a –CHO group. In the other, (87 lower), a similar line of atoms has bent around and the –CHO group has reacted with an –OH group near the far end of the molecule to form a closed six-membered ring, with one of the atoms in the ring being an oxygen atom. A solution of glucose should be pictured as consisting of a writhing, continuously interchanging collection of open molecules, six-membered rings, and five-membered rings like the one shown for fructose (88).

Some of the properties of glucose should be obvious from the structure of the molecule. In particular, the oxygen atoms bring it properties that are quite different from those of its parent hydrocarbon *hexane*, in which each of the six carbon atoms carries only hydrogen atoms (as in octane, 20). These numerous oxygen atoms result in glucose being highly soluble in water, since they can form strong hydrogen bonds to water molecules. Note particularly that, in the six-membered ring (87 lower), all the –OH groups are arranged along the perimeter like teeth on a gear wheel; this high degree of exposure allows water molecules to form numerous strong hydrogen bonds with them. Consequently, the molecule readily slips into solution.

Glucose resembles hexane in being a fuel, because its carbon atoms are ripe for conversion into carbon dioxide (3) by oxygen; but here we see a compromise: hexane carries no oxygen atoms, and many new and strong carbon–oxygen bonds are formed when it burns. Glucose is already partly oxidized, and fewer new carbon–oxygen bonds are formed. However, hexane is not soluble in water, and so cannot be transported to cells by blood. Glucose, on the other hand, can be regarded as a strip of water-soluble carbon atoms, less efficient perhaps than hexane but much more readily transportable through the body.

Glucose is also known as *dextrose* (because solutions of glucose rotate the plane of polarized light to the right). It occurs in ripe fruits, the nectar of flowers, leaves, saps, and blood and is variously called *starch sugar*, *blood sugar*, *grape sugar*, and *corn sugar*. It is the primary fuel for biological cells, and more complex sugars and starches are broken down into individual glucose units when they are digested. The glucose ingested when we eat and not used immediately is stored temporarily in muscle and liver as glycogen, a polymer of glucose similar to amylopectin (92). Typically, an adult stores about 350 grams of glycogen, which can fuel us for about a day. If we don't draw on this fuel, the body converts it into fats, which can be stored more economically, if less aesthetically, until we need them. Once glucose molecules are in solution, their energy is immediately available for any demands that a metabolic process may make. The hydrocarbons stored in adipose tissue (in the obese) are available only much more slowly, since they must first be made soluble, then transported, and only then can be used.

Glucose is so soluble in water that it grips water molecules to itself and forms a *syrup* when the concentration of glucose is high. *Corn syrup* is partly degraded starch (92) consisting of individual glucose molecules as well as short chains of glucose molecules still joined together. It is formed by the action of enzymes in the bacterium *Aspergillus oryzae*, which breaks down the starch molecules; its viscosity is due to the entangling of the chains and their attraction to the surrounding water molecules. (*A. oryzae* is also used to prepare rice starches for fermentation into ethanol in the production of *sake*.) The syrup-forming ability of glucose is used in the manufacture of hard candy – a flavoured and coloured solution held together as a glassy solid by water-attracting glucose (and sucrose) molecules that do not allow the little water that remains to drip away.

88 FRUCTOSE $C_6H_{12}O_6$

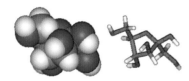

Fructose is another sugar that, like glucose, consists of a single chain of six carbon atoms. In its open-chain form it differs from glucose in the location of the doubly bonded oxygen atom, which is attached to the next-to-last carbon atom of its chain. Like glucose, it also forms six- and five-membered rings (the latter is shown in the illustration). Fructose is also known as *laevulose* and *fruit sugar*. The former name comes from its effect on polarized light: it behaves oppositely to glucose (dextrose), rotating the plane of polarization to the left. The name 'fruit sugar' comes from its widespread occurrence in fruits and vegetables. Fructose is the major sugar in many forms of *honey*, for nectar (the fluid exuded by plants, usually flowers, perhaps to control the osmotic pressure of their fluids and also to attract insects) contains a high proportion of the sugar.

Two features of fructose account for its properties and uses: fructose is about 50 per cent sweeter than sucrose (89) and more soluble than both sucrose and glucose. The first property makes it useful in low-calorie diets, for with it the same amount of sweetening can be achieved with a smaller mass of carbohydrate. It may also, depending on its price, be more economical to use than sucrose. That fructose is more soluble than sucrose may account for the conversion of any sucrose in nectar into fructose, for then higher concentrations of carbohydrate may be achieved in a given mass of water. This greater solubility accounts for the softness of *brown sugar* compared with pure, white sucrose; brown sugar consists of sucrose crystals coated with the glucose and fructose that remain in molasses after most sucrose has been removed by repeated crystallization. It also provides a neat way of making soft-centred chocolates: semisolid sucrose filling can be injected into the hollow centre with enzymes (obtained from yeast) that convert sucrose to glucose and fructose; as the latter is formed, it dissolves in the little water remaining, and gives a soft, creamy texture to the filling. Since fructose retains water better than sucrose, it is used in jams and candies to reduce the chance of crystallization.

Commercial fructose is prepared from the glucose in corn syrup by enzymatic action (by *Streptomyces*), which rearranges the atoms of the glucose molecule into the marginally different fructose molecule. As well as providing a sweetener that can be used in smaller concentrations, this process has the additional economic advantage of using a readily available raw material.

Fructose is also the sugar that powers the motion of sperm. Men synthesize it in their seminal vesicles, and it is incorporated into semen and used by the sperm as fuel for their brief but portentous journey.

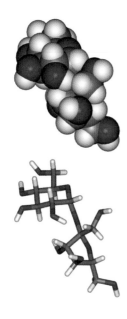

89 SUCROSE C$_{12}$H$_{22}$O$_{11}$

Sucrose is an example of a *disaccharide*, a molecule consisting of two glucose-like units (which are themselves *monosaccharides*) linked together. In sucrose, one of the units is a six-membered glucose ring (87b), and the other is a five-membered fructose ring (88). The two parts are linked by a chemical bond (so that sucrose is not a mixture, but a definite compound) at the points where glucose and fructose themselves would break open to give their open-chain forms. The sucrose molecule is therefore pinned together into a double-ring form, and it does not pop open in solution.

Sucrose is our common table sugar and one of the purest of every-day compounds. It occurs in most plant materials but is particularly abundant in sugar cane (hybrids of *Saccharum officinarum*) and sugar beet (*Beta vulgaris*), from which it is extracted by slicing, leaching, and refining. Beets provide the additional advantage to the farmer

A stem of suger cane (*Saccharum officinarum*), the principal source of sucrose.

of having such deep roots that they mine nourishment from well below the surface, causing a circulation of nutrients. *Maple syrup*, the concentrated sap of the maple tree (*Acer*, especially the sugar maple, *A. saccharum*), is a solution of sugars, about 65 per cent being sucrose, with small amounts of glucose and fructose. The sap is gathered from holes bored in the bark of the tree in early spring, being pushed out when carbon dioxide, produced by metabolic activity and dissolved in water in the trunk, comes out of solution as the tree grows warmer in spring sunlight. (Most gases are less soluble in warm water than in cold, for the water molecules effectively shake the gas molecules out of solution as they begin to move more rapidly.) The brown colour of maple syrup is the result of a reaction between the various sugars and the amino acids that are also present.

When the enzyme *invertase* is added to a sucrose solution, the two parts of the sugar molecule are snipped apart, giving a solution of equal parts of glucose and fructose known as *invert sugar*. (The name reflects the effect of the sugar solution on polarized light: sucrose rotates the plane of polarization of light in one direction and invert sugar in the other.) This mixture is sweeter than the parent sucrose because of the fructose.

When sugar is heated, a complex series of decompositions occurs. Each molecule already has many oxygen atoms (so that in a sense it is already partly burned), and their rearrangement is encouraged by the heat. The molecules break up, and the smaller fragments (including acrolein, 113) either vaporize immediately or dissolve and remain trapped in the complex solid residue called *caramel*, adding to its flavour.

Several other common sugars are disaccharides. Maltose, the sugar present in malted barley, is a combination of two glucose units; and lactose, which occurs in milk, is a combination of glucose and a similar sugar, galactose.

90 RAFFINOSE $C_{18}H_{32}O_{16}$

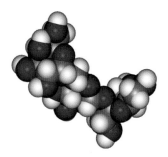

Raffinose is an example of an *oligosaccharide*, a sugar consisting of several glucose-like rings. The raffinose molecule is made up of a fructose ring (on the right in the illustration), a glucose ring, and the six-membered ring of a sugar called *galactose* (which also occurs in milk).

When sucrose is ingested, it is broken down into its component sugars by enzymes in the body. However, raffinose, which is found in

peas and beans, is too much for our enzymes, and it passes undigested into the large intestine. There it is pounced on by hungry intestinal flora (including the bacterium *Escherichia coli*). They break it down and, in the process, release large amounts of gas – typically hydrogen, carbon dioxide (3), and methane (15) – in proportions that depend upon the person. Thus, meals of beans can result in considerable social inconvenience.

A similar failure to digest even disaccharides occurs in many individuals who, for genetic reasons, lack the enzyme *lactase*, which breaks down the disaccharide *lactose* that occurs in abundance in milk. In fact, production of this enzyme in adults appears to be the exception rather than the rule and is confined primarily to descendants of northern Europeans. It may have arisen in connection with calcium uptake and the whitening of the skin, which led to an increase in production of vitamin D in the sunlight-poor north. Vitamin D, or cholecalciferol, is a precursor of a substance that transports calcium and phosphate ions through cell membranes and helps to build bones: a deficiency leads to *rickets*. 'Abnormal people', that is, people of northern-European stock, can comfortably consume milk even after they have been weaned; the 'normal people' of Africa and the orient simply feed their colonic bacteria and suffer the pains of indigestion, which are often the pains of gas pressure in the colon. Hardly

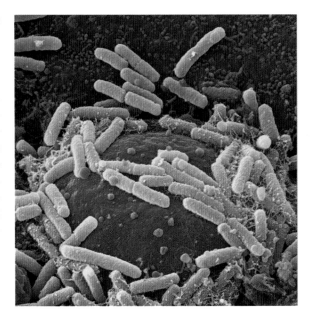

Our intestines swarm with *Escherichia coli*, some of which (from human bladder cells) are shown here as a coloured scanning electron micrograph. They break down foods that we cannot digest and synthesize amino acids and vitamins, particularly vitamin K, which is necessary for blood clotting, If they invade the blood, however, they can induce disease, urinary tract infections, and septicaemia. Some strains have arisen that cause gastroenteritis in young children.

anyone has difficulty digesting cheese and yogurt because bacterial action eliminates much of their lactose during fermentation.

91 AMYLOSE

92 AMYLOPECTIN

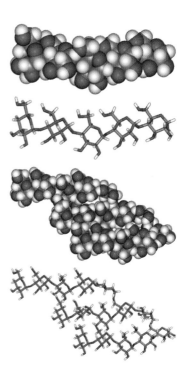

Now we shall see the magic that nature conjures from the subtle deployment of meagre resources – how, in a profound example of elegant economy, major components of organisms can be spun from the almost endless repetition of glucose (87) rings, and how a deft, genetically controlled twist of a bond can transform food into structure. We shall see how a single molecular mutation (and hence its endless repetition, for that is the nature of life) can represent an opportunity.

Starch is an example of a *polysaccharide*, a molecule consisting of many glucose-like units linked together into a polymer. It is a digestible polymer that is consumed in huge amounts worldwide in cereal grains and potatoes. It is present in such abundance in seeds (accounting for 75 per cent of the weight of wheat flour) because the developing plant embryo requires a compact supply of energy. The first step in the digestion of starch is to cut the polymer chains into individual glucose molecules, which are then oxidized in cells and used to power growth, action, and thought.

Starch consists of two varieties of glucose polymer, *amylose* and *amylopectin*; the latter is the major component in most plants, comprising about three-quarters of the total starch in wheat flour. Amylose consists of long chains of glucose monomers, with very little branching. Amylopectin, on the other hand, consists of amylose-like chains with, occasionally, a different linkage to a glucose unit that results in a branch. This gives rise to a tree-like structure, as shown in the illustration. Amylopectin molecules normally contain many more glucose units than do amylose molecules. Both are nature's solution to the need for glucose for energy and the problem that glucose itself is so soluble. The solution is to link numerous glucose units together chemically, so that they are anchored to each other, yet in such a way that they can be snipped off by enzymes when a metabolic requirement arises.

Hydrogen bonding is part of the reason why starches are used in cooking. In natural starch, the molecules are bound tightly to each other by hydrogen bonds, and they form compact solids that are encased in a membrane that is almost impenetrable to water. However, when starch is heated in water, the water penetrates the granules and the hydrogen bonds between the molecules break down at about 65 °C; then water molecules flood into the solid and adhere all over the starch molecules by forming their own hydrogen bonds with the innumerable –OH groups. The starch molecules suddenly swell as the water molecules penetrate into them. In addition, their entanglement with each other and the adhesion of the water molecules together result in a sudden surge of viscosity. That is, the starch molecules thicken watery solutions, which can include gravy and sauces. However, on standing, the amylose chains exude and liberate water in a process called *retrogradation*; as they do so, they partially crystallize by sticking to each other again.

Cereal starches are essential to bread-making. When dough is kneaded, the starch granules of the flour are broken down, and the enzymes that are present in the granules cut some of the polymers into sugar molecules. These are fermented by the added yeast (which is usually brewer's yeast, *Saccharomyces cerevisiae*), which converts the sugar molecules into alcohols and carbon dioxide, the former flavouring and the latter leavening the dough. The staling of bread is a form of retrogradation. Bread crumbs become hardened and bread becomes stale as the amylose chains and the linear amylose-like branches of amylopectin trees align with each other and crystallize. Some bread ends up as alcohol: the Russian drink *kvass* is a low-alcohol beverage made by the incomplete fermentation of stale bread and cereals.

Animals do not store their glucose as starch, but some is stored in muscle and in the liver as *glycogen*, which is closely related to amylopectin. The anaerobic metabolism of these reserves comes into play when the oxygen supply is reduced, as in sport, or terminated, as in death; in each case the metabolic product is lactic acid (32). In athletes, the resulting increase in acidity leads to cramp. In death, it leads to termination of the enzymatic action that metabolizes the glycogen, consequent cessation of the transport of calcium ions, and hence the more permanent cramp of *rigor mortis* – the locking of muscles in their current state of extension or contraction. Glycogen levels are depleted less if animals are tranquil before they are slaughtered. The levels of lactic acid are then higher and the meat is protected during storage by the mild preservative action of the acid.

Starches are used to encourage the biodegradability of polymers. One of the most convenient forms of rubbish disposal is the landfill. Whereas organic products decompose, modern synthetic materials are largely resistant to degradation. That, of course, is an advantage while they are still in use, but not when they are discarded. A biodegradable polymer is one that is susceptible to bacterial or enzymatic action. In short, microorganisms must be induced to eat the material once it has fulfilled its function and produce metabolites that are not toxic to the environment. One procedure is to incorporate digestible, sugar-like components into the polymer chains or to blend together starch and synthetic products. For instance, the product sold as *Mater-Bi* is a blend of starch and synthetics, and has been used to make composting bags, pens, and cutlery. It is a mixture of 'destructured' starch, obtained by extracting the amylose and amylopectin polymers from starch granules, and poly(vinyl alcohol), which is like PVC (52) but has a –OH group where PVC has a chlorine atom. This synthetic material, and others that are used, are also biodegradable on account of the –OH groups, which make them digestible, but suffer from being soluble in water (for the same reason). Other approaches include grafting these degradable synthetics onto the starch molecules themselves, to give modified starches.

93 CELLULOSE

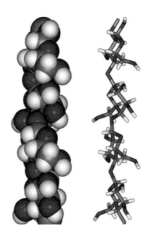

Although the monomers of cellulose are the same as those of starch, neighbouring glucose units are linked differently. This results in long, flat, ribbon-like chains that are supported by hydrogen bonds between neighbouring units. These flat ribbons pack together, and the hydrogen bonds between them stabilize the structure into a solid, rigid mass. The difference between starch the fuel and cellulose the scaffolding – a simple twist of a link – shows unconscious nature at its most brilliant.

Cellulose, like starch, is a polysaccharide, but it is not digestible by humans. Ruminants need several stomachs to digest grass, and even then they must draw on the cooperation of specialized fungi and other microorganisms in their gut, which cut up the chains and convert the glucose units into butanoic acid (39) and various other compounds. Although rabbits have only one stomach, they have developed the anatomically economical, but to us socially unacceptable,

The tunicates are one of the few animals to produce cellulose – their outer wall (the tunic that gives them their name) is made up of *tunicin*, a form of cellulose. This is a colonial (as distinct from solitary) sea squirt (*Clavellina lepadiformis*) found in Scotland.

solution of multipass digestion, in which they eat some of their own excrement.

Natural cellulose is always found embedded in a matrix of *lignocellulose*, an amorphous mixture of *hemicellulose* (a polysaccharide formed from many different sugar molecules, with extensive branching into side chains) and *lignin* (a cross-linked polymer of aromatic molecules). In paper-making and the production of cellulose fibres, the lignocellulose must be removed. Nature has done this almost completely in cotton, which has only a thin covering of lignocellulose. Wood, however, is heavily embedded, and wood pulp must be treated thoroughly. The lignocellulose is also responsible for the dark colour of paper pulp, which must therefore be bleached before it can be used. The cheapest method, which is used for the pulp destined for newsprint, involves the reduction (the opposite of oxidation) of the compounds responsible for the colour. That is why newsprint turns yellow soon after it is exposed to air and light: atmospheric oxygen reoxidizes the material, thereby undoing the bleaching accomplished by the reduction.

Purified wood cellulose can be converted into fibres if the majority of the hydrogen bonds between the chains can be eliminated. One approach is to convert many of the –OH groups into acetate groups ($-O-CO-CH_3$) by esterification with acetic acid (31). This removes the crucial hydrogen atoms, eliminates hydrogen bonding, and hence insulates each chain from its neighbours. The product is *cellulose acetate*, one of the cellulose derivatives used for textiles and photographic film. Celluloid, which is still used for table-tennis balls, is *cellulose nitrate*. Celluloid is made by the reaction of cotton with a mixture of nitric and sulfuric acids, which results in two $-NO_2$ groups becoming attached to each glucose ring. If the nitration continues, to attach three groups to each ring, then the product is the explosive *gun cotton*.

Rayon (also known as *viscose*) is cellulose that has been dissolved from organic sources (typically wood pulp) and then spun into yarn. The solvent has to be rather aggressive, and the pulp is heated with sodium hydroxide (caustic soda) and the foul-smelling liquid carbon disulfide (to be fair to carbon disulfide, its stink may be due to impurities). When the solution is passed through fine nozzles, the product is rayon thread; when the solution is extruded into sheets it is *cellophane*. When you press on a cellophane sheet or try to tear it, you are experiencing yet again the strength of hydrogen bonds. The modern version of the production of rayon involves dissolving the cellulose in a complicated organic solvent (*N*-methylmorpholine oxide) and then extruding the solution into water. The product is marketed as *Tencel*, the brand-name version of the generic name *lyocell*. The fabric has excellent draping qualities as a result of the spaces created between the yarns during processing, which allows movement within the fabric. The surface of the fibres can also be roughened into microfibrils, to give a fabric with a very soft touch and a texture like the skin of a peach.

94 CHITIN

Nature's economical and elegant deftness with limited resources is shown in *chitin*, the structural material of the flexible inner parts of the exoskeletons of arthropods, including scorpions and crabs. Chitin also occurs in the beautiful, colourful, thin sheets we see as insect wings. It is identical to cellulose, except for the replacement of one $-OH$ group on each glucose unit by an $-NH(CO)CH_3$ group. The nitrogen and sugar of chitin are valuable commodities, and some newly emerged insects postpone the taxing process of hunting by eating their discarded exoskeletons.

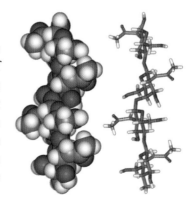

The cell walls of fungi are also of chitin, rather than cellulose. The adaptation here may be that chitin is less liable to undergo microbial

Insect wings and the flexible inner parts of their skeletons are made of chitin. These are the wings of a cicada.

degradation than is cellulose. The advantage would stem from fungi having very high surface-to-volume ratios and hence being in almost total contact with their environment. This contact makes up for their immobility, for the fungi can scavenge for food by spreading through their surroundings. Some can do this enormously quickly, producing more than a kilometre of new mycelium (the collection of the individual filaments, or *hypae*, that make up the fungus) in a day. The rigidity of chitin means that a fungus cannot engulf its prey but must attack it with enzymes and absorb its components. These enzymes account for the decay wrought by fungi, but sometimes that decay is advantageous, for it includes fermentation by yeast and the production of antibiotics.

4

Taste, smell, and pain

This is the first of two chapters that deal with the chemistry of sensation. Here we see how molecules can act as messengers from the external world to the internal universe of consciousness within our heads. All sensation is ultimately chemical, for all neuronal activity in our brains depends on the transport of molecules and ions from one location to another and the reactions in which they participate. However, some messages from the outside world involve the direct perception of molecules that act as messengers. This direct perception of our surroundings is involved in taste, for which the sensors are in the tongue, smell, for which the sensors inhabit the nose, and certain varieties of pain, for which the receptors pervade the skin and organs within the body. Flavour is an enormously complex phenomenon, since the molecules responsible for it are released sequentially. Moreover, olfaction plays a central role, either as we sniff the food as it goes into our mouths or because, once food is inside the mouth, its volatile components spread up into the nose from behind. The release of these odours may also be sequential. Thus, when we eat a strawberry, breath analysis shows that volatile esters are released with the first breath; the second breath is rich in ethers; and the third breath contains an aldehyde.

In this chapter we see that odour is so closely related anatomically to emotion that molecules can charm us into a recollection or a mood. We shall also see many of the molecules that act as stimulators of taste, smell, and pain. These are the molecules to have in mind when we are sensing a flavour or a perfume, the molecules that, once we know them, may enhance the delight we find in a dinner or a person. These are the molecules of pleasure, warning, corruption, and communication.

Taste and smell are examples of *chemoreception.* Taste in mammals is confined to the damp region inside the mouth, but some insects taste through their feet, and the bodies of fish are covered with chemoreceptors. Within the mouths of humans, chemoreceptors are largely confined to the mobile slab of muscular tissue called the tongue. Adult tongues are about 10 centimetres long and carry about 9000 *taste buds*, which are groups of 50–100 adapted epithelial cells that are innervated by a smaller number of nerve endings. In this regard, taste is distinct from olfaction, for which the sensors are the actual nerve endings themselves. In the adult, the taste buds are largely confined to the perimeter of the tongue, and their number declines with age, particularly after age 45. The tongues of children are covered by taste buds. As shown in the illustration, different regions of the tongue respond to what some regard as the four basic tastes: sweet, salt, sour, and bitter (but see later for a fifth elementary taste).

SWEETNESS

Sweetness, the taste considered in this section, is detected at the front of the tongue. Cats have very few 'sweet receptors' and are among the few animals that do not prefer a sweet taste. Preference for sweetness and dislike of bitterness may be an evolutionary adaptation, since many ripe fruits are sweet (with ripeness, there is a decline of acidity, and the sweetness of sugars becomes more apparent) whereas many vegetable poisons are bitter.

Molecules that evoke taste are called *sapid* (from the Latin word *sapere* meaning 'to taste'). One criterion of sapidity is solubility, because a substance must dissolve in water before it can penetrate into the taste buds. Particular tastes are evoked by molecules with groups of atoms in characteristic arrangements called *saporous units.* The

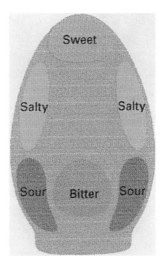

saporous unit responsible for sweetness is called a *glucophore*; the structure of the glucophore presumably matches the structure of a protein in a taste receptor in a taste bud near the front of the tongue. When the molecule binds, perhaps by forming hydrogen bonds with the protein, a signal is sent to the brain.

Several models have been proposed for glucophores. The problem is to identify a group of atoms in a particular geometrical arrangement that, if it is present in a molecule and the rest of the molecule is not too bulky to allow it to approach the receptor protein molecule closely, can bind to the protein molecule and result in the sensation of sweetness. The glucophore is thus a kind of molecular key and the receptor protein a molecular lock. One model is shown in the illustration overleaf. The red atoms labelled A and B must be electron-attracting atoms (usually oxygen, but sometimes nitrogen) that can participate in hydrogen bonding. Sweetness may also depend on the presence of a hydrocarbon group near the A and B atoms.

There are many odd features of taste and sweetness in particular. For instance a substance of unknown structure in a fruit known as agbayun (*Synsepalum dulcificum*) modifies the sweet receptor mechanism so that it will respond to hydrogen ions, which are normally the cause of sourness. Eating the fruit causes sour substances to taste sweet for about an hour.

Fructose (88) is the sweetest sugar (when it is in six-membered-ring form). Sucrose is generally experienced as being about one and a half times as sweet as glucose, which may be because it contains two glucophores in the right arrangement for them to fit two sites

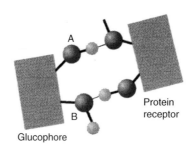

on neighbouring proteins. However, starch (91, 92) is not sweet, even though it is rich in glucophores, probably because its individual glucose units are held back from fitting into the receptor sites by the rest of the chain. Some small molecules, including glycols such as ethylene glycol (72) and glycerol (33), are sweet. The α-amino acids, such as glycine (80), are sweet, but amino acids with the $-NH_2$ group further removed from the carboxyl group are not: the neurotransmitter GABA (27) is tasteless. The tastes of mirror-image α-amino acids differ; D-amino acids (the ones that do not occur naturally) are all sweet, but the corresponding L-amino acids may be sweet, bitter, or tasteless. This emphasizes how important the shape of the molecule is in determining its ability to attach to the receptor protein molecule: the mirror image of a molecular key might not fit the molecular lock. Some solutions of metal ions are sweet; for example, beryllium salts evoke a sweet taste, as do some lead salts (including lead acetate). Even solutions of common salt (sodium chloride) taste slightly sweet if they are extremely dilute.

95 SACCHARIN $C_7H_5O_3NS$

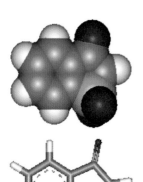

Saccharin was discovered in 1879 by a dirty, careless chemist who failed to wash his hands after a session in the laboratory. With its discovery came an opportunity for the food industry to satisfy the lust for sweetness without the penalty of obesity, for saccharin is not metabolized by the body but instead is excreted unchanged. Saccharin went into commercial production in 1900 and is available as, for instance, Sweet 'N Low (the 'Low' referring to its low calorific value). It is approximately 300 times sweeter than sugar but has a bitter, metallic aftertaste that is difficult to mask. Its sweetness is lost if the hydrogen atom attached to the nitrogen atom is replaced by a $-CH_3$ group, which indicates the involvement of the hydrogen atom in the saporous unit. Since saccharin itself is not very soluble, it is normally used in the form of the sodium or calcium salt, either of which is more soluble than saccharin. Incidentally, saccharin does not fool bees or butterflies into treating it as sugar.

Saccharin is under suspicion as a carcinogen, since it has caused cancer in rat bladders. However, it is not metabolized by rats (or humans), whereas most carcinogens are. It appears that the sodium form of saccharin combines with rat urine, creates stones in the animal's bladder, and induces cancer. Human urine, it is claimed, is so different from rat urine that it does not react in the same way. Moreover, because rats concentrate their urine more than humans do, it remains in their bladders for an unusually long time. The matter remains highly controversial.

96 CYCLAMATE $C_6H_{12}O_3NS$

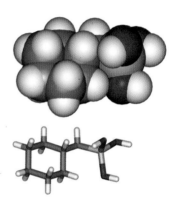

Cyclamate is used (in countries that still allow it) as the sodium salt (or as the calcium salt in low-sodium diets) and, as such, is about 30 times sweeter than sucrose. It was discovered in 1937 – like saccharin, by accident – in this case, by a careless chemist smoking a cigarette that had picked up some of the compound. Cyclamates were banned in the USA in 1969, perhaps unnecessarily, when it was found that massive doses of a mixture of drugs that included cyclamates led to the formation of bladder tumours in rats. This was traced to the formation of the known carcinogen cyclohexylamine as a result of the action of intestinal flora. Cyclohexylamine, a cyclohexane molecule composed of a ring of six $-CH_2-$ groups with one hydrogen atom replaced by an $-NH_2$ group, is formed by loss of the $-SO_3^-$ group from the cyclamate ion. Result of subsequent studies have suggested that the fear that it is a carcinogen is unfounded, but there remains controversy about its possible role as a co-carcinogen (a compound that enhances the carcinogenicity of other compounds).

One advantage of cyclamate over aspartame (99) is that the former is heat stable, so it can be used in foods that require heat treatment. Cyclamate and saccharin are commonly employed jointly in the ratio of 10 : 1, which is found to give a better aftertaste and a synergistic sweetening effect.

The cyclohexane ring provides the hydrocarbon group of the glucophore, and the sweetness is lost if the $-NH$ hydrogen atom is replaced or the cyclohexane ring is modified. The water-repelling character of the cyclohexane ring is overcome by the ability of the $-OSO_2$ and $-NH-$ groups to form hydrogen bonds with water molecules, so the cyclamates are soluble in water.

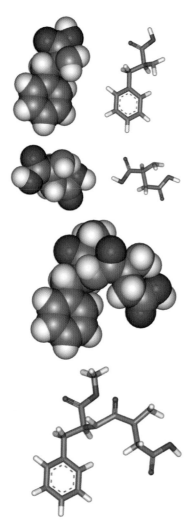

97 PHENYLALANINE $C_9H_{11}O_2N$

98 ASPARTIC ACID $C_4H_7O_4N$

99 ASPARTAME $C_{14}H_{18}O_5N_2$

The aspartame molecule, a dipeptide sold as Nutrasweet and Equal, is a combination of two naturally occurring amino acids, aspartic acid and phenylalanine, with a slight modification of the remaining acid group of aspartic acid, to turn it into the methyl ester in order to avoid the bitter contribution of the acid itself. It may be considered a tiny protein, because proteins are polypeptides (Chapter 3). Aspartic acid is almost tasteless and phenylalanine is bitter, but their esterified dipeptide is quite different. It tastes 100–200 times sweeter than sucrose and lacks the unpleasant aftertaste of saccharin. However, as is typical of proteins, it is sensitive to heat and cannot be used in cooked foods. Aspartame also decomposes slowly in liquids, so soft drinks sweetened with it have a limited shelf life. A mixture of saccharin and aspartame is sweeter and more stable than either substance on its own.

Aspartame is a white crystalline solid that was also discovered (in 1965) by accident – once again confirming that carelessness can be profitable as well as dangerous. In this case the careless chemist, who happened to be working on a cure for gastric ulcers and was synthesizing a four-amino-acid compound normally found in the stomach using aspartame as an intermediate, licked his dirty fingers and tasted sweetness. Because aspartame is a kind of protein, it is metabolized in the body like the other proteins we ingest and is a source of amino acids, but people who suffer from phenylketonuria lack the enzyme that converts phenylalanine into tyrosine should be cautious. Because aspartame is much sweeter than sucrose, less has to be incorporated into food and it is therefore less fattening than sucrose. Some people prefer its taste to that of sucrose.

SOURNESS AND BITTERNESS

Now we consider the sides and rear of the tongue, the sites where we taste sourness and bitterness. Sourness is due to the presence of free hydrogen ions (H^+), which are released by acids such as the acetic acid (31) of vinegar, the phosphoric acid added to some cola drinks to enhance their zest, and the carbonic acid (3) of soda water.

It is speculated that the taste buds on the sides of the tongue contain protein molecules rich in carboxylate groups ($-CO_2^-$), which are carboxyl groups that have lost a hydrogen ion; in an acid medium, they are converted back into carboxyl groups and consequently cause a change in the shape of the protein molecules, which triggers impulses to the brain. It is quite easy for manufacturers to achieve the sensation of sourness simply by adding small concentrations of acid to their products.

Bitterness is often associated with the presence of organic compounds of nitrogen known as *alkaloids* whch are widely present in the *angiosperms* (the flowering plants). Many of these alkaloids (which include strychnine, nicotine, 184, and caffeine, 179) are poisonous, and the ability to detect them by taste may have arisen as an adaptation for survival. There is even speculation (it is no more than that) that the very limited ability of reptiles to detect bitterness may have contributed to the demise of the dinosaurs, which occurred at about the same time as the emergence of the angiosperms. That the detection of bitterness is an avoidance signal is supported by the observation that in only very few cases – quinine (103) and caffeine (179) among them – is bitterness sought for pleasure, and then only after training. The inclusion of some bitter principles in aperitifs may be a distant, highly domesticated echo of our ancestors' coping with survival, for they stimulate the secretion of saliva. For us as modern sophisticates that is the prelude to a meal. For those who once needed to survive environments harsher even than cocktail parties, increasing salivation may have been a last line of defence against poisons.

Perhaps because the food industry, always true to the echoes of our past, is more often concerned with the achievement of sweetness than bitterness, less is known about the receptors for bitterness than about those for sweetness. Some patterns have been discerned, including solubility (so as to pass through the saliva into the taste buds) and the presence of several $-NO_2$ groups in a molecule. One group of particular interest relates to the glucophore mentioned in the previous section. There, we saw that, for sweetness, the distance between the AH and B groups should be about 3×10^{-8} centimetres if the molecular key is to fit the sweetness lock. In some bitter compounds, the saporous unit is very similar to that of a glucophore but the AH-to-B distance is half that in the glucophore. If this is true, then we have an example of how a tiny change in a molecule can be the difference between pleasure and displeasure.

Some substances have keys that match both sweetness and bitterness receptors and can therefore evoke both tastes. One such molecule is found in the woody nightshade (*Solanum dulcamara*), which is also appropriately known as 'bittersweet'. Others include ionic compounds and acids, where the cation may stimulate one response and the anion another. This is the case with salicylic acid (169): the hydrogen ion it releases in solution stimulates sourness receptors, but the accompanying salicylate anion stimulates a sense of sweetness that swamps the sourness; as a result, salicylic acid is sweet.

100 OXALIC ACID $C_2H_2O_4$

101 MALIC ACID $C_4H_6O_5$

102 CITRIC ACID $C_6H_8O_7$

Oxalic acid occurs in appreciable concentrations in many leafy green plants, including rhubarb (which gets its name from the Greeks, who called it the vegetable of the barbarians beyond the river Rha, the Volga) and spinach. The toxicity of rhubarb leaves was once ascribed to the presence of oxalic acid; however, spinach is rich in salts of the acid but is not toxic. It seems that the compound responsible for rhubarb-leaf toxicity has not yet been identified.

Apples (the fruit of trees of the genus *Malus*) are rich in *malic acid*, which is closely related to oxalic acid but has a –CH(OH) group separating the two carboxylic-acid groups; when you taste their sharpness, you should have this molecule in mind. The molecule comes in left- and right-handed forms, one the mirror image of the other; the L-isomer, as shown in the illustration, is the naturally occurring form.

Citrus fruits are particularly rich in citric acid, lemon being the most concentrated, then grapefruit, and finally oranges. Citric acid is added to lemonade, and the discerning drinker can taste both the hydrogen ions it provides, which account for its sharpness, and the anions, which stimulate the sweetness receptors. Citric acid is widely used to acidify foodstuffs and is found as the acid agent in effervescent powders and tablets: its ability to function as an acid is switched on when it dissolves in water and can donate one of its –COOH hydrogen ions. Citric acid is known as *sour salt* in cookery. It is also used to clean metal surfaces (where, as an acid, it reacts with oxides) and, as an iron salt, to make blueprint paper.

103 QUININE $C_{20}H_{24}O_2N_2$

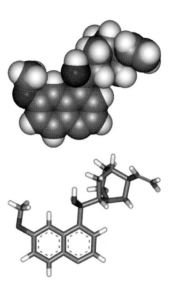

Quinine, a white, crystalline solid, is extracted from the bark of the *cinchona* trees of South America (*quina* is the Spanish rendering of the Inca word for 'bark'). In countries unaffected by malaria (literally 'bad air'), its bitter taste is most familiar in the form of the *tonic water* used in, among other things, gin and tonic drinks (and in which its taste is usually enhanced by a little citric acid, 102). It is also a contributor to the taste of Dubonnet.

Quinine's more serious application, the abatement of malaria, depends on the ability of the quinine molecule to bind to DNA of the *Plasmodium* parasite transmitted by the female malaria mosquito *Anopheles balabacensis*, and hence to inhibit its replication. It affects only infected cells because they absorb the molecule in higher concentrations than do unaffected cells. Quinine has been replaced to a large extent by synthetic drugs (for the treatment of malaria, not in G&Ts) such as mepacrine proguanil (Paludrine), chloroquine, and mefloquine, but the resistance of mosquitos to these synthetics is increasing and quinine itself is being more widely used again. Quinine also has a mildly analgesic action. In doses stronger than that of a gin and tonic, it causes contractions of the uterine muscles and induces abortion.

104 HUMULONE $C_{21}H_{30}O_5$

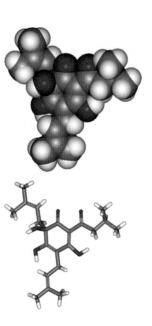

The yeast fermentation of cereal carbohydrates is the basis of the production of *beer*. In this process, starch molecules (91, 92) from the kernels of barley are cut into individual monosaccharide and disaccharide (89) units by the action of enzymes that are naturally present in the sprouted grain. This is the process of *malting* (the term comes from an old word for 'softening', as in melting), and occurs naturally to provide the sprouting seedling with an energy supply in the form of glucose. Much the same initial treatment is needed for the rice starch used in the production of *sake* (which is also technically a beer, although it is commonly called a 'rice wine'), but in this case the predigestion is due to the *Aspergillus oryzae* that grows on the rice. The germination is terminated by dehydrating the malt at a suitable stage. This 'kilning' process also results in browning reactions that contribute to the beer's final colour, with the more heavily roasted malt resulting in darker beers, such as stout and porter. Some malts

are bleached with sulfur dioxide to give a lighter beer. The product of this initial process is the *mash*, a concentrated liquid that is then separated into a solid part and a liquid part known as the *wort*. At this stage the hop resin is added to the wort.

Hop resin is a viscous yellow substance contained in glands at the base of the blossoms of the female hop plant (*Humulus lupulus*), a relative of the cannabis plant (180); it is rich in bitter components, including humulone and its relative lupulone, which are extracted by the wort and serve to offset the insipidity and sweetness of un-hopped, sugary beer. Hop leaves were probably added early in the history of beer for their bactericidal action, to act as a preservative; now they are added to provide flavour. The bitter molecules probably undergo a slight molecular rearrangement in the mixture in which the six-membered ring breaks open and reforms as a five-membered ring, producing isomeric molecules that are even more bitter than humulone and lupulone themselves. Yeast is then added to the wort; it converts the sugars into ethanol (26) by fermentation, releasing carbon dioxide in the process. The technique of fermentation may be either 'top fermentation' or 'bottom fermentation'. These techniques differ with regard to the region where the used yeast cells accumulate. The former is used to produce English beers. These are more acidic and stronger than other beers partly because of the access that the yeast cells have to air. Bottom fermentation, the conventional procedure in the USA, is used to produce the lighter *lager*. The name (from the German word for 'store') reflects the process used to clear the beer, that of storing it at just above the freezing point.

When beer is left exposed to sunlight, a photochemical reaction occurs, in which the humulone molecules react with other molecules that contain sulfur atoms, resulting in the formation of several sulfur-containing products. One of these is 3-methylbutane-1-thiol (146), a molecule manufactured even more liberally by the skunk.

HOT, SPICY, AND COOL

Hot, spicy, and cool tastes are chemical stimulations of pain. There are two types of pain nerve. Class A nerves are slender fibres that carry signals rapidly (at about 20 metres per second); class C nerves are thicker and carry signals more slowly (at about 1 metre per second). Their signals are referred to as *fast pain* and *slow pain*, respectively. Fast pain is the response to injury and is often sharply localized. Slow pain is often a dull, aching sensation that is usually less sharply localized.

Both types of nerve fibre enter the spinal cord, together with nerves responsible for sensation of temperature; there, they stimulate neurons that lead to the brain and their signals undergo some local processing. An important feature of pain nerves is the interaction of the two types in a gelatinous part of the spinal cord called the *substantia gelatinosa*. Signals arriving along the A fibres excite cells of the *substantia gelatinosa*, but those arriving along the C fibres inhibit them. The net effect can be to inhibit the cells that are responsible for transmitting A and C signals to their processing centre in the brain (the thalamus). Hence there is a complex interplay between the signals arriving initially as fast and slow pain (a point that will be illustrated in what follows). Moreover, in response to pain signals, the brain can secrete its own analgesics – the *endorphins* and *encephalins*. Both are polypeptides, with the endorphins having long chains and the encephalins short chains, that affect the transmission of nerve signals, and both are mimicked by opiates (173). The pain receptors that initiate all this complex signalling are highly branched nerve endings themselves: there are no specific innervated pain receptors.

There are, however, receptors that respond to thermal stimulation. They are essentially of two types, one of which responds to hot and the other to cold; the latter are more numerous by a factor of about ten. Their signals, like pain signals, are carried by class A and class C nerve fibres, ultimately to the thalamus, so that intense thermal stimulation can be interpreted as pain. Many of the spices used in curries and other foods stimulate pain-detecting nerve endings in the mouth (and elsewhere), but the relation between molecular structure and response is not known. 'Noxious heat' (above 52 °C) stimulates one receptor, moderate heat (42 °C) and capsaicin (106) stimulate another receptor, and coolness (below 22 °C) and menthol (108) stimulate a third receptor. The missing receptor for the range 22–42 °C was identified only in 2002, and is found in the skin, on the tongue, and in brain and nerve tissue. It is speculated that such widespread occurrence of the receptor is related to its role in response to injury and inflammation.

105 PIPERINE $C_{17}H_{19}O_3N$

106 CAPSAICIN $C_{18}H_{27}O_3N$

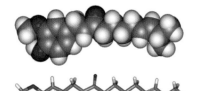

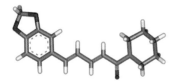

Piperine is the active component of white and black pepper (the berries of the tropical vine *Piper nigrum*). Black pepper is obtained by allowing the unripe fruit to ferment through the action of the fungus *Glomerella cingulata* and then drying it. White pepper is obtained by removing the skins and pulp of the ripe berries and drying the seeds. Piperine is an alkaloid, and its presence presumably is an adaptation for the protection of the plant.

Another 'hot' spice is capsaicin, the pungent component of various species of *Capsicum*, including red and green chilli (chile) peppers and especially *C. annum* and the small, virulent *C. frutescens*. It is the active component of *paprika*. Capsaicin molecules – more generally the dozen or so members of the capsaicinoid family of molecules – are located in sacs along the fruit's inner wall, or placenta; these sacs break open when the fruit is cut open and the capsaicinoids spread onto the seeds.

The action of capsaicin (and perhaps piperine) seems to have several components. It stimulates the excretion of saliva, which aids digestion. It also stimulates movement of the colon and helps to encourage the passage of the remains of food. At the anus, it can cause itching (*pruritis*) and a pleasing sensation of warmth during defaecation. The sense of well-being one gets after a meal of hot spices

Red chillis (*Capsicum*) are pungent on account of the capsaicinoids they contain. One of the most abundant capsaicinoid molecules is capsaicin itself.

has been ascribed to the ability of these pain-producing compounds to stimulate the release of soothing endorphins in the brain. The production of capsaicin in plants is probably a deterrent mechanism, which dissuades mammalian predators from eating the plant; birds apparently do not respond to the burning sensation and help spread the seeds with impunity. To quench the fire of chilli, take a drink of milk, for milk and milk-based products such as milk chocolate (and beans and nuts) contain casein, a protein that expels capsaicin from nerve receptors on the taste buds.

The pungency of chilli peppers has traditionally been measured in 'Scoville units'. On this scale, the bell pepper ranks a paltry zero, whereas the fiery Habanero is ranked at 300 000. Cayenne and tabasco, which also contain capsaicin, come in at about 30 000 units. Pure capsaicin registers a fearsome 16 million units. Chilli growers like to flaunt their hotness, so one has to take their claims about their peppers with a pinch of salt, but Naga Jolokia (*Capsicum frutescens*), which grows in Tezpur in Assam, is perhaps the hottest of all, weighing in at about 700 000 units.

107 ZINGERONE $C_{11}H_{14}O_3$

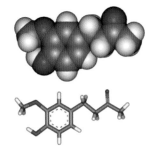

Although the relation between molecular structure and pain response is not known, it almost certainly has something to do with the shape of the molecule, which acts to fit a protein in the wall of the pain-nerve ending: the key fits the lock, the protein changes shape, and a signal is on its way. Some confirmation of this idea comes from a comparison of the shapes of the capsaicin molecule (106) and the zingerone molecule shown here. Notice how closely they resemble each other: the zingerone molecule lacks a hydrocarbon tail and a nitrogen atom, but is otherwise very similar to capsaicin.

Zingerone is the pungent, hot component of *ginger*, the rhizome (or underground stem) of *Zingiber officinale*. The name of the plant derives ultimately from Sanskrit *shringavera*, meaning 'shaped like a deer's antlers'. Closely related compounds are also present in ginger; they differ mainly in the length of the chain that replaces one of the hydrogen atoms of the terminal $-CH_3$ group (on the right in the illustration). Different forms of ginger have different proportions of these related compounds. When green root ginger is dried and powdered, it loses not merely its free water but also one H and an OH unit (as H_2O) from neighbouring atoms of a side chain, leaving a double bond between the two atoms and a tail with a different shape. This loss affects the flavour slightly because it changes the composition of the zingerone-like mixture.

Ginger beer is a soft drink made from sugar, ginger extract, and carbonated water.

108 MENTHOL $C_{10}H_{20}O$

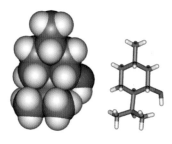

Menthol, from *oil of mint*, has a characteristically cool taste. It is extracted from the Japanese peppermint (*Mentha arvensis*) by cutting the plant when it blooms, curing it like hay, and then distilling the oil in steam. Menthol is also prepared synthetically from turpentine (137). It is present in the common mint herb (*Mentha piperita*) and is used in cigarettes, soaps, and perfumes for the pleasant odour that accompanies its cooling taste. Menthol comes as left- and right-handed forms: the isomer shown is L-menthol; this is the desired product since, although both isomers have a mint-like odour, L-menthol has a much greater cooling effect.

The cooling taste arises from menthol's effect on cold-temperature receptors in the skin. When menthol is present, the sensors responsible for signalling 'cold' become active at a higher temperature than normal. Hence, an environment that is actually warm (such as the mouth) may be interpreted as being cool. The cold receptors are very similar structurally to hot receptors and to proteins that allow ions to penetrate cell membranes. Their similarity suggests a mechanism for the fact that in some cases cold may feel like a burn and that a burning pain is sometimes experienced when one touches warm and cool surfaces simultaneously.

In ancient Greece, students taking examinations would braid rosemary into their hair, believing that it enhanced their intellectual vigour. In an echo of that ritual, it has been found that a peppermint fragrance in the air does in fact increase the speed and accuracy of typists.

Spearmint (*M. spicata*) is the common form of mint in Britain, often as an accompaniment of lamb. Peppermint is a hybrid of water mint (*M. aquatica*) and spearmint. 'Black Mitcham' has unusually dark leaves as a result of the presence of anthocyanine pigments (162); varieties free of anthocyanines are known as 'white peppermint'.

MEATINESS AND BARBECUES

Most meat is muscle tissue and consists largely of proteins, so its principal components are strings of amino acids (80). Hence many of the properties of meat are the properties of its amino acids. The workhorse proteins of muscle are *actin* and *myosin,* which lie in layers between each other and slide past each other when the muscle is stimulated to contract. Contraction is maintained by temporary chemical bonds that form between the two proteins. The muscle fibrils that contain the actin and myosin molecules and the muscle fibres that are formed by groups of fibrils are encased in connective tissue, which is principally the protein *collagen.* A collagen-like protein is the principal structural component of commercial sponges (members of the class *Demospongiae* of the phylum *Porifera*), which makes them inedibly tough and gristly; fish muscles have very little collagen and are correspondingly very tender. When collagen is heated in boiling water, it forms *gelatin*.

Muscle contraction must be fuelled as well as stimulated. The energy resource used depends on the type of muscle and affects the appearance of meat, especially its colour. There are two broad classes of muscle fibre: fast and slow. Fast muscle fibre (which is also called *white fibre*) is used for rapid motion and uses as its fuel carbohydrates, particularly glucose (87). This is readily available in the blood and as glycogen and can be used (albeit inefficiently) even without oxygen. Slow muscle fibre (which is also called *red fibre*) is used for steadier sustained motion. Its energy supply is stored as fat (35), for which oxygen is essential. Hence the oxygen supplied by the blood needs to be stored in slow muscle fibre, so that it can continue to function at least briefly even when the oxygen supply is insufficient to meet the demands of sustained action. The storage molecule is another protein, *myoglobin*, which is closely related to the oxygen-carrier haemoglobin (84) and contains an iron atom at its heart. Myoglobin is red when it is oxygenated, which is why slow muscle fibre is red whereas fast muscle fibre, which does not need this storage molecule, is pale.

Fish (especially cod, which spend most of the time lazing around on the sea floor) have skeletal muscles that are built predominantly of fast muscle fibre. Hence they have pale, even white, flesh. Poultry, which are largely earthbound, have unexercised wings and hence white breast meat but pinker legs; the legs are greasier than breast meat because fat is the fuel of their red muscle fibre. Game birds fly frequently, require pectoral muscles that can sustain prolonged activity, and hence are myoglobined throughout. The muscle of most domesticated animals is slow and hence red. Diving animals, especially whales, require extensive oxygen reserves, and their flesh is very dark.

When myoglobin loses its oxygen after death, it becomes pale purple. When it is cooked it turns brown, as the iron at its centre is oxidized and, in its new form, absorbs light of a different wavelength. When meat is salted, as in the preparation of ham and some cooked meats, the myoglobin molecule picks up nitrite ions (NO_2^-), and its colour changes to pink. The different tastes and odours of animal muscles are modified by cooking, which breaks proteins and other components into smaller fragments. Some of these fragments are small enough to be volatile and thus add to our perception of the flavour. Some originate from molecules of the substance called ATP, with which we begin our description.

The savoury taste of meat is now recognized as a fifth type of taste (together with sweetness, bitterness, sourness, and saltiness), called *umami*. Umami, a Japanese word, was first identified by investigating the distinctive flavour of seaweed broth. The taste appears to be specific to glutamates (111) and is responsible for the distinctive taste of meat, especially aged beef. The taste of ham, tomatoes, asparagus, and cheese and the sweet taste attributed to many shellfish is also umami. The ability to detect umami is perhaps an evolutionary response to the need to identify high-protein foods.

109 ADENOSINE TRIPHOSPHATE $C_{10}H_{16}O_{13}N_5P_3$

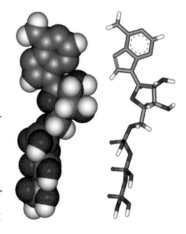

The adenosine triphosphate (ATP) molecule looks complicated, but it can be pictured as being formed from three components. The central component is a sugar molecule, *ribose*, in the form of a five-membered ring resembling one form of the fructose molecule (88). Attached to this are the two other components. One is a group consisting of linked five- and six-membered rings of carbon and nitrogen atoms. Such groups are called *bases*, and this particular one is *adenine* (199). The combination of the ribose and the base makes up a unit called a *nucleoside*. Another atom of the ribose ring is attached to a string of three phosphate groups. As far as we are concerned here, this string is where the action is.

ATP is one of the most important molecules in living things, and it is abundant in muscles. Its role is as an immediate source of energy for powering biochemical reactions, such as may be needed for the contraction of muscle cells when lifting a load. It is also used much more broadly: for example, in the metabolism of foods, in the construction of proteins from the DNA template, and in powering the processes that allow you to see these words and reflect on their significance. Indeed, where there is life, there is ATP. Its ubiquity explains why phosphates are so important in the diets of plants and animals. The phosphate-fertilizer industry, in essence, converts old bones into an assimilable form of phosphorus that can be used to maintain the presence of ATP in growing, living plant cells.

ATP performs its role by releasing the terminal phosphate group at the demand of an enzyme; in so doing it releases energy to drive some other reaction, which may be the construction of a protein from amino acids or the contraction of a muscle. The liberated phosphate group is then reattached to the molecule using energy derived from ingested food. After death, the ATP is no longer reconstructed, and the muscles stiffen into *rigor mortis*. Rigor mortis sets in more quickly if the ATP supplies have been depleted in a fight, a struggle, or even a state of anxiety before death.

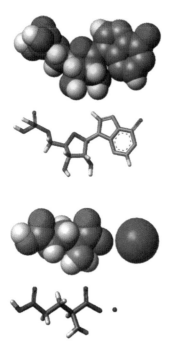

110 INOSINE MONOPHOSPHATE $C_{10}H_{13}O_8N_4P$

111 MONOSODIUM GLUTAMATE $C_5H_8O_4NNa$

After death, when the reconstruction of ATP ceases, its decomposition goes beyond the loss of a single phosphate group. After losing two phosphates from its tail, to form adenosine monophosphate, it loses one of its nitrogen atoms, which is replaced by an oxygen atom. This sequence of events produces a molecule of inosine monophosphate (IMP), a substance with a slight meaty flavour.

The sodium salt of the naturally occurring amino acid glutamic acid is *monosodium glutamate*, or MSG. It occurs in meat as the meat ages and its proteins decompose. Like IMP, MSG does not have a very pronounced meaty flavour on its own; but the two together are strongly meaty and are the principal compounds responsible for meat's taste. MSG is the cheaper and more readily available of the two; it is added by producers of meat foods to bring out the flavour of their product. It appears to enhance the sensitivity of the salt and bitter taste receptors of the tongue, but its precise mode of action remains obscure. The savoury taste of meat stimulated by glutamates is referred to as *umami* and appears to have a specific receptor.

Different meats contain different proportions of IMP and MSG, and their flavours differ accordingly. Beef has twice as much MSG as pork, but approximately the same amount of IMP. Mushrooms are also rich in proteins composed of glutamic acid, which accounts for their slightly meaty flavour and their ability to enhance the flavour of most dishes. The part of the mushroom we eat is the *basidiocarp*, the fleshy, spore-producing body.

Although most microorganisms conserve their precious nitrogen atoms, there are several strains of bacteria (*Micrococcus* and *Brevibacterium* among them) that excrete glutamic acid when they are fed a

These oyster mushrooms (*Pleurotus ostreatus*) have a slightly meaty flavour, partly because, like meat, they contain proteins rich in glutamic acid.

diet rich in ammonia (6). This 'fermentation' of ammonia is now the standard commercial method for producing MSG.

112 HYDROGEN SULFIDE H₂S

The hydrogen sulfide molecule is the sulfur analogue of water (5), with a central sulfur atom in place of the central oxygen atom. Some striking differences arise from this simple replacement.

To begin with, hydrogen sulfide is a foul-smelling poisonous gas. It is a gas at normal temperatures because sulfur is a less aggressive seeker after electrons than oxygen is (it is a bigger atom, and the attracting nucleus is buried under clouds of electrons). Consequently, the nuclei of its hydrogen atoms are less exposed than are those in the water molecule and they are much less able to take part in hydrogen-bond formation. Hydrogen sulfide molecules therefore interact much less strongly with each other and are able to move freely as a gas.

Why hydrogen sulfide and many volatile sulfur compounds are foul-smelling is less easy to explain. We should note here that one's perception of the gas declines after a brief exposure, which is very dangerous because hydrogen sulfide is more poisonous than hydrogen cyanide (115). In a sense this resembles a similar response to water: because we are pervaded by water from conception to cremation, its smell is undetectable to us.

Hydrogen sulfide forms when sulfur-containing proteins decompose, either after death or during cooking. It has the characteristic smell of rotten eggs (and, in smaller quantities, the pleasant smell of a freshly boiled egg) and is indeed formed by eggs as their sulfur-rich albumin protein molecules decompose. The pale green colouration where the white of a boiled egg meets the yolk is another sign of its presence. The green is a precipitate of iron sulfide that forms where the hydrogen sulfide released from the albumin meets the iron-rich proteins of the yolk. Hydrogen sulfide reacts with many of the other compounds produced during the thermal degradation induced by cooking, and it gives rise to numerous other odorous molecules. It is particularly important in producing molecules that contribute to the flavours of cooked chicken.

Another fate for hydrogen sulfide is its conversion into elemental sulfur, just as water was the source of atmospheric oxygen. In much the same way as cyanobacteria learned to snip hydrogen atoms off water molecules and excrete the oxygen, which wafted off as a gas of O_2 molecules, so purple sulfur bacteria (*Rhodospirillum rubrum*) have learned how to snip hydrogen atoms off hydrogen sulfide molecules. What they excrete is solid elemental sulfur, which lies around in great piles and does not waft away. These bacterial dung hills are the deposits once quarried for sulfur.

The Black Sea (in ancient times, the Euxine), is so called because the hydrogen sulfide formed by the sulfate-reducing bacteria that inhabit its stagnant depths stains the mud at the bottom black by the formation of sulfides. The dissolved gas preserves the euxinic (as such stagnant sulfide-rich bodies of water are called) character of the lower levels of the sea because it reacts with any dissolved oxygen that diffuses down from the aerated upper layers. Some Norwegian fiords are euxinic, and sometimes the deep, stagnant waters well up to the surface and release their noxious fumes.

113 ACROLEIN C_3H_4O

Acrolein, an aldehyde, is a colourless volatile liquid with an acrid smell. It is formed when the fatty acids (34) present in meat break down under the stress of heating: the long-chain fatty acids break off their glycerol anchor in the storm of thermal motion caused by the heat, and the glycerol molecule itself loses two molecules of water, forming acrolein. Acrolein's acrid smell adds considerably to what some regard as the pleasure of a barbecue, where it is easily detected in the smoke. It also contributes to the flavour of the caramel prepared by heating and partially decomposing sucrose (89).

Wood smoke also includes the simplest aldehyde of all – formaldehyde (28). Its attack on the eyes is partly responsible for the tears elicited by smoke. Its attack on bacteria is responsible for the preservation of smoked meats. Other substances in smoke include phenols; they act as antioxidants helping to protect fats from oxidation by diverting the attack of atmospheric oxygen.

FRUITS AND FOODS

The flavour of food is a combined response to two chemical senses, taste and odour. Taste has already been considered and here we concentrate on odour. This is the more sensitive sense, even in humans, and it is the major contributor to the perception of flavour. Whether or not a molecule has an odour depends on whether it can excite the olfactory nerve endings. In humans, these nerve endings (some 50 million or so) occupy an area of yellow–brown epithelium of area about 5 square centimetres in the nasal cavity approximately level with the eyes; except in sniffing, it is eddy currents (rather than direct drafts) that carry molecules to this area.

Smell is peculiar because the olfactory epithelium consists of bare nerve endings. This is unlike any other sensation (except pain), for which normally some kind of transducer acts as a buffer between the outside world and the nervous system. With smell, the nervous system is in direct contact with the outside world. In essence, the brain is exposed in the nose. This suggests that olfaction is one of the oldest and most primitive senses. Smell is also closely linked, at least in terms of the physical proximity of their processing centres, with one of the more primitive parts of the brain – the *limbic system*, the seat of the control of emotions. This location may account for the powerful, and sometimes unconscious, impact of an odour. That amphibious animal the frog has noses for all seasons, or at least different chambers within its nose: one pair of chambers is used for when it is breathing air and detecting volatile molecules and the other pair for when it is submerged and detecting dissolved molecules, with a valve that switches between them; tadpoles appropriately have only the wet nose; the dry nose develops as they undergo metamorphosis.

Smelly molecules are called either *odorivectors* or *osmophores* (from the Greek word for 'smell'). However, the relation between their molecular structures and the sensations they excite remains obscure. One criterion for smell is that an odorivector must be volatile, for otherwise it would never reach the nose (unless you are a frog). A second is that it should be at least slightly soluble in water, for otherwise it would not dissolve in the *mucus*, an aqueous solution of proteins and carbohydrates that is exuded by cells in the olfactory epithelium and coats the nerve endings there. However, this is a less clear-cut criterion, because organic molecules in the mucus could act as detergents and carry insoluble molecules through the water and into receptor sites. An odorivector must also, presumably, interact with a protein molecule in the olfactory nerve endings, modify its shape, and hence stimulate the nerve cell to send a message to the brain.

It seems plausible that the interaction between an odorivector and a nerve-cell protein depends on the same kind of lock-and-key mechanism as that which activates taste: a molecule of a particular shape can attach to a given protein molecule so long as it matches its shape in some respects. The difficulty with the testing and development of this idea is that there are so many different odours, and hence so many different protein locks, and – the killer problem – no way of measuring the subjective response to smell quantitatively or even objectively. About 30 different types of anosmia – partial odour blindness – have been identified, which suggests that there are at least 30 different locks that can be opened, but that is probably a considerable simplification, especially when we consider olfaction by other species. There have been naive attempts to propose the shapes of these receptors, but none can be taken seriously. There are also rival proposals that olfaction is due not to shape but to the vibrational characteristics of molecules. However, even vibration is an aspect of structure, so it is not clear whether such speculations are independent of correlations with shape. The field is wide open, and enormously complex: the bottom line is that no one really knows how we respond to this most primitive of senses. That is the problem: late-developed senses, such as vision, are highly sophisticated; primitive senses like olfaction are the whores of sensation, responding to any old molecule that happens to drop by.

Pungency and putridity are less specific than other odours; they may be responses that are signalled by cells other than the ones described here. Lurking in the olfactory epithelium, among the mucus-exuding cells, are cells that are part of the system that innervates the face (the *trigeminal nerve*). It is suspected that pungent and putrid molecules penetrate into them, interact with their proteins, and stimulate them to fire. Thus there may be two types of olfaction, what might be called 'first smell', the ordinary variety for specific odours, and 'second smell', for non-specific pungency and putridity.

The molecules discussed in this section are related to typical fruits and foods. The essential oils of plants and parts of bodies are also rich sources of olfactory stimulation, for good or for ill, and are described in the following two sections.

114 BENZALDEHYDE C$_7$H$_6$O

115 HYDROGEN CYANIDE HCN

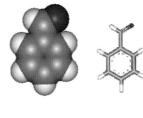

Benzaldehyde is a colourless liquid that smells of bitter almonds. A closely related molecule, phenylethanal, is obtained conceptually by inserting a –CH$_2$– group between the benzene ring and the –CHO group. The latter molecule fits a floral receptor better than does benzaldehyde itself. It smells of hyacinth and is used in perfumes under the name *hyacinthin.*

Hydrogen cyanide is an almond-smelling, colourless, poisonous gas with an odour that fades on prolonged exposure. Hydrogen cyanide interferes with the transport of oxygen by the red blood corpuscles and with the action of the energy-source molecule ATP (109), so that it brings the body to a stop.

The aroma of cherries and almonds is due to benzaldehyde, but the hydrogen cyanide in cherries also contributes somewhat. Benzaldehyde and hydrogen cyanide both occur quite widely in drupes and pomes (multiple-seed and single-pip fruits), especially apricots and peaches. They are released when the pips are crushed and enzymes can get to work. This much the Romans and Egyptians also knew, for they ground peach kernels to make poisons. The source of both molecules is the compound *amygdalin*, which is a modified disaccharide (specifically, a *glycoside*) consisting of two glucose (87) units linked together, with one of the –OH groups changed to a benzaldehyde-like group with a –CN group attached. The enzyme *emulsin* can disrupt this molecule, releasing two glucose molecules, a benzaldehyde molecule, and a hydrogen cyanide molecule. Some cyanide appears in fruit jams that contain pip extracts, such as quince. The cyanide is normally in too small a concentration to be dangerous.

116 ISOAMYL ACETATE $C_7H_{14}O_2$

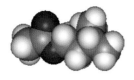

117 ETHYL 2-METHYLBUTANOATE $C_7H_{14}O_2$

With these two molecules we see nature building different compounds in similar ways and from the same kit. The isoamyl acetate molecule is an ester formed from acetic acid (31) and an alcohol, isoamyl alcohol. The ethyl 2-methylbutanoate molecule is also an ester with the same numbers of carbon, hydrogen, and oxygen atoms, but they are bonded in a different pattern.

Both compounds grow in prominence as apples ripen and, as their concentration increases, they mask the characteristic flavour of the unripe fruit. Esters with about seven carbon atoms have characteristic fruity smells, occur widely in fruits, and result from the breakdown of long-chain fatty acids (34) as the cell membranes are oxidized during the ripening process.

118 2-HEPTANONE $C_7H_{14}O$

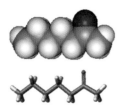

This compound is a liquid with a clove-like odour (it occurs in oil of clove). Its oxygen atom allows it to be transported through the mucus of the olfactory epithelium, and it stimulates receptors that are characteristic of woody–fruity smells. The presence of 2-heptanone accounts for the odours of many fruits and dairy products. The molecule is also responsible for the aroma of blue cheese, which is formed by inoculating the curing mixture with moulds such as *Penicillium roquefortii*. Like butanedione (40), which gives butter, buttermilk, sour cream, and cottage cheese their characteristic odour, 2-heptanone is a ketone.

119 3-(*para*-HYDROXYPHENOL)-2-BUTANONE $C_{10}H_{12}O_2$

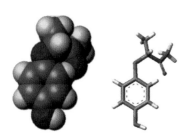

120 β-IONONE $C_{13}H_{20}O$

These two ketones are included here because they show how two molecules in a mixture can conspire to produce an aroma that may evoke memories of others. 3-(*para*-Hydroxyphenyl)-2-butanone is the molecule chiefly responsible for the smell of ripe raspberries, and it is included in the recipe for the synthetic flavour. The fresh smell of the newly picked fruit is due partly to β-ionone, which is also responsible for the odours of sun-dried hay and violets.

β-Ionone is the fragrant component of oil of violets, which is obtained (by extraction in solvents) from the flowers of the blue and purple varieties of *Viola odorata*. Natural oil of violet is too expensive as a source, and most of the 2-ionone molecules used in the perfume and food-additive industries are made synthetically.

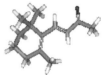

121 METHYL 2-PYRIDYL KETONE C$_7$H$_7$ON

122 2-METHOXY-5-METHYL PYRAZINE C$_6$H$_8$ON$_2$

Molecules containing a benzene-like ring but with one or more of the carbon atoms replaced by nitrogen atoms play an important role in the aroma of heat-treated foods. The pyridyl molecule (121) is responsible for the pervasive odour of popcorn. The pyrazine molecule (122) is responsible for the odour of peanuts. Pyrazines also contribute to the aromas of crusty bread, rum, whisky, chocolate, and some uncooked vegetables, including peppers. When you take delight from these aromas, you are responding to molecules like the

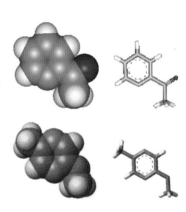

That fresh raspberries (here, *Rubus idaeus*) can evoke the smell of new-mown hay is no accident: they contain the same molecule (ionone) as hay.

Roasted coffee beans. Their colour is largely due to the browning reaction (page 177) that occurs when organic substances containing nitrogen are heated. Temporarily trapped within the beans are the molecules responsible for coffee's flavour and stimulating effect.

ones shown here, as they stimulate the part of your brain that ends in your nose.

123 2-FURYLMETHANETHIOL C$_5$H$_6$OS

This molecule is one of those responsible for the aroma of *coffee*, the roasted beans of *Coffea arabica* (the plant first cultivated near Mocha in the Yemen) or of the more climatically tolerant *C. canephora*. (*Mocha* now means a mixture of coffee and chocolate.) The stimulating action of coffee is due to caffeine (179). See the description of that molecule for more details.

124 DIALLYL DISULFIDE C$_6$H$_{10}$S$_2$

125 ALLYL PROPYL DISULFIDE C$_6$H$_{12}$S$_2$

126 PROPANETHIAL *S*-OXIDE C$_3$H$_6$OS

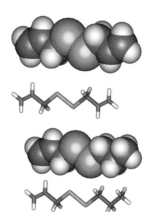

Sulfur compounds are responsible for the pungent odour of members of the genus *Allium*, including garlic (*A. sativum*) and onion (*A. cepa*). All these plants contain a high concentration of amino acids (80), particularly *cysteine*, that include sulfur atoms on the side chains. Garlic and onion are odourless until crushed and chopped; then the damage done to the cells allows enzymes to reach their contents and

Garlic cloves (*Allium sativum*). Pungent odours, including those of garlic and onion, are often due to molecules that contain sulfur atoms.

to convert nitrogen- and sulfur-containing amino acids into volatile compounds, including ammonia (6) and the compounds shown here. The smell of garlic powder is different from that of fresh garlic because it was acted on by enzymes in the past, and the most volatile compounds have had time to evaporate and to oxidize.

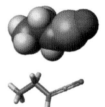

The aromas of species of *Allium* are due largely to disulfides, in which two sulfur atoms act as a bridge between two hydro-carbon fragments. Diallyl disulfide is responsible for the odour of garlic, and allyl propyl disulfide for that of the onion. Enzymatic action on the onion amino acids also results in the formation of the thiopropionaldehyde-S-oxide, which is the lachrymatory component of chopped raw onions.

FLOWERS AND ESSENTIAL OILS

The *essential oil* of a plant is the essence of its fragrance. More pro-saically, it is the volatile material that can be isolated from a single species of plant. It is often obtained by heating the leaves or petals in steam and is collected as the oily fraction. Some essential oils are pressed out of the plant, and others (particularly those from flowers) are extracted using solvents. About 3000 essential oils have been identified, and several hundred are available commercially. The subtlety of nature shows up here, however, for a natural essential oil may involve several hundred different types of molecules, and its impact may be changed by removing those in even very minute concentration.

A major use for essential oils is in the blending of perfumes, but they are also used to flavour foods. When next you wonder how a perfume can have so strong an impact, or bring back a memory, remember that olfactory signals are processed near the limbic system, the part of the brain that is closely associated with emotion. Most molecules that occur in essential oils contain up to about a dozen carbon atoms. This allows them to be moderately volatile but to have a wide range of distinctive structures.

Flowers work a seven-hour day. At least, snapdragon (*Antirrhinum majus*) is known to produce its volatile emissions most abundantly during the hours that coincide with the peak foraging activity of bees. The emissions are weak when the flower first opens and reach a peak between five and eight days later when the reproductive apparatus of the plant has been developed fully. Some plants change the composition of the odour they produce once they have been pollinated, to dissuade bees from returning and perhaps causing damage.

Just as some smells attract us and others repel us, the same is true in the symbiotic world of plants and insects. Thus, bees and flies are attracted by 'sweet' smells whereas beetles favour musty, spicy, and fruity odours. In perfumery, ingredients are said to contribute different 'notes' to the mixture: 'high notes' are typically the most volatile, with citrus, herbal, and green (for instance, cut foliage) odours; the 'middle notes' are typically floral, woody, and spicy; the 'end notes'

The common feature of terpenes is that they are constructed by stringing together five-carbon-atom units based on isoprene.

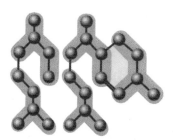

are the heavier tones of musk, ambergris, and vanilla. A successful perfume is an evolving symphony (a synosmia) of these sequential notes.

What follows is just a quick tour through the types of molecules encountered in different oils. Several of the molecules shown are members of the class of organic compounds called *terpenes*. These are compounds based on the isoprene unit (67), as shown in the illustration; in a sense, they are very tiny, fragrant fragments of rubber (68), for rubber is a longer string of isoprene units. The smallest terpene unit consists of two isoprene units, but bigger versions include the carotenoids (158) that are responsible for many of nature's colours, and the steroids (195). All terpenes are formed as a result of the metabolism of glucose (87).

127 BENZYL ACETATE $C_9H_{10}O_2$

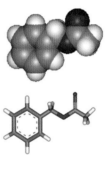

This molecule represents the class of essential oils that are esters; benzyl acetate is an ester of acetic acid (31) and benzyl alcohol. The latter is a modification of methanol (25) in which a hydrogen atom from the –CH_3 group has been replaced by a benzene ring (22). Benzyl acetate is one active component of *oil of jasmine*; another is *linalool*. Although jasmine (genus *Jasminum*) is a source of this compound, it is generally synthesized directly from acetic acid and benzyl alcohol. Since it is cheap and readily available, jasmine is a common perfume for toiletries.

128 CARVONE $C_{10}H_{14}O$

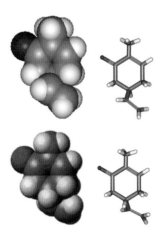

Carvone is representative of the terpenes. It is the main active component of *oil of spearmint* and is distilled from the leaves of the spearmint plant (*Mentha viridis*), a relative of common mint. Traditional chewing gum is flavoured with spearmint. The gum part was originally *chicle*, the coagulated latex of the sapodilla tree (*Achras sapota*), but it is now made predominantly from a synthetic styrene–butadiene copolymer (69). The outer casing is dried sugar, held together by hydrogen bonds until those bonds are replaced by water molecules from saliva.

A clever trick of molecular structure transforms spearmint flavour into dill or caraway. Both molecules have the same atoms joined together into the same network. They differ in that one is the mirror image of the other, as a left hand is to a right. That a left-handed molecule (in a technical sense) should smell differently from a right-handed version of the same molecule is consistent with the view that the olfactory receptors are sensitive to shape. One molecule is a key for particular receptor proteins; its mirror-image molecule may unlock different doors and hence different odours.

129 CINNEMALDEHYDE C_9H_8O

130 2-METHYLUNDECANAL $C_{12}H_{24}O$

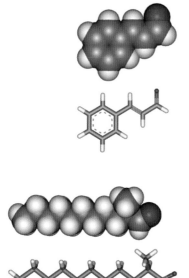

The cinnamaldehyde molecule is an example of a fragrant aldehyde. It occurs in *oil of cinnamon* and is obtained by steam distillation of the bark of the cinnamon tree (*Cinnamomum zeylanicum*) and the leaves of the Chinese cinnamon. Cinnamon stick or powder is the dried inner bark. As well as for its fragrance, cinnamon has a reputation for its *carminative* action, its ability to release the gases hydrogen sulfide (112), methane (15), and hydrogen from the intestine and the stomach in one direction (belching) or the other (politely, flatus), perhaps as a result of its irritant action.

2-Methylundecanal, which is like cinnemaldehyde but with the benzene ring unrolled into a linear chain, has been included because it is the dominant molecule of the fragrance of *Chanel No. 5*, which was introduced in 1921 by Coco Channel. This fragrance established synthetic materials in the perfume industry and brought perfumes to the masses.

131 EUGENOL $C_{10}H_{12}O_2$

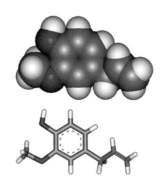

Eugenol occurs in *oil of bay* and is extracted from bay leaves – originally from the tall Mediterranean *Laurus nobilis* but now more commonly from the Californian *Umbellularia californica*. Eugenol is also an active component of *oil of cloves*, which is obtained from the dried nail-like (*clavus* is Latin for nail) flower buds of *Eugenia aromatica*. It is also the principal component of *allspice* (a spice so called because it combines the flavours of cinnamon, nutmeg, and cloves), also known as *Jamaican pepper*, which is the dried unripe berry of *Pimenta dioica*, a type of myrtle. The odour of carnations is in part due to eugenol, and perfumiers incorporate it in carnation-like fragrances. It is, like cinnamaldehyde (129), a carminative compound.

Isoeugenol differs from eugenol only in the location of the double bond in the hydrocarbon side chain: this change shifts the odour

A growing nutmeg (*Mystirica fragrans*) in its red coat, growing in Grenada, the 'spice island' that produces about a third of the world's nutmeg. The coat when dried and ground is mace.

from clove to nutmeg (*Myristica fragrans*), one of the most important of the traditional spices. Nutmeg is derived from the ground seeds of the tree, and *mace* from the seed coatings.

132 GERANIOL C$_{10}$H$_{18}$O

133 2-PHENYLETHANOL C$_8$H$_{10}$O

Geraniol occurs in many essential oils, such as the *citronella* obtained from Javanese citronella grass, and (as esters) in the leaves of the geranium. Its most delicate origin, though, is the rose, for, together with 2-phenylethanol, it is responsible for their fragrance and is extracted from their flowers. Now, when you smell the fragrance of a rose, you will know what molecules are triggering a signal to your brain; but what goes on in your head, how your emotions respond, and what the nature of enjoyment is, are still far from being understood.

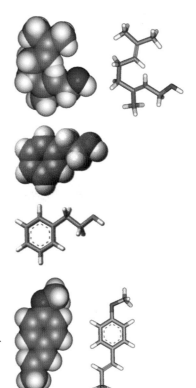

134 ANETHOLE C$_{10}$H$_{12}$O

Anethole is the active component of *oil of aniseed* (*Pimpinella anisum*). It also contributes to the flavours of fennel and tarragon. A close relative of anethole, anethole trithione, in which the hydrocarbon side chain is completed to form a ring with sulfur atoms, is one of perhaps a hundred substances that taste bitter to some European and Indian people and are tasteless to others; African, Native American,

Chinese, and Japanese people do not seem to display a taste blindness of this kind.

135 CAMPHOR C$_{10}$H$_{16}$O

Camphor, a white solid, is obtained by steam-distilling the trunk, roots, and branches of the camphor tree (*Cinnamomum camphora*) which grows in China and Japan. It is used as a counterirritant in medicine, like oil of wintergreen, and as an *antipruritic* (anti-itching agent), perhaps because, like menthol (108), it selectively stimulates cold sensors. It was once used as an *analeptic* (from the Greek word for 'restore'), a drug that stimulates the respiratory system, for it provokes deep breathing and raises the blood pressure. However, in larger doses camphor can lead to convulsions and respiratory collapse.

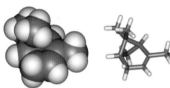

136 α-PINENE C$_{10}$H$_{16}$

137 α-TERPENEOL C$_{10}$H$_{18}$O

138 DIHYDROMYRCENOL C$_{10}$H$_{20}$O

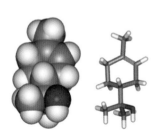

The α-pinene molecule is closely related to the camphor molecule (135), but it has no oxygen atom: α-pinene is a hydrocarbon. If we think of the awkward central 'handle' of a α-pinene molecule bursting open, then we have almost envisaged the α-terpineol molecule. It needs another –CH$_3$ group on the ring in place of a hydrogen atom, and an –OH group at one end of the broken handle.

Oil of turpentine is obtained by steam-distilling the resin exuded by various species of conifer trees, particularly (in the USA) longleaf pine (*Pinus palustris*) and slash pine (*P. elliottii*). Its major component is the fragrant hydrocarbon α-pinene. Although *pine oil* can be extracted naturally, it is often prepared by treating α-pinene with acid, which converts it into α-terpineol; the latter is the major component of pine oil and the molecule responsible for the fragrance of the juniper. Pine oil is the perfume and bactericide in many domestic cleaners.

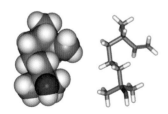

Open the ring of α-terpineol, and we get dihydromyrcenol (138), an alcohol with a fresh floral note (more specifically floral-muguet, where muguet is lilly-of-the-valley, *Convallaria majalis*). This alcohol is the dominant fragrance of the after-shave *Drakkar Noir*.

139 VANILLIN $C_8H_8O_3$

140 2,6-DIMETHYLPYRAZINE $C_6H_8N_2$

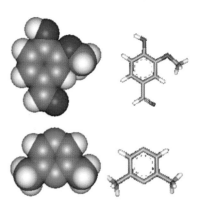

Vanillin is the essential component of *oil of vanilla*, which is extracted from the dried, fermented seed pods of the vanilla orchid (*Vanilla fragrans*), which is grown principally in Madagascar, Mexico, and Tahiti. During the curing process, the vanillin molecule is released from its glycoside, a compound in which the molecule is attached to a molecule of sugar. Vanillin is one of the most widely used flavour and odour compounds, and natural supplies are inadequate. It is therefore synthesized on a large scale; one method is by the oxidation of eugenol (131).

Vanillin can be detected in extremely low concentrations, yet the strength of its perception does not increase greatly as its concentration is raised (such is the oddness of olfaction). It is used in perfumery,

The fermented seeds of the cocoa plant (*Theobroma cacao*) and a variety of chocolate products made from them. The cocoa beans are roasted and ground to produce the chocolate liquor, which is then solidified and sweetened to make dark chocolate, or milk solids added to make milk chocolate. Some of the liquor is pressed to release the fat known as cocoa butter. Baking chocolate is pure chocolate liquor without sweetening; sweet chocolate is chocolate liquor with added cocoa butter, sugar, and vanilla. In milk chocolate some of the chocolate liquor is replaced by milk solids. For chocolate coatings some of the chocolate liquor is replaced by cocoa powder and cocoa butter is replaced by vegetable fats of higher melting points. White chocolate contains none of the solids from the cocoa bean that give chocolate its dark colour and rich flavour, and is principally cocoa butter with added sugar, milk solids, and perhaps some vanilla and lecithin, which emulsifies the ingredients to give a smoother consistency.

in confectionary (chocolate is a blend of vanillin and cacao), and for masking the smell of some manufactured goods. Vanillin molecules are leached out of the oak barrels (made from the European *Quercus robur* or *Q. sessilis*, not the American *Q. alba*) used to age wine, and contribute to the wine's 'finish'. One problem with use of vanillin in white soaps made from vegetable oils is that it undergoes a browning reaction, not unlike that responsible for bruising of apples, and discolours the product, making it look like, but not taste like, chocolate. One way round this problem is to modify the –OH group or the –CHO group, to make the molecule less reactive.

Such is the appetite for the taste of chocolate since its inception as *chocalatl*, the presumed aphrodisiacal drink of nobles in the Aztec court, that commerce has been stimulated to synthesize its flavour. One recipe depends on a mixture of vanillin (139) with an organic sulfide and 2,6-dimethylpyrazine (140).

Chocolate itself is obtained from the fermented seeds of *Theobroma cacao* (the name comes from the Greek words for 'food of the gods'). After fermentation, they are roasted to remove volatile matter, including butanoic acid (39) and tannins (see 162), then ground, and finally mixed with sugar.

The sharp melting point of chocolate and its cool taste are due to its predominance of fats that have very similar hydrocarbon-chain lengths, so its softening resembles the melting of a pure solid rather than a mixture (see 36). Cocoa differs from chocolate in that it is made from cacao beans that have been pressed to remove some of the fats; the beverage made from it then does not acquire a fatty layer on top. The main stimulant in chocolate is *theobromine*, which differs from caffeine (179) only in the replacement of one N–CH$_3$ group by N–H.

Synthetic vanillin is much cheaper than natural vanillin, and crooks are keen to represent the former as the latter. One technique for detecting the difference is to note that natural vanillin is weakly radioactive as it contains some radioactive carbon-14 atoms captured from the atmosphere during photosynthesis (carbon-14 is present as a result of the impact of cosmic rays on the atmosphere). Synthetic vanillin is made from coal tar, from which the radioactivity has long decayed. Chemically astute crooks then adjust the radioactivity artificially; but chemically more astute analysts know how the various isotopes of carbon and hydrogen are distributed in the natural molecule, and can use techniques to detect their locations. It looks as though fraudsters will have to resort to growing their synthetic vanillin if they really want to go undetected.

141 LIMONENE $C_{10}H_{16}$

142 CITRAL $C_{10}H_{16}O$

One really important essential oil is lemon oil, which is obtained by pressing the skin of the fruit of the lemon tree (*Citrus limonum*), a tree introduced into the western Mediterranean region from northern India about a thousand years ago. Lemon oil is widely used to evoke a sense of freshness, and is common in men's fragrances, such as *Eau Sauvage* and in a variety of *eaux de Cologne*. The oil itself is a complex mixture of several hundred terpenes, of which limonene and citral are characteristic. One problem with using lemon oil and its connotations of cleanliness and freshness is that citral is susceptible to oxidation and therefore cannot be used in bleaches, which act by oxidation. The industry has circumvented this problem by replacing the –CHO group by –CN, to give the compound geranyl nitrile, which is more stable but has a similar smell.

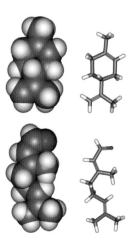

ANIMAL SMELLS

Animals have a rich system of chemical signalling in which a substance called a *pheromone* is emitted by one and causes a response in another. Once again, the compulsion to respond may be due to the proximity of the region in the brain where olfactory sensations are processed and the limbic system, which is associated with emotion. Here chemistry has its most immediate impact on our emotions and behaviour.

Many animals, including humans, continually emit a variety of molecules from their skin and sundry orifices. Indeed, over three hundred different compounds have been detected in human effluvia, one of the more abundant being, oddly, isoprene (67); but perhaps it is not so odd when we realize that isoprene units are part of the vitamin A molecules involved in vision (155).

The most obvious emanation from humans, other than their exhaled breath, is due to flatulence. People have to rid themselves of about half a litre of intestinal gas each day, about half of which is nitrogen (2) from the air, gulped down with food. Most of the rest is carbon dioxide (3), the product of the metabolism of organic matter in the intestine by the bacteria that inhabit it (among them, *Escherichia coli*), together with a little methane (15) and molecular hydrogen (H_2). So far, so good, for none of these gases smells. However, nitrogen atoms are abundant in proteins (80), and usually sulfur atoms are as well. The fate of these atoms, once bacteria get to work on the amino acids of these protein molecules, includes the production of trace amounts of the pungent gas ammonia (6) and the foul-smelling gas hydrogen sulfide (112). They flavour the flatulence, and, although one's own limbic system may respond favourably, that need not be true of one's neighbour's. The volume of gas may be increased if the food cannot be digested before it reaches the colon, as is the case with some oligosaccharides (90) found in beans. The odour may be worsened if the food is rich in sulfur-containing amino acids and other sulfur compounds, as are varieties of the cabbage (*Brassica oleracea*), especially Brussels sprouts.

The axillary region of the body (that is, the armpit) is a warm, moist region that is often a fount of odour. Once again bacteria are the cause, for the *Streptococcus albus* that inhabit the skin excrete lactic acid (32), thus increasing the acidity of their environment. This encourages other bacteria present to digest the organic components of sweat. The sweat itself is largely odourless, but the bacteria spice it with ammonia, hydrogen sulfide, and related compounds.

Another component of male underarm sweat provides an engaging story. This component is a hormone molecule that closely resembles one secreted in the saliva or by a male pig encouraging mating behaviour in a sow, but that is not all. The same pheromone is also secreted by the fungus we know as the *truffle*, the fruiting body of the *Tuberales*. Because truffles do not appear above ground, they must be sought out from their symbiotic cohabitation among the roots of certain trees, which pigs do, only to be frustrated, and which dogs and goats can be trained to do. Whether our enjoyment of truffles is related to our perhaps unconscious enjoyment of our own underarm sweat is a matter of conjecture.

Rather oddly, and therefore not necessarily believably, results from a series of experiments suggest that women prefer the smell of more symmetrical men, especially mid-way through their menstrual cycle.

There is no evidence that men prefer the smell of more symmetrical women.

143 CIVETONE $C_{17}H_{30}O$

144 MUSCONE $C_{15}H_{28}O$

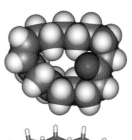

The civetone molecule (143) does not lie in a perfect circle but can twist and turn (except at the carbon–carbon double bond) and adopt many different shapes. That perhaps allows it to slip into many different olfactory receptors.

Civetone is responsible for the sweet odour of *civet*, the soft, fatty secretion of the perineal gland of the civet cat (*Viverra civetta*). The civet, which is initially unpleasantly pungent, may be collected. The perfume glands are largest in the male and are located between the anus and the genitalia. Civetone has been used in perfumery for centuries, and most of it is now prepared synthetically. The musk deer (*Moschus moschiferus*) of central Asia has a small sac in the skin of its abdomen in which a secretion collects as a viscous, brown, strong-smelling oil, especially during rutting. The secretion is complex and includes cholesterol (38), long-chain fatty-acid esters (35), and a small amount of muscone (144), a molecule closely resembling civetone and largely responsible for its musky odour.

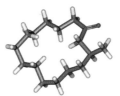

Musk is used in two ways in perfumery. It is used for its odour, as a component of heavy, musky, oriental perfumes. It is also used as a *fixative*, sometimes in concentrations so small that its own odour is masked. That is, it is added to more volatile fragrances to retard their evaporation, so that they are experienced as a symphony of odours rather than as a sequence in order of decreasing volatility. Several totally unrelated compounds have musk-like odours and are easier to synthesize; they are widely used in household products. One musk odour compound is

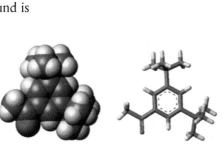

The civet cat (*Viverra civetta*) which inhabits Africa and Malaya. It is closely related to the hyena and is the source of civetone.

However, to emphasize the extraordinary specificity of odour, the following three closely related molecules are odourless:

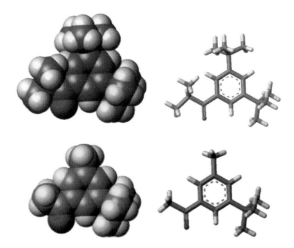

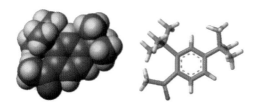

145 AMBERGRIS C$_{16}$H$_{28}$O

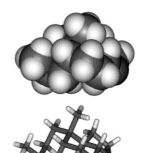

Perfumiers will go to all lengths to achieve their ends, including scavenging the beaches for ambergris (from the French *ambre gris*, grey amber), a substance produced, perhaps in response to an irritation, in the intestinal tract of the sperm whale (*Physeter catodon*) and excreted in multi-kilogram lumps. The actual material deposited is a complex mixture, of which the terpenoid ambreine is a major component. The substance degrades in the presence of sunlight, air, and salt water, to give the composite ambergris. A major component is the molecule shown here.

Ambergris, which it has been made illegal to import into a number of countries in an attempt to protect the whale, is a grey–black waxy mass. It is used as a fixative in perfumery and has been prized for the animal notes of its fragrance. A synthetic form of ambergris is now available to replace its forbidden (and, where it is not forbidden, scarce) natural source.

146 3-METHYLBUTANE-1-THIOL C$_5$H$_{12}$S

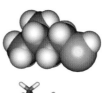

This molecule can be thought of as being like an alcohol (26), but with a sulfur atom in place of the oxygen atom. Such compounds are called *thiols* and were once called *mercaptans* (because they captured mercury atoms). This thiol bears out the reputation of sulfur compounds in general (as expressed in garlic and onions, 124) for having fierce odours: 3-methylbutane-1-thiol is the molecule squirted out in abundance by the striped skunk (*Mephitis mephitis*). In the skunk, the anal sacs that contain the secretion flavoured by this molecule are embedded in the muscle that erects the tail; thus, in a neatly economical piece of biological engineering, tail erection can be accompanied by an immediate squirt towards the adversary.

Sulfur compounds like the one shown, but with an –S–CH$_3$ group in place of –S–H, contribute to the odour of the urine of the red fox. Similarly, the odours of mink, stoat, and ferret are produced by cyclic compounds in which the hydrocarbon chain has bent around so that the carbon atom at its other end has replaced the hydrogen atom of the –S–H group.

147 UREA CH$_4$ON$_2$

148 BILIRUBIN C$_{33}$H$_{36}$O$_6$N$_4$

149 STERCOBILIN C$_{32}$H$_{44}$O$_6$N$_4$

Urea is the major organic component of human urine, the end product of the breakdown of the strings of amino acids that constitute proteins. An adult excretes about 25 grams of urea each day. As urine goes stale, microbial action converts it into ammonia (6), with its typically pungent smell.

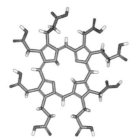

The cloudiness of the urine of cattle, horses, and some vegetarians is due to the precipitation of calcium and magnesium phosphates in the more alkaline solution that their diets produce. The golden yellow of urine is due to several molecules, specifically stercobilin (149) and the related compound uroporphyrin. Urine is said to fluoresce in ultraviolet light – which is perhaps an as yet unexploited source of entertainment in men's rooms. The African bird *Turacos*, which is remarkable for its plumage, uses a version of uroporphyrin in which a copper atom lies between the four nitrogen atoms and provides the red colour of its feathers.

The colour of faeces is largely due to stercobilin, a molecule that results from a long series of changes. First is the decomposition of some blood, then its conversion in the liver into the pigment bilirubin (148) of bile, then the conversion of bilirubin into a colourless compound, and finally the oxidation of that compound to give the yellow–orange stercobilin.

150 PUTRESCINE C$_4$H$_{12}$N$_2$

151 CADAVERINE C$_5$H$_{14}$N$_2$

The names of putrescine and cadaverine speak for themselves; little more need be said to describe their odour or to indicate their origin in rotting flesh. However, they also contribute to the odours of living

animals and are partly responsible for the smell of semen. Both add to the odour of urine and are present in bad breath (exhaled air that, sometimes in the lungs, has picked up volatile compounds from the blood). Putrescine is a poisonous solid and cadaverine a poisonous, syrupy liquid; both have disgusting odours.

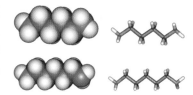

We dress in the odour of death. One of the monomers in nylon-6,6 is hexamethylenediamine (78), a diamine with six $-CH_2-$ groups. However, we cannot smell death in a nylon garment, partly because the amine molecules are anchored to their neighbours and cannot reach the nose.

152 TRIMETHYLAMINE C_3H_9N

The trimethylamine molecule is a derivative of the ammonia molecule (6) in which there are three $-CH_3$ groups in place of ammonia's three hydrogen atoms. Trimethylamine is a gas (but condenses to a liquid at $2\,°C$). It has the odour of rotten fish. Indeed, when you smell rotting fish, this is the molecule you can think of as being inside your nose, for it is given off as it is formed by enzymatic and microbial attack on fish proteins. The enzymes released during gutting are particularly potent at carving proteins into trimethylamine. This amine is also secreted by the coyote and the domestic dog, which also sometimes smell of rotten fish. Trimethylamine is also responsible for the odour of herring brine.

The distressing disease known as trimethylaminurea, in which the victim – more often women than men – exudes a persistent fishy odour (like Caliban in *The Tempest*) of trimethylamine is due to a defective enzyme that is responsible for breaking down proteins and eliminating possibly toxic compounds from the gut. There is some evidence that the same enzyme breaks down drugs such as anti-depressants, thereby making them less effective.

5

Sight and colour

Communication between an object and an eye is physical, for the message is carried by a ray of light, a stream of photons. However, molecules are often the agents of colour in an object and also act as sensors that respond to light once it has entered our eyes.

Many natural colours depend on the presence of particular molecules, and the following pages show some that inhabit petals, leaves, and skin and are responsible for the colours of flowers, vegetables, and people. When you look upon a rose with the information described here, you will know the molecules that contribute to its shades and hues. You will discover what changes give rise to the colours of a leaf in the autumn, and you will comprehend the ebb and flow of its colour. You will also understand the events that take place in your eye when you perceive these colours. You will see that, in vision, molecule speaks to molecule, and that the activated molecule in the eye triggers a signal that, in the deep unknown of the head, is interpreted as a colour. You will also learn what molecules contribute to the colours of hair, meat, and fat, and how some people do away with their colouration while others seek to modify theirs naturally or artificially.

The incandescence of the surface layers of the Sun results in the mixture of wavelengths that we are accustomed to interpret as 'white' light. If any colour is removed from white light, then the light takes on a hue. Filtering out orange light, for instance, results in

Complementary colours (indicated in units of nanometres) are diametrically opposite each other on an artist's colour wheel. Thus, if a substance absorbs red light from incident white light, the reflected light is the complementary colour of red, which is green.

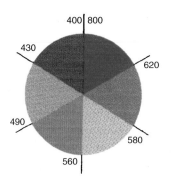

blue–green (cyan) light; filtering out cyan results in orange light. The colour resulting from the removal of a colour from white light is the latter's *complementary* colour. A traditional way of demonstrating the relation between colours is in terms of an artist's *colour wheel*, on which complementary colours are opposite each other along a diameter.

Substances absorb characteristic wavelengths of light when the incoming radiation excites electrons into different locations within the molecule. The difference in energy between the new and initial locations determines what wavelengths are absorbed. If only a little energy is needed to relocate the electrons, then the substance can absorb red light, for photons of red light have a relatively low energy. If the incident light is white, then the light reflected by such substances is perceived as green. If a lot of energy is needed to change the arrangement of the electrons, then the substance absorbs only high-energy, short-wavelength light. If the substance absorbs blue–indigo light (which is short-wavelength light), then the light it reflects is perceived as orange.

The colour green is interesting in this regard. On the colour wheel, it lies opposite both red and violet, where the two ends of the visible spectrum meet. Hence green can arise when low-energy, long-wavelength red light is removed from white light, when high-energy, short-wavelength violet light is removed, or when both are removed.

VISION

Although the receptors in the eye are sensitive to a physical stimulation – light of various wavelengths – once light has struck, all the detection, processing, and transmission of the neural signal is chemical.

The retina of the eye consists of two kinds of receptors, called *rods* and *cones*. The billion or so rods in the human eye function in dim light but do not distinguish between colours (light of different wavelengths). The three million or so cones function in bright light and do distinguish between colours. In fact, there are three kinds of cones, each of which absorbs red, green, or blue light and sends signals to the brain accordingly. Each receptor contains molecules that are sensitive to light, and their response to illumination is the trigger for a message to the brain.

153 11-*cis*-RETINAL $C_{20}H_{28}O$

The retinal molecule has a shape that will recur as a structural theme in this and the next section, for reasons that will soon become clear. The molecule is largely a hydrocarbon, and its most important feature is that it consists of an alternating chain of carbon–carbon single and double bonds. That chain terminates in the –CHO group typical of aldehydes (hence the *-al* in the molecule's name). The –CHO group is important insofar as it is quite reactive and can combine with (and anchor the retinal molecule to) other molecules in its vicinity – particularly protein molecules. Retinal is formed from vitamin A (155) and β-carotene (158) in the diet.

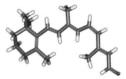

The alternating double and single bonds result in two properties. One is the rigidity of the tail: the hydrocarbon chain is not free to wind up into a coil like a singly bonded hydrocarbon (20), but sticks out rigidly from the six-membered ring. The other consequence is that the electrons in the chain are only loosely held and can be moved to new locations reasonably easily. Hence, the molecule can absorb energy from light that falls on it, storing that energy as a shift of its electrons into a new arrangement.

The 11-*cis*-retinal molecule is the one that absorbs incident light in the rods and cones of the eye. In the rods, the retinal molecule is linked to a protein, *opsin* (156), to give the light-sensitive *rhodopsin*. Rhodopsin is also called *visual purple*. In the cones, retinal molecules are linked to three slightly different opsins that change the wavelength of the light they can absorb to red, blue, and green. Retinal can also be excited by ultraviolet radiation, but this dangerously destructive radiation is normally filtered out by a yellow pigment in the cornea, and we are confined to seeing by longer-wavelength 'visible' light. Some people who have had cataracts removed, however, can read by ultraviolet light.

The different colour-absorption characteristics of retinal in combination with the three opsins stem from the modification of its energy levels in the different environments provided by the opsin molecules. The red and green opsins are most alike, and their genes reside on the X chromosome. The gene for blue resides on another chromosome (the seventh) and that for rhodopsin on yet another (the third). The occurrence of the red and green genes on the X chromosome accounts for the observation that red–green colour blindness is largely confined to males (who have only one X chromosome) and is rare among females (who are XX, and a defect in one X chromosome may be dominated by the other). Dichromatic colour blindness occurs when one of the red or green opsins is defective and cannot bind a retinal molecule. If instead a mutation causes a change in the environment of the retinal, but still allows it to bind, then the individual will have abnormal trichromatic vision.

It should not be surprising that the light-absorbing ability of retinal and its protein combination rhodopsin are used to extract energy from sunlight, not merely for vision. The purple, non-sulfur bacterium *Halobacterium halobium*, which is commonly found in sunny, concentrated brine that is about seven times more salty than sea water (as at the edges of salt pools), uses it in the form of bacteriorhodopsin as space stations use a solar cell. The halobacterium uses the energy the retinal captures to empty its cell of hydrogen ions, which leads to the formation of ATP (109) from ADP, which in turn powers the activities that constitute its little life.

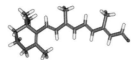

154 All-*trans*-RETINAL $C_{20}H_{28}O$

155 RETINOL $C_{20}H_{30}O$

The all-*trans*-retinal molecule is identical to the 11-*cis*-retinal molecule with the exception that one half of the molecule has been twisted into a new rigid arrangement around one of the double bonds, so the side chain is fully extended (and still rigid).

When light strikes a rod or cone, it is absorbed by the *cis*-retinal in that receptor. As in all molecules, this absorption of light causes a shift in the location of an electron. In this case, the immediate outcome is that a double bond is broken open, leaving a single bond. Single bonds are like hinges and one end of the molecule is now free to rotate relative to the other. As a result, *cis*-retinal suddenly changes into *trans*-retinal. Once it has done so, the double bond can snap back into position and freeze the molecule into its new shape. This massive change of shape affects the shape of the opsin protein (156) and causes a signal to be sent along the optic nerve to the brain. After that has happened, the *trans*-retinal breaks away from its location on the opsin, where it no longer fits, is converted elsewhere back to *cis*-retinal, and then reattaches to an opsin to await its next illumination.

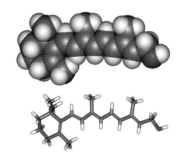

All image-resolving eyes (which are found only in the molluscs, arthropods, and vertebrates) have adopted retinal as their light-sensitive component, even though they may have undergone separate evolution. The halobacteria mentioned above, although they are of a different kingdom, also have a vestigial light-sensing ability based on retinal. All these creatures have adopted the molecule because it is ideal for its purpose. It undergoes a large change of shape when it changes from *cis* to *trans*, and its absorption wavelength is readily and widely modified by its protein anchor. Moreover, the *cis* form is structurally stable and does not readily convert to the *trans* form in the dark (so that we are not misled and confused by false signals). It can also be synthesized from readily available precursors, including the carotenes (158).

The closely related retinol molecule, (155), differs from all-*trans*-retinal only by having an –OH group at the end of its tail, so formally it is an alcohol and is a metabolic product of β-carotene (158). Retinol is Vitamin A. Good sources of it are liver, egg yolk, butter, and whole milk. Not only is retinol a precursor of retinal, and hence essential to the maintenance of vision, but also it is involved in cell growth and the formation of mucus and, in a related process, the maintenance of healthy epithelial tissue. Retinol interferes with the formation of hard forms of proteins and hence inhibits the formation of dry, hard skin. For skin-care preparations, the retinol molecule (an alcohol, remember) is combined with a long-chain fatty acid (typically palmitic acid), which helps to decrease its rate of oxidation. Variations on the theme of the retinol molecule are used to treat acne and psoriasis.

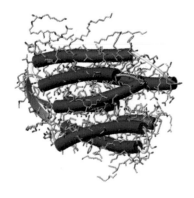

156 OPSIN

As remarked in the discussion of retinal (153, 154), a retinal molecule is attached to a protein, opsin (from the Greek for 'to see'), to form rhodopsin, which acts as a miniature camera. Rhodopsin consists of seven helices (represented by the red cylinders) that are strung together by the rest of the polypeptide chain, and lies embedded in the membrane of a cell. The retinal molecule is attached, through its –CHO group, to a lysine amino acid on the seventh of these helices, about half way down the barrel they form. A human rod cell contains about 40 million rhodopsin molecules distributed over about a thousand flattened discs of the cell membrane. When light is absorbed by the retinal molecule, it undergoes a change of shape that causes the opsin molecule to modify its shape. That in turn triggers a cascade of events that send a signal into the visual cortex. Meanwhile, the retinal molecule is pushed out of its barrel-like cavity and is converted back into its active form elsewhere.

LEAVES, CARROTS, AND FLAMINGOS

The bright reds, yellows, and blues of flowers are due to a class of compounds treated in the next section, but many of the yellow and orange hues of nature are due to molecules called *carotenoids*, which we consider here. As we examine their structures, it will be good to keep in mind their common structural unit, for carotenoids (and parts of chlorophyll too) can all be regarded as mainly strings of isoprene units (67). It should hardly be surprising that rubber is exuded by trees, because rubber is a polyisoprene (68), or that humans emit it, for vitamin A (155) is a chain of isoprene units. Many plant odours (particularly the terpenes) are also built from isoprene units – emphasizing again the economical elegance, or penny-pinching, of nature. Here, though, we are concerned with colour and carotenoids. Carotenoids can be passed from an individual to its predator, and that, as we shall see, leads to some interesting relations.

157 CHLOROPHYLL $C_{55}H_{72}MgN_4O_5$

The ubiquitous green of leaves is due to *chlorophyll*. The atom at the centre of the molecule, lying there like an eye, is magnesium. This hugely important molecule absorbs both high-energy violet and low-energy red light, which is exactly what is required for the reflected light to appear green. Hence vegetation is green. The light absorbed provides the energy for the task of *photosynthesis*, in which carbon dioxide (3) and water (5) are combined to form carbohydrates (93). The chlorophyll molecule is therefore the antenna that green plants use to harvest the Sun and open the way to the processes of life.

Chlorophyll absorbs light very strongly and can mask many other colours, some of which are unmasked when the fragile chlorophyll molecule decays in the autumn (158). When vegetation is cooked, the central magnesium atom is replaced with hydrogen ions. This changes the energy needed to excite the electrons in the rest of the molecule, so cooked leaves change colour – sometimes becoming an insipid green. This insipid colour is also characteristic of some processed food, such as processed peas. The colour can be restored, or brilliant green added to confectionary, by replacing the central magnesium ion by a copper ion; canned peas and mint jelly, though, are normally coloured with a coal-tar dye (Green S).

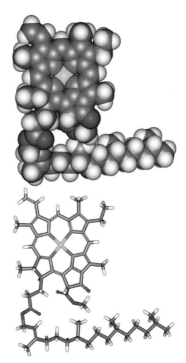

158 β-CAROTENE $C_{40}H_{56}$

The β-carotene molecule is a hydrocarbon and a string of eight iso-prene (67) units. We take it as the starting point for all the other molecules in this section. The most striking feature of the molecule is its chain of alternating carbon–carbon single and double bonds. As explained in the discussion of the closely related molecule reti-nal (153), this arrangement has two consequences. One is that the molecule is stiff and inflexible. The other is that the electrons in the extended chain are only loosely held and can readily be ex-cited. β-Carotene itself can absorb blue–indigo light and hence looks orange. Another feature of the molecule relevant to its occurrence in the world is that, being a hydrocarbon, it is soluble in fats, which provide a similar oily environment, but not in water.

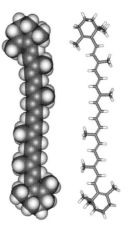

β-Carotene occurs in carrots, but their yellow core is due to slightly oxidized versions of carotene known as *xanthophylls*. In fact, our familiar orange carrots are a fairly recent innovation, for their pre-cursors were purple, owing to the presence of molecules like those described in the next section. However, a mutation led to a variety in which that pigment did not develop, leaving the golden yellow β-carotene molecule dominant. β-Carotene is also partly responsi-ble for the colour of the mango and the persimmon. The pale cream of milk and the yellow of butter are due largely to the presence of molecules related to β-carotene. Such molecules are added to

The persimmon owes its colour to carotene and its astrigency to tannins. The latter become insoluble as the fruit ripens and the bitterness lessens. The persimmon tree is a member of the ebony family (*Ebenaceae*); the friut shown here is that of the Japanese persimmon (*Diospyros khaki*).

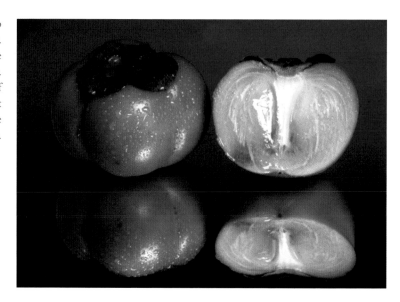

margarine to emulate the colour of butter. The fat of meats is often tinged slightly yellow by β-carotene that the animal ingested, which, by virtue of its hydrocarbon nature, dissolved in the fat.

β-Carotene accompanies chlorophyll in photosynthetic organisms. Its role is partly to harvest some sunlight that is not absorbed by chlorophyll, as well as to react with energetic oxygen molecules so as to protect the cells from degradation. There is typically about one carotene-like molecule to every three chlorophyll molecules in a leaf, so the darker the green of the leaf the greater the concentration of carotenes. The yellow–orange of β-carotene remains masked by the chlorophyll until the autumn, when the chlorophyll molecule decays and is not replaced; that leaves the sturdier β-carotene molecule to exhibit its powers of light absorption, and the leaves turn yellow.

Retinol (Vitamin A, 155) is a metabolic product of β-carotene and both it and β-carotene itself are believed to be active against some forms of cancer, possibly as a result of their role in the development of cells but also because they act as antioxidants by scavenging free radicals.

Leaves that contain carotene, such as the leaves of birch and hickory, change from green to bright yellow as the chlorophyll disappears. In some trees with a high sugar concentration, anthocyanins (162) are formed and cause the yellowing leaves to turn red. When grass is cut and bleached in the sun, its carotene molecules are broken down into molecules of β-ionone (120), which has the characteristic odour of hay.

159 LYCOPENE $C_{40}H_{56}$

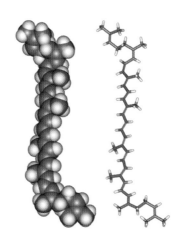

Lycopene is like a carotene molecule with its two rings broken open. It is a deeper red than carotene and is responsible for the colour of tomatoes (*Lycopersicon esculentum*). As a green tomato ripens, the chlorophyll of the unripe fruit decays, the increasing amount of lycopene is unmasked, and the fruit turns red. Lycopene and carotene join forces to contribute colour to apricots. Tomatoes, when they were first cultivated by the Aztecs, were yellow, but selective breeding has turned them red.

Dietary studies have focussed on the excellent antioxidant properties of lycopene, and it has been found that more is absorbed by the body from tomato paste than from the raw fruit, perhaps on account of the cell breakdown that has occurred when the tomatoes are mashed. Tomato juice turns out to be a poor source, perhaps because it lacks the fats that are needed to aid transport in the body. Lycopene is relatively rare as a colourant, but it is also responsible for the colour of pink grapefruit, fresh papaya, raw guava, watermelon, apricots, and rosehips.

A shorter version of the lycopene chain, with each end partially oxidized to become a carboxyl group, –COOH, is the molecule *crocetin*. This molecule is responsible for the colour of *saffron* (from the Arabic word *za'faran*, meaning 'yellow') and is used, among other applications, to colour saffron rice. Saffron is obtained from the stigma of the eastern crocus (*Crocus sativus*). A close relative of crocetin, differing from it by way of a slight lengthening of the chain and the conversion of one carboxyl group into its methyl ester, is *bixin*. Bixin is the molecule responsible for the red colour of *annatto*, which is obtained from the pulpy parts of the seed of the tropical shrub *Bixa orellana* and used to colour cheeses such as red Leicester.

160 ZEAXANTHIN $C_{40}H_{56}O_2$

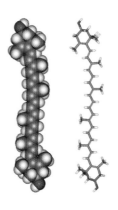

With zeaxanthin we move to the oxygen-containing *xanthophylls* and to the corn belt, for it and its relatives occur widely and make great stretches of the world golden. Zeaxanthin and carotene jointly colour corn (*Zea mays*) and contribute to the colour of the mango. Zeaxanthin is also a partner in the colouration of egg yolk and orange juice, where it is joined by *lutein*, a molecule that differs from zeaxanthin only in the location of the last double bond on the right. Both kinds of molecule are present in deeply coloured vegetables, such as

The yellow corn (*Zea mays*), or maize, take its colour from zeaxanthin. A corn crop is said to mature somewhere in the world every month of the year; half the world's total is produced in the USA.

spinach. When they are ingested, lutein and zeaxanthin dissolve in hydrocarbons (because they are largely hydrocarbon themselves) and contribute to the yellowish tint of animal fats.

Both zeaxanthin and lutein collect in the macula region of the retina (the *macula lutea*, or 'yellow spot'), where their two-fold function is to filter out some of the blue light and ultraviolet radiation, to protect the retina against its constant exposure to light, and to scavenge free radicals. Macular degeneration is the most common form of blindness in the elderly, and zeaxanthin is nature's version of internal sun-glasses.

161 ASTAXANTHIN $C_{40}H_{52}O_4$

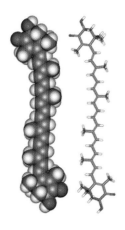

The presence of four oxygen atoms in astaxanthin changes the energy needed to move its electrons, so its colour is different from that of carotene. Astaxanthin is, in fact, pink (despite its name, which comes from the Greek words for 'yellow flower') and is responsible for the colour of salmon, trout, and red seabream. Astaxanthin, which is often extracted from krill or made synthetically, is used in fish-farms to enhance the colour of the fish which otherwise would have a diet deficient in it. Astaxanthin also occurs in the carapaces of shellfish, including lobsters and shrimps; however, its pink colour is

A prawn in its shell (prawn is the name given to the larger species of shrimp). The common European shrimp (*Crangon vulgaris*) is dark brown or grey but becomes pink when cooked. The change in colour is a result of the relase of astaxanthin from a combination with a protein. Crabs, crayfish, and lobsters are related crustaceans.

not apparent in the live animals because the molecule is wrapped in a protein, giving a blackish hue. When lobsters and shrimps are boiled, the protein chain uncoils, liberating the astaxanthin molecule, and the lobsters and shrimps turn pink.

The detailed structure of astaxanthin provides a way of distinguishing farmed from natural salmon. The –OH groups in astaxanthin may be both below the plane of the rings at each end of the molecule, both above, or one above and the other below. The first of these is the form found in the wild Pacific and Atlantic salmon. Synthetic astaxanthin is principally the one-above-one-below isomer, and farmed salmon fed synthetic astaxanthin are unable to convert it into its natural form.

An astaxanthin molecule that has lost its two –OH groups is called *canthaxanthin*. That molecule is responsible for the colour of the American flamingo, which owes its colour to its diet: flamingos in captivity lose their pink colour if they are not supplied with adequate amounts of carotenoid-containing shrimp.

FLOWERS, FRUIT, AND WINE

Many of the bright colours of the world in spring, summer, and autumn are due to a single class of compounds, called *flavonoids*. They have in common a basic framework:

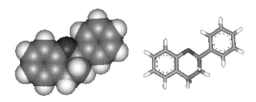

Most of them occur in combination with a sugar molecule, producing a glycoside, but the sugar component will not be shown in the models. Flavonoids occur in leaves as well as in petals. Their function in leaves is to absorb the ultraviolet radiation that could be so destructive to the genetic material and the proteins in the cell.

Nature mixes its palette of colours from the framework of flavonoids by attaching different groups to different places around them, linking different types of sugar molecules to them, and changing the acidity of their environment.

Eyes other than our own can see a richer palette. Bees, for instance, have eyes that are sensitive to ultraviolet radiation, so they see a richer range of colours where we might see only one. An example is the (to us) plain yellow of the fleabane (*Pulicaria dysenterica*). To the bee, which notices both the ultraviolet and the yellow that are reflected, these flowers appear a colour called 'bee's purple'.

Absorption of ultraviolet light can occur accidentally in animals that have digested flower pigments and can induce distress. When cattle feed in pastures that includes the yellow St John's wort (genus *Hypericum*), the colouring pigment finds its way to the surface of the skin, where ultraviolet radiation is absorbed and initiates a chemical reaction that inflames the skin and, in severe cases, can lead to death.

162 PELARGONIDIN $C_{15}H_{11}O_5$

Pelargonidin is the simplest of the *anthocyanidins*, the principal class of flavonoids discussed here. In combination with sugar molecules such as glucose (87), anthocyanidins become *anthocyanins* (from the Greek words for 'blue flower'). Anthocyanins are responsible for many

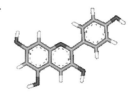

of the red, purple, and blue colours of nature (but not the red of either beet or *bougainvillea*, which is due to other types of compounds called *betacyanins*). Pelargonidin itself is responsible for the red of the common geranium (genus *Pelargonium*) and contributes to the colour of ripe raspberries and strawberries. *Cyanidin*, an anthocyanidin with two –OH groups on the right-hand benzene ring, provides the violet on the palette; it is responsible for the colour of ripe blackberries and contributes to the colours of blackcurrants, raspberries, strawberries, apple skins, and cherries. Its colour is responsive to the acidity of its surroundings: in acid solution it is red, but in alkali blue.

Anthocyanins are formed by a reaction involving sugars in sap that occurs once the concentration of sugar in the sap is high and in the presence of light. The need for light is the reason why apples often turn red on the side exposed to the Sun but remain green on the side in shadow. Red maples, red oaks, and sumac produce high concentrations of anthocyanins and turn bright reds and purple in the autumn. Chlorophyll is destroyed by low temperatures and bright sunshine but the formation of anthocyanins is enhanced by both these factors. The concentration of anthocyanins is also increased by dry weather, for that increases the concentration of sugar in sap. The brightest autumn colours are therefore the result of dry, sunny days followed by cool, dry nights.

Here nature again shows its economy, for the strikingly different colours of the blue cornflower (*Centaurea cyanus*) and the red poppy (*Papaver rhoeas*) actually arise from the same molecule. In the cornflower the sap is alkaline, and the cyanidin molecule loses a hydrogen ion and turns blue. In the poppy the sap is acid; because the molecule is in an environment rich in hydrogen ions, it acquires one and turns red. Rhubarb, which is rich in oxalic acid (100), is coloured red by the acid form of cyanidin. Red cabbage retains its colour, which is due to cyanidin, if it is kept acid as it is cooked. Flowers sometimes modify the acidity of their sap and change colour after pollination so that they become less conspicuous to insects.

Replacing an –OH group of the cyanidin molecule with an –OCH$_3$ group brings us to the *peonidin* molecule and the colour of the peony (genus *Camelia*). Peonidin also contributes to the colour of cherries and grapes. Some flowers change their acidity as they open. For instance, the Japanese Morning Glory (*Ipomoea hederacea*) has red–purple buds but blue petals when it is fully opened. The change, which is due to a decrease in acidity of the sap, is due to a protein that extracts hydrogen ions and replaces them by sodium ions as the flower opens.

The colours of spring, summer, and autumn are found in a glass of wine. Red wine acquires its colour from anthocyanidins. As it matures, the anthocyanidins react with other largely colourless but bitter flavonoids that are also present and are known collectively as a type of *tannin*. This reaction removes the tannins and improves the taste of the wine. As red wine grows even older, the reaction between the anthocyanidins and the tannins results in the removal of the red anthocyanidins and leaves the brown tannins visible. The colour of white wines is due in part to quercetin (163), which turns deeper brown as more is oxidized with age. Initially, a young white wine may have a greenish hue from the chlorophyll (157) molecules that survive fermentation.

163 QUERCETIN $C_{15}H_{10}O_7$

The quercetin molecule is a flavonoid, but it is a *flavonol* rather than an anthocyanidin (the extra oxygen on the ring makes the difference). This molecule is yellow and is responsible for the colour of Dyer's oak (*Quercus tinctoria*). Quercetin also occurs in many leaves, but its colour is masked by chlorophyll (157) until the latter decays. Some flavonols are colourless (to our eyes) and are present in leaves to prevent damage by absorbing ultraviolet radiation. They do not absorb blue and red wavelengths and hence do not intrude on the energy-harvesting

The colours of the carotenes, anthocyanins, and flavonols in leaves are masked by chlorophyll, but when that decays in the autumn their yellows, oranges, and reds are revealed.

function of chlorophyll. In the autumn, when the chlorophyll decays, these colourless flavonols are stripped of the oxygen atom attached to the ring and thus are converted into anthocyanidins such as the scarlet pelargonidin (162). This switch of compound, induced by no more than the loss of a single oxygen atom, is what we perceive as the majestic blaze of autumn glory.

If the quercetin molecule is modified by removing the –OH group on the heterocyclic ring, the *luteolin* molecule is obtained. Luteolin is an example of a *flavone*, another common type of pigment molecule. It is the yellow dye of the chrysanthemum.

BROWNS, BRUISES, AND TANS

Not everything is brightly coloured, and some things that are fresh and white become brown when they are bruised. Many people are brown even before they are bruised. Hair, unless it courts fashion deliberately and synthetically, is rarely brightly coloured. Some browns are formed by reactions between carbohydrates (87) and the amino acids (80) of proteins. Such a combination, which is called a *Maillard reaction*, generally occurs only when the two substances are heated, and it results in a very complex mixture of products. At the same time as the brown colour develops, molecules that cause taste and odour are formed, so browning induced by heating contributes to flavour. Maillard reactions occur on the hot, dry crust of bread while it is being baked, and some of the more mobile flavour molecules diffuse into the interior of the loaf. Browning reactions also occur when maple syrup is boiled, as the sugar and amino acids in the sap combine: the longer it is boiled, the more extensive the reaction and the deeper the colour of the syrup. The colour of beer also darkens during brewing, as the unfermented sugar reacts with the amino acids present. Roasting coffee and cocoa beans and nuts of various kinds also brings out the flavour as well as the colour by stimulating Maillard reactions between their carbohydrate and protein components.

Another familiar brown is that of *caramel*, which is formed when sucrose (89) is heated. This is not a Maillard reaction, for no amino acids are present. Instead, it is the result of very complex decomposition and recombination reactions caused by the heat. Small, tasty, smelly molecules, including acrolein (113), are also produced by the decomposition and contribute to the flavour of caramel. Many of the browns and blacks of nature can be traced to a single molecule – melanin – the one we consider next.

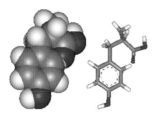

164 TYROSINE C₉H₁₁O₃N

165 MELANIN

Tyrosine is a naturally occurring amino acid that is part of many proteins. It is also the starting point for the synthesis, in the body, of some of the molecules responsible for transmitting signals from one nerve cell to another. For our purposes, though, it is the start of the trail that leads to melanin. Under the action of enzymes, tyrosine molecules polymerize to give the melanin molecule, a part of which is shown here. The electrons in the alternating double and single bonds are only loosely held and can be moved around by the energy of light of any wavelength. Hence, all light is absorbed by melanin and an object that contains it looks black. Melanin molecules become attached to protein molecules and accumulate in granules that range in colour from yellow through brown to black.

Melanin contributes to the pigmentation of skin and hair, except that of redheads, where the iron-rich pigment trichosiderin is dominant (85), and of those who have destroyed the double bonds by oxidation with hydrogen peroxide (11). It also provides a dark

This octopus (*Octopus cuanea*) from the Red Sea is one animal that deploys melanin to change its colour.

background that makes colours due to scattering of light more obvious, as in the iris of the eye. The number of melanin-producing cells is much the same in dark- and light-skinned people, but they are more active in dark-skinned people.

There are various versions of melanin in hair, including the black pigment eumelanin and the red and yellow pigment pheomelanin. In black and dark brown hair there is a large number of big granules of eumelanin; brown hair contains a smaller number of granules; in blonde hair the granules are much smaller and more widely dispersed, but still contain mainly eumelanin. The pheomelanin granules in red hair are much less dense. Granules containing both eumelanin and pheomelanin are believed to give rise to golden brown or auburn hair. The formation of pigment in hair and skin is controlled by numerous genes: in mice, 147 genes have been implicated. The condition of albinism in humans stems from the genetic corruption of the enzyme tyrosinase, which leads to the failure of the body to produce melanin from tyrosine.

Melanin is a part of the colour-change mechanism of the chameleon, in which it is shipped around through channels in the skin and used to cover the brighter pigments below. Animals that darken themselves, including the octopus and its relatives, achieve the change by dispersing granules of melanin; the skin lightens when the melanin granules aggregate again.

Melanins with slightly different compositions are also formed when fruit is bruised; their formation follows damage to cell walls that allows an enzyme, *phenol oxidase*, to act on the material within. This enzyme is absent from citrus fruits, melons, and tomatoes, which do not brown so rapidly when they are bruised. However, the precursors of melanin in fruit are not quite the same as those in animals, for, although they contain, like tyrosine, an –OH group attached to a benzene ring (that is, they are phenols), they are not necessarily amino acids. The dark colours of tea are due to similar melanin-like polymerized phenols.

In addition to reddening of the skin, the ultraviolet radiation in sunshine causes two types of tanning. One form occurs immediately (and continues in corpses) but disappears within a few hours; it is due to damage to the molecules from which melanin is formed. A longer-lasting tan (which depends on biochemical processes and does not occur in the dead) is induced by ultraviolet radiation in a narrower range of wavelengths. This tan is due to an increase in activity of the melanin-producing cells deeper in the skin; their product becomes apparent when the melanin has had time to reach the surface, normally within a day or so. The melanin acts like the flavonoids of plants (162), protecting the DNA in cells from damage.

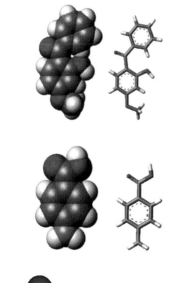

166 2-HYDROXY-4-METHOXYBENZOPHENONE $C_{14}H_{12}O_3$

167 *para*-AMINOBENZOIC ACID $C_7H_7O_2N$

168 1,3-DIHYDROXYPROPANONE $C_3H_6O_3$

There are three types of solar ultraviolet radiation. Ultraviolet A (UV-A) penetrates the top layer of skin, the epidermis, and damages the lower layer, the dermis. As a result, it causes ageing and wrinkling. It can also contribute to the formation of skin cancers. Ultraviolet B (UV-B) has a shorter wavelength and its photons are more energetic. As a result, they do more damage and cause sunburn and skin cancers. Window glass blocks UV-B but allows UV-A to pass through. Ultraviolet C (UV-C) has a still shorter wavelength and is blocked by the ozone layer (4); it is a serious hazard now that the ozone layer is becoming depleted, and is also a hazard in welding, where the arc generates this damaging radiation.

A suntan cannot be achieved without sunburn, because sunburn triggers production of melanin. Sunscreens act by lowering the dose of damaging radiation received by the skin, so that production of melanin has a chance to catch up with sunburn. The benzophenone molecule shown here (166) is a common constituent of broad-spectrum formulations designed to reduce the effects of UV-A and UV-B: it screens out UV-A. The benzoic acid derivative (167) is known as PABA; it and modifications of it are typical of the molecules contained in sunscreens aimed at reducing the effects of UV-B radiation. The early sunblocks used finely divided titanium dioxide or zinc oxide – essentially, white paint – to shield the skin. Modern versions use much smaller particles: these microparticles do not scatter visible light but do scatter ultraviolet radiation. Therefore, they screen out ultraviolet radiation but are invisible.

Quick-tanning lotions cut out all the radiation that causes sunburn (and hence true tanning), permitting only the immediate-response, broad-spectrum, destructive tanning. Their effect is often enhanced by the inclusion of a skin dye such as dihydroxypropanone (168), which reacts with the amino acids in the upper layers of the skin and colours them brown through a version of the Maillard reaction used for browning foods. This tan is ephemeral, for it is lost as the stained cells are rubbed off.

6

The light
and the
dark

One of chemistry's major contributions to the welfare of humanity has been its provision of pharmaceuticals. We know how some
of these drugs act, and it is possible to discern a relation between
the structure of their molecules and the influence they have on our
bodies. Some of them quell pain. Others induce calm or expunge
a depression. Still other molecules do the opposite and induce a
sense of euphoria, sometimes simultaneously inducing addiction and,
through addiction, ruination.

Unhappily, chemists achieve evil as well as good, sometimes by
accident but sometimes by intention. It would be improper to conceal
this dark face of their activity, to show their benevolent creations but
not their malevolent ones, so a few of their pernicious contrivances
are described here too.

ANALGESICS AND CURES

Analgesics can be classed into two broad categories. One consists of
those that act peripherally, at the site of the pain, to interfere with

the pain signals at their source. The other group acts centrally, on the central nervous system, to modify the brain's processing of the signals it receives along its pain nerves. Centrally acting analgesics often produce disadvantageous side effects, such as addiction and mood modification. Peripherally acting analgesics are not addictive; and they do not modify mood directly.

169 SALICYLIC ACID $C_7H_6O_3$

170 ACETYLSALICYLIC ACID $C_9H_8O_4$

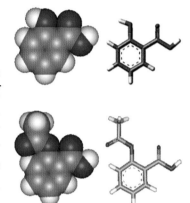

Several of the milder non-addictive analgesics on the market are aspirin-like. They are often derivatives of salicylic acid, the name of which is derived from the Latin name of the white willow, *Salix alba.* The acid appears in the bark of that tree, in combination with a sugar molecule as the glycoside *salicin.* Extracts of the bark were originally used to relieve pain, as in *oil of wintergreen.* Extracts of the leaves and bark of a number of trees and shrubs have similar medicinal properties that can be traced to the same substance or close relatives.

Salicylic acid, taken as its sodium salt, had uncomfortable residual effects in the stomach. A great improvement in tolerance was achieved when the analgesic property of acetylsalicylic acid was discovered in 1897 by Dr Felix Hoffmann, then working for Bayer AG, who apparently was motivated by the wish to find a treatment for his father, a sufferer from rheumatoid arthritis. The compound went on sale in 1899 under the tradename Aspirin: this name is still protected in some countries (Germany, for instance), but is a generic name in most others. Its mode of action remained unknown for a long time. It is now thought to interfere with the synthesis of compounds called prostaglandins by inhibiting the action of an enzyme, *prostaglandin cyclooxygenase.*

Prostaglandins are locally acting hormones that are involved in numerous processes in the body; among them is the modification of signals transmitted across synapses (the connections between nerves), particularly pain signals. Prostaglandins may also be involved in the dilation of blood vessels that causes the pain experienced as a headache (if the vessels are intracranial) or as a migraine (if they are external to the skull). In these cases, aspirin and other local analgesics may act by inhibiting the synthesis of the prostaglandins that might cause the pain as well as help transmit it. The zigzag lines often seen by victims of migraine are caused by constriction of the blood vessels in the region of the brain responsible for vision. Since they have a different cause from the headache, they can be treated independently with small doses of substances (including amyl nitrite) that can dilate the vessels.

The applications of aspirin seem endless. It reduces the risk of recurrence of heart attacks and strokes by reducing the tendency of platelets to clot and allowing the blood to flow more freely. Franz Kafka, somewhat unromantically, once explained to his fiancée Felice Bauer that aspirin was one of the few things that eased the unbearable pain of being.

The proprietary formulation *Alka-Seltzer* consists of aspirin, anhydrous citric acid (102), and sodium bicarbonate ($NaHCO_3$). When the mixture is added to water, the aspirin (an acid) is converted into its more soluble sodium salt, and the citric acid releases carbon dioxide (3) from the bicarbonate ions, giving the fizz as well as a pleasant taste.

171 *N*-ACETYL-*para*-AMINOPHENOL $C_8H_9O_2N$

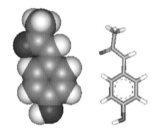

The analgesic properties of drugs such as that shown were discovered by accident when some *acetanilide* (like the molecule in the illustration, but without the –OH group) was added by mistake to a patient's prescription. Acetanilide can, however, be toxic, and less harmful compounds were sought. One of them is *N*-acetyl-*para*-aminophenol, better known as *paracetamol* and sold under the trade-name *Tylenol*. Acetanilide is, in fact, converted into paracetamol in the body, which gives it its analgesic properties; but some is also converted into *aniline*, a benzene molecule with one hydrogen atom replaced by an –NH$_2$ group, which gives it its toxicity.

The marked similarity between this molecule and acetylsalicylic acid (aspirin, 171) should be noted. Although they are built up from different atoms, they have similar shapes, with an –NH group in the aminophenol taking the place of an –O in aspirin. Because of their similarity they are both recognized by the same enzyme, the one responsible for the biosynthesis of prostaglandins, and paracetamol also acts by inhibiting production of prostaglandins.

172 DIAZEPAM $C_{16}H_{13}ON_2Cl$

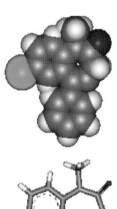

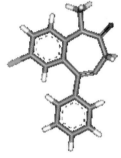

Much insomnia is due to anxiety, which is associated with heightened neuronal activity in the limbic system, the region of the brain associated with emotion. This activity can spread to the brain stem, the most primitive part of the brain, and maintain wakefulness. The molecules called *benzodiazepines*, of which diazepam is an example, bind to protein molecules in the junctions between nerves and enhance the ability of the neurotransmitter GABA (27) to bind to neighbouring sites on the same molecule. As explained in the discussion of the depressant action of alcohol, GABA inhibits the firing of nerve cells, so the presence of the benzodiazepine molecule encourages this inhibition. The sites to which benzodiazepines bind are particularly rich within the limbic system, so ingesting a benzodiazepine suppresses the abnormal activity there that we experience as uneasiness and generalized fear and know as anxiety. That is, benzodiazepines are specifically anti-anxiety drugs that (in larger doses) act indirectly as sedatives. They are used for the treatment of the effects that accompany alcohol withdrawal, such as tremors, delirium, seizures, and hallucinations.

Benzodiazepines are slightly addictive but not particularly lethal. However, ethanol and the benzodiazepines bind at neighbouring sites on the same protein, and the two together can so change the shape of the protein that there is a strongly increased propensity for GABA to bind. This results in a massive inhibition of nervous activity and may lead to death. The benzodiazepines include the diazepam molecule illustrated here, which is widely sold under the trade name *Valium* and used as a muscle relaxant as well as a tranquilizer. If the chlorine atom on the left of the diazepam molecule is replaced by $-NO_2$, and the $-NCH_3$ by $-NH$, the result is *nitrazepam*, which is sold as *Mogadon*. Another tranquilizer, *chlordiazepoxide* has a very similar structure and is sold as *Librium*.

173 MORPHINE $C_{17}H_{19}O_3N$

174 DIACETYLMORPHINE $C_{21}H_{23}O_5N$

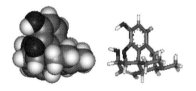

Morphine is the principal component of *opium* (from the Greek word *opion*, for 'poppy juice'), which is obtained as the milky juice that exudes from unripe poppy seed capsules (*Papaver somniferum*). Unlike

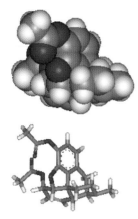

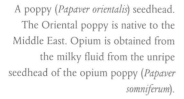

the aspirin analgesics, morphine and related compounds act on the central nervous system and can induce addiction. The specific action of morphine and its relatives appears to be related to the ability of the molecule to fit into and block a specific receptor site on a nerve cell. The shape of the receptor that has been proposed is shown in the illustration. The benzene group of the morphine molecule fits snugly against the flat part, and the neighbouring group of carbon atoms is at the correct distance and orientation to fit into the groove. Beyond the groove is a group with a negative charge, which can attract the positive charge of the nitrogen atom. By fitting the shape of the receptor and binding to it, the incoming morphine molecule eliminates its action. In this respect the molecule mimics the body's natural pain killers, the encephalins.

Morphine acts on the deep, aching pain described earlier as *slow pain* (see Chapter 4) but has no effect on fast pain. Large accumulations of morphine-receptor sites are found in the *substantia gelatinosa*, the region of the spinal column where pain signals are first processed; there, morphine acts by raising the threshold at which slow pain is experienced. Morphine-receptor sites are also abundant in the medial region of the thalamus, the part of the brain that acts as an input region for slow-pain signals. High concentrations are also found in the limbic system, which, as we have noted, is a region closely associated with emotion. This point will become relevant when we consider the

A poppy (*Papaver orientalis*) seedhead. The Oriental poppy is native to the Middle East. Opium is obtained from the milky fluid from the unripe seedhead of the opium poppy (*Papaver somniferum*).

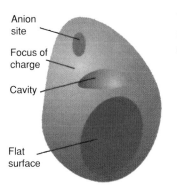

Anion site

Focus of charge

Cavity

Flat surface

The supposed shape of the protein that acts as the receptor site for morphine and related molecules.

stimulation of euphoria in the next section. A close relative of morphine is *codeine*, which is obtained when the –OH group on the left in the illustration is replaced with –OCH₃. In the body, the –OCH₃ group is replaced with –OH, so that codeine is converted back into morphine.

A diacetylmorphine molecule (174) is a morphine molecule in which the hydrogen atoms of two –OH groups have been replaced with acetyl groups (–COCH₃). Other names for diacetylmorphine are *heroin* and *diamorphine*. The replacement of the hydrogen-bonding –OH groups results in heroin being less soluble in water than morphine but more soluble among the hydrocarbon chains of fats (35). Therefore, although it must be injected directly into the blood, it passes more rapidly through the blood–brain barrier, the barrier that prevents water-soluble and large molecules from passing between the two. As a result, it is more potent than morphine (more 'heroic'), but its effect does not last as long. Once heroin has been absorbed into the body, the acetyl groups are removed, forming morphine, which provides its analgesic and euphoric action.

175 THALIDOMIDE C₁₃H₁₀O₄N₂

Thalidomide (which was marketed under many different names) became a popular sedative and mild hypnotic when it was first introduced onto the market in 1956; even massive overdoses were not lethal. In the early 1960s, however, an increasing number of congenitally deformed (teratogenic) babies were born to mothers who had taken the drug during the first three months of pregnancy, and the drug was soon withdrawn from sale in Europe. (It had never been sold in the USA, because of certain neurological

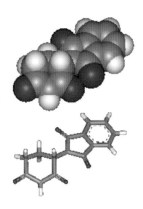

side effects that had been reported.) The specific teratogenicity was phocomelia, in which hands and feet are attached directly to the body.

We need to recall that some compounds are chiral and exist in left-handed and right-handed forms (32); we show both forms here. It is now known that the right-handed form is responsible for the anti-inflammatory action of thalidomide but the teratogenicity is due to the left-handed form. The commercial product was a mixture of left- and right-handed forms. However, taking just the right-handed form does not avoid the problem, for an enzyme in the liver converts it into the left-handed form.

The action of thalidomide is to inhibit angiogenesis, the formation of new blood vessels. Its action is thought to stem from the similarity of its structure to the bases adenine (199) and guanine (200) that are part of the genetic code carried by DNA (203). Hence the thalido-mide molecule attaches to regions of DNA that are rich in these bases (and specifically rich in guanine). There it inhibits the formation of proteins that are crucial to angiogenesis.

The inhibition of angiogenesis, though of tragic consequences for embryos, is of crucial importance for the treatment of some kinds of disease, and it has been found effective for the treatment of some kinds of cancer and leprosy. Now the search is on for analogues of the thalidomide molecule that are more potent treatments but avoid the risk of its side effects. This is a molecule that went from the light, when it was the wonder sleeping pill, to the dark, when it caused awful defects, and has now returned to the light again.

176 PACLITAXEL $C_{47}H_{51}O_{14}N$

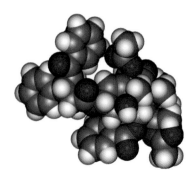

Paclitaxel is sold, and better known, under the trade-name *Taxol*. It is obtained from the pacific yew tree (*Taxus brevifolia*) and is a treatment for cancers of the ovary, breast, head, and neck. It is an antiproliferative in the sense that it inhibits the growth of cancerous cells. Its mechanism of action is based on the need for malignant cells to undergo rapid division, which requires the rapid dis-assembly and re-assembly of the cytoskeleton, the supporting framework of the cell formed by microtubules made from the protein tubulin. Taxol promotes the formation of stable bundles of microtubules, which prevents the dis-assembly of the cytoskeleton and hence impedes the proliferation of the malignant cells.

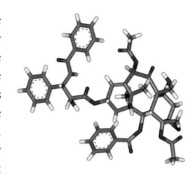

Taxol can be synthesized from a precursor that is found in the leaves of the European yew (*Taxus baccata*). It is also produced by a fungus found in the bark of the yew, probably because a gene has been transferred to it from its host, and it continues to produce the compound even when it is separated from the tree. Taxol has been synthesized from much simpler compounds, but such routes are currently more elegant artistry than economically viable technology. Alternative sources of Taxol are necessary because the pacific yew is one of the slowest growing trees, and it takes the product of six century-old trees to treat just one patient.

A drawback of Taxol is that it is only very slightly soluble in water, and, to make it into a formulation suitable for injection, it needs to be mixed with a detergent-like solubilizing agent. The solubilizing agent causes adverse reactions, which have to be countered by the administration of steroids and anti-histamines.

STIMULANTS

Many drugs are used for non-medical purposes in ways that vary in their degrees of social acceptability. This section describes a few that are used in one way or another to bring about a sense of euphoria. Ethanol, a depressant (26), has already been discussed.

As we have noted in several previous sections, the region of the brain associated with emotion is the limbic system: stimulation of the neurons of the limbic system, either by enhancing the ability of neurotransmitters to communicate between them at synapses or by depressing the ability of the inhibitory transmitter GABA (27) to bind, can result in enhancements of emotional activity. Now we need to bring the story slightly more into focus.

Many of the connections in the limbic system involve highly branched neurons that originate in a tiny blue region of the brain stem called the *locus coeruleus*. The chemically significant feature of these neurons is that their neurotransmitter is a molecule called *nor-adrenaline* (*norepinephrine* in the USA); this neurotransmitter is also abundant in the limbic system itself. The neurophysiologically significant feature of these neurons is their branching, which results in numerous synaptic connections with other neurons, both in the limbic system and in the higher regions of the brain. It seems reasonable to assume (but is wholly speculative) that such highly branched, polysynaptic neurons correlate with emotional responses, whereas less branched, even monosynaptic, neurons correlate with more logical, rational functions. If so, then any modification of the response of these highly branched clusters of neurons can be expected to modify emotional responses. The key to that modification may be their neurotransmitter, the molecule noradrenaline.

177 ADRENALINE $C_9H_{13}O_3N$

178 AMPHETAMINE $C_9H_{13}N$

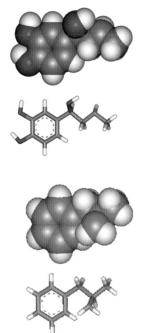

Not all stimulants need to be administered from outside; some, including adrenaline, are produced within the body and participate in the transmission of signals from one nerve cell to another. Adrenaline is produced in the renal gland of the kidney. It is also called *epinephrine* (in the USA), a Greek version of its name (*ren* is the Latin word for 'kidney', *nephros* the Greek). Adrenaline circulates in the blood and affects the autonomic nervous system (the nerves over which we do not have voluntary control, as distinct from the central nervous system). It acts on the heart muscles to increase their force of contraction, it dilates the pupils of the eye, and it stimulates the secretion of sweat and saliva. However, the hydrogen-bonding –OH and –NH groups in the molecule keep it from passing across the hydrocarbon-rich curtain of the blood–brain barrier.

Nevertheless, adrenaline and closely related compounds can be made in the brain. One of these related compounds is the noradrenaline mentioned in the introduction to this section. Its precursor is the amino acid tyrosine (164), which has the same basic ground plan as adrenaline and is converted into noradrenaline by enzymes. This conversion occurs abundantly in the neurons of the *locus coeruleus* and limbic system and, in a sense, feeds our emotions.

Now consider the amphetamine molecule. Amphetamine, which was once widely available under the trade name *Benzedrine*, is a stimulant. The similarity of its shape to that of noradrenaline suggests that it can act in a similar manner, to stimulate the limbic system and the *locus coeruleus*. Because the latter is also connected to the higher centres (particularly the cerebral cortex), amphetamine may also stimulate the higher, cognitive functions and lead to greater alertness and a feeling of prowess.

The action of amphetamine is not simply to take the role of noradrenaline as a transmitter. It appears that amphetamine molecules can mimic noradrenaline molecules so closely that the former take the place of the latter in storage sites inside the presynaptic (upstream) neuron. This results in noradrenaline molecules being displaced from storage and entering the space between the presynaptic and postsynaptic (downstream) neurons. Because of their increase in concentration there, more noradrenaline molecules attach to protein molecules in the wall of the postsynaptic neuron, change the shape of the protein molecules, and trigger a signal. Hence, the activity of the neurons is enhanced, and sensations of euphoria and increased alertness are experienced.

The amphetamine molecule can exist in two forms, one being the mirror image of the other (like lactic acid, 32, and other chiral molecules). The form that rotates plane-polarized light to the right is called *dexedrine* and is much more potent than its mirror-image molecule. The two differ in shape, like hands, and only one of them readily fits a specific protein 'glove'.

The *methamphetamine* molecule differs from that of amphetamine in having a –CH$_3$ group in place of one of the hydrogen atoms of the –NH$_2$ group. This stimulating molecule is sold as *Methedrine* and more commonly known as *speed*.

179 CAFFEINE C$_8$H$_{10}$O$_2$N$_4$

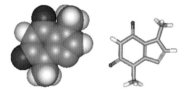

Caffeine is the component of coffee and tea that stimulates the cerebral cortex; it does so by inhibiting an enzyme (*phosphodiesterase*) that in turn inactivates a certain form (called 'cyclic AMP') of the energy-supply molecule ATP (109). That caffeine can pass itself off as ATP should not be too surprising: a comparison of their structures shows that both include two fused heterocyclic rings (rings containing atoms of other elements as well as of carbon).

A typical cup of coffee or tea contains about a tenth of a gram of caffeine. Coffee is obtained from the roasted seeds of *Coffea arabica*, and tea from the fermented leaves of *Camellia thea*. Caffeine also occurs in the seeds of the West African kola plants (*Cola acuminata* and *C. nitida*), which are now grown extensively in South America. Extracts from these plants are used to flavour (and add stimulant action to) cola drinks, in place of the cocaine (183) they originally contained. *Theobromine* is a close relative of caffeine; its molecule differs from that of caffeine only in the replacement of an –NCH$_3$ group by –NH. It is the stimulant in chocolate (140).

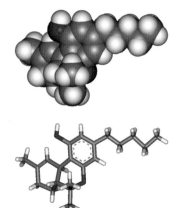

180 TETRAHYDROCANNABINOL $C_{21}H_{30}O_2$

Tetrahydrocannabinol is the active component of *cannabis*, which is obtained from *Cannabis sativa*, a plant in the same family as the hop plant used in beer-making (104). *C. sativa* has been grown widely for its fibres called *hemp*; canvas takes its name from cannabis, for it was woven of fibres obtained from the plant. The plant exudes a resin that covers its flowers and nearby leaves, and this resin is incorporated into some preparations that humans take to consume cannabis. *Marijuana* denotes the dried, crushed leaves and flowers of the plant; *hashish* is more precisely the resin. The mode of action of this structurally highly complicated molecule has not yet been determined.

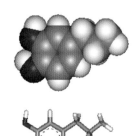

181 DOPAMINE $C_8H_{11}O_2N$

182 3,4-DIHYDROXYTYROSINE $C_9H_{11}O_4N$

Dopamine is an important neurotransmitter. Its distribution in the brain is highly non-uniform, and is concentrated in the *corpus striatum*, which is a region of the brain involved in the coordination of movement, and the limbic system. A decline in the production of dopamine is responsible for the onset of Parkinson's disease, of which the principal symptoms are rigidity, tremors, and hypokinesia (slowness in initiating movement). An excess of dopamine is associated with the onset of Huntington's chorea, a genetic disorder in

which the symptoms are severe involuntary movements. Dopamine is also implicated as a neurotransmitter in the brain pathways that control reward and pleasure: see cocaine (183) and nicotine (184).

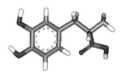

The molecule 3,4-dihydroxytyrosine is better known as L-DOPA, the acronym coming from its alternative chemical name dihydroxy-phenylalanine and the L, for *laevo*, indicating a particular ('left-handed') stereochemical form. Dopamine is synthesized, via L-DOPA, from tyrosine (164) in the *substantia nigra*. The colour of this region is due to an accumulation of melanin (165), which we saw is also formed from tyrosine. As the cells in the *substantia nigra* die, less dopamine is produced and the symptoms intensify. Temporary alleviation of the condition is achieved by supplying L-DOPA, but the cells continue to die. Dopamine itself does not cross the blood–brain barrier, and so is ineffective when it is supplied intravenously but a little (about 1 per cent) of the L-DOPA ingested does manage to scrape through, and, once it has passed inside the brain, is converted into dopamine by the loss of the –COOH group. Because so little L-DOPA reaches the brain, large, frequent doses are needed, but its effectiveness decreases as the disease advances.

The symptoms of Parkinson's disease can be brought on by folly: compounds present as contaminants in some supplies of illicit heroin kill cells in the *substantia nigra* and result in Parkinsonism.

183 COCAINE $C_{17}H_{21}O_4N$

Cocaine (alias coke, crack, dust, snow, blow, etc.) is an alkaloid obtained from the leaves of the coca plant (*Erythroxylon coca*), which is native to the eastern slopes of the Andes and grown widely in Bolivia, Peru, Ecuador, and Columbia. The plant has evolved the production of the alkaloid to serve as a pesticide: cocaine inhibits the re-uptake of the insect-specific neurotransmitter octopamine. In humans, it inhibits the re-uptake of the neurotransmitters dopamine (181), noradrenaline, and serotonin (185) in the brain. These neurotransmitters stay in the synaptic cleft for longer; cocaine can also stimulate the release of more dopamine. The immediate effect is of euphoria, garrulousness, and increased motor activity; but the feeling soon passes and is succeeded by a feeling of depression and dysphoria, which the user feels can be relieved only by more cocaine.

The usual route of administration of cocaine in the form of its hydrochloride salt (cocaine in combination with hydrochloric acid) is by nasal inhalation; however, the resulting prolonged constriction of blood vessels can lead to necrosis of the nasal septum. *Crack cocaine* is cocaine hydrochloride from which the acid has been removed, commonly by heating the compound in a solution of baking soda (sodium bicarbonate, $NaHCO_3$) until the water evaporates. The 'free base' obtained in this way makes a cracking sound when it is heated as a result of the release of carbon dioxide from the hot bicarbonate; hence the name 'crack'. Because the free base is molecular rather than ionic (as in the hydrochloride), it vaporizes at a low temperature, and hence can easily be inhaled through a heated pipe.

Cocaine has no medicinal uses except as a mild anaesthetic in ophthalmology.

184 NICOTINE $C_{10}H_{14}N_2$

Nicotine is found in the leaves of the tobacco plant (*Nicotiana tabacum*, named after an early importer of tobacco, the French ambassador to

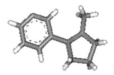

Portugal, Jean Nicot) in combination with citric acid (102) and malic acid (101). Alkaloids commonly occur in plants to deter predators by their sharp taste. When the tobacco crop is harvested, the leaves are dried, which changes the protein, carbohydrate, and alkaloid composition. Cigarette tobacco is largely flue-cured, a process in which the leaves are hung in a ventilated building with flues and a heat source. After about a week the green leaves become yellow ('bright tobacco') as the carotenes (158) become dominant. The leaves have a high concentration of sugars and nicotine. Cigar tobacco is air-cured, being hung in the open for several weeks. The resulting product is low in sugar and variable in nicotine. Fire-cured tobacco, produced by exposing the green leaves to the smoke and heat of wood fires, is used largely for pipe tobacco; it is low in sugar and nicotine. After curing, the leaves are fermented by storing in a moist environment: the flavour is changed and the leaves become rehydrated, which makes them less brittle.

A typical cigarette contains about 10 milligrams of nicotine, of which about 1 milligram is absorbed by the smoker, principally through the lungs to which it is carried attached to tar particles. Cigar and pipe smoke is less acidic than cigarette smoke, so the nicotine in it is more likely to be in its free-base form; as a result, more nicotine is absorbed in the mouth and nasal cavity. Cigarette smoking gives a more rapid rise in concentration of nicotine in the blood; cigar and pipe smoking gives a slower rise but a greater duration.

Nicotine has complicated and extensive pharmacological effects, some of which are virtually identical to those of cocaine (183), and the outcome often depends on the dose. Thus, in some cases smokers report a greater alertness and in other cases a relaxation. Nicotine, like cocaine, heroin, and marijuana, increases the level of the neurotransmitter dopamine (181), which affects the brain pathways that control reward and pleasure. Furthermore, because it causes a discharge of adrenaline (see the introduction to this section) from the adrenal cortex, the ingestion of nicotine results in an almost immediate 'kick'. The adrenaline, through the endocrine system, then stimulates the sudden release of glucose, with a feeling of alertness. This stimulation is then followed by depression and fatigue. The hormone corticosterone which is released during episodes of stress reduces the effects of nicotine and hence leads to a greater consumption to achieve the same effect.

Nicotine is addictive, and cigarettes without nicotine are unacceptable to most smokers. 'Low-tar' cigarettes are also low in nicotine, but smokers puff harder, inhale more, and smoke more often. As a result, the net effect of switching to low-tar cigarettes is often to increase the intake of carbon monoxide, with a consequently greater risk of cardiovascular disease.

185 SEROTONIN $C_{10}H_{12}ON_2$

186 FLUOXETINE $C_{17}H_{18}ONF_3$

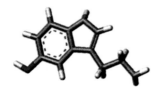

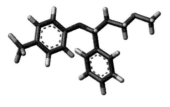

Serotonin (also known as 5-HT from its alternative chemical name, 5-hydroxytryptamine), a neurotransmitter, is a slightly modified version of an essential amino acid, tryptophan, used to construct proteins. Although food contains some serotonin, most is formed inside the body by the action of enzymes on the amino acid tryptophan. Serotonin itself participates in a wide range of activities, especially, but not only, in the central nervous system and modulates mood, emotion, sleep, and appetite. The cerebrospinal fluid of depressed people contains significantly lower concentrations of substances derived from serotonin, and there is also less in the brains of many suicides. These low concentrations may be due to the slower than normal transport of tryptophan across the blood–brain barrier into the brain.

Serotonin is a powerful vasoconstrictor. When it acts on the smooth muscle of veins it impedes the flow of blood from the capillaries and gives rise to blushing. The action of serotonin is affected by certain hallucinogenic drugs, most famously lysergic acid diethylamide (LSD), but the precise – probably multifaceted – mode of action remains unknown.

Melatonin is closely related to serotonin: the –OH group on the benzene ring is replaced by $-OCH_3$ and the $-NH_2$ group on the dangling side chain is converted into $-NHCOCH_3$. Melatonin is produced in the pineal gland, a structure resembling a third eye, which Descartes thought to be the seat of the soul, and on the vascular side of the blood–brain barrier. Melatonin is thought to control the diurnal variation in levels of hormones, and is taken by some as a way of overcoming jet-lag. Sunlight has an effect on the body through the eyes by stimulating the pineal gland to release serotonin and melatonin. An excess of melatonin in the body induces sleep, drowsiness, and lethargy.

Seasonal affective disorder (SAD) is a condition, the 'winter blues', experienced when people are exposed to low levels of light for extended periods, such as in winter at high latitudes. No one really knows the cause of SAD, but there are several hypotheses. Possible causes are that the low levels of light signal the body to increase the production of melatonin or cause low levels of serotonin.

Fluoxetine (186), in combination with hydrogen chloride to give its fluoride salt, is sold as the anti-depressant *Prozac*. In a depression, there is an imbalance of serotonin at a synapse; for instance, there may be insufficient serotonin present to stimulate the transmission of a signal in the postsynaptic neuron. It appears that the role of Prozac is to inhibit the serotonin-uptake pump that helps to scavenge the serotonin molecules from the synapse and return them to the presynaptic (upstream) neuron. As a result, serotonin remains in the synapse at a higher concentration and can attach to the postsynaptic (downstream) neuron, which responds by firing off its signal, as in an undepressed person. Prozac appears to be specific to serotonin receptors, and therefore it has fewer side effects than do drugs that interact more generally.

187 METHYLENEDIOXYMETHAMPHETAMINE $C_{12}H_{17}O_2N$

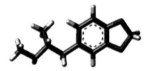

This little monster, known for short as MDMA, is known on the street as *Ecstasy* (literally: to stand outside oneself). Its chemical structure is similar to those of methamphetamine, methylenedioxyamphetamine (MDA), and mescaline, all of which cause brain damage. It appears that MDMA affects neurons that use serotonin (185) as a neurotransmitter and consequently influences mood, aggression, sexual activity, sleep, and sensitivity to pain. The drug was originally developed as a treatment for obesity. Long-term (six or seven years) depletion of serotonin receptors occurs with extended use of the drug, with loss of concentration and cognitive ability. In high doses it causes a rise in body temperature, which can result in muscle breakdown and the failure of the kidney and cardiovascular system. At that point, one does stand outside oneself, permanently.

NASTY COMPOUNDS

It cannot be denied that some chemicals have done evil – sometimes by design and sometimes by accident. Evil chemicals, like evil deeds, capture bigger headlines than do the good ones. Chemicals that poison, which for some are the only public manifestation of chemistry and chemicals, also often poison the image of chemistry in the public's eye. This section describes a ragbag of molecules that have hit the headlines on various occasions by virtue of their unpleasant effects on people and other animals – which include poisoning them, deforming them, or simply tearing them apart. I found it rather depressing to write about them, and hope that you will remember instead the better lives of other molecules.

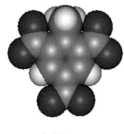

188 TRINITROTOLUENE $C_7H_5O_6N_3$

The trinitrotoluene molecule is a toluene molecule (23) to which have been attached three *nitro groups* ($-NO_2$). Trinitrotoluene is the high explosive *TNT*. It is explosively unstable because it is an assembly of carbon atoms that are on the brink of oxidation: its oxygen atoms have only to be nudged into slightly different locations for them to be able to swoop down on the carbon and hydrogen atoms of the benzene-like ring and carry them off as carbon dioxide (3) and water (5), leaving the nitrogen atoms to fall together and move off as gaseous nitrogen (2). In an instant, therefore, the compact molecule can be converted into a voluminous cloud of gas, and the pressure wave of its expansion is the destructive shock of the explosion. That is the general function of high explosives: the sudden creation of a large volume of gas from a small volume of liquid or solid. Typically, 1 gram of explosive suddenly produces about 1 litre of gas, corresponding to a thousand-fold increase in volume. There is not enough oxygen in a TNT molecule to oxidize it completely, so its explosion is marked by a good deal of black smoke.

The rearrangement of the TNT molecule is achieved by hammering it with a sharp pressure wave from another and more easily induced explosion in a detonator. A typical detonator is the solid *lead azide*, $Pb(N_3)_2$, which contains the azide ion (N_3^-). It is more sensitive to shock than TNT is; the azide ion shakes itself apart into nitrogen gas when lead azide is struck or exposed to the shock of an electric discharge. One advantage of TNT, to munitions manufacturers if not to humanity as a whole, is that it melts at a low temperature (80 °C). It can thus be poured into shells and bombs.

189 NITROGLYCERINE $C_3H_5O_9N_3$

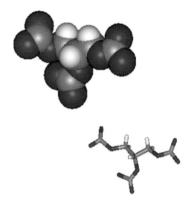

The nitroglycerine molecule is derived from the glycerole (alias glycerine) molecule (33) by replacing the hydrogen atom of each –OH group with a nitro group ($-NO_2$). Nitroglycerine is an oily colourless liquid. In it the nitroglycerine molecules contain all the seeds of their own destruction, for the carbon and hydrogen atoms can be converted into carbon dioxide and gaseous water, and the nitrogen atoms can pair up without any input of oxygen from outside. A mechanical shock will so distort the nitroglycerine molecule that its atoms can change partners and, as with TNT (188), generate a rapidly expanding cloud of gas.

Nitroglycerine is very unstable and very sensitive to shock, impact, and friction, as may be familiar from the film *The Wages of Fear*. It is so unstable that it is normally dissolved in an absorbent material, a process that gives *dynamite*. This invention, together with the oilfields in Russia that he owned, provided Alfred Nobel with his fortune. Originally the absorbent material was *kieselguhr*, a clay, but modern dynamites use a mixture of wood flour, ammonium nitrate, sulfur, and sodium nitrate.

190 *bis*(2-CHLOROETHYL)THIOETHER $C_4H_8SCl_2$

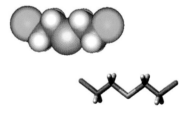

This sinister little molecule is *mustard gas*, which was used first at Ypres in 1917 and then stockpiled in large quantities during World War II. Mustard gas, which is in fact a volatile liquid, is odourless and therefore is not immediately detectable by smell. Where it touches the skin and is inhaled, it forms blisters. Those who do not die at once, or from the infections that follow the blistering, suffer a generalized poisoning that renders them ill for the rest of their lives. 'Improvements' to the molecule that reflect the progress of civilization since Ypres include the lengthening of the hydrocarbon chain so that the molecule can pass itself off as a hydrocarbon and more easily penetrate protective clothing.

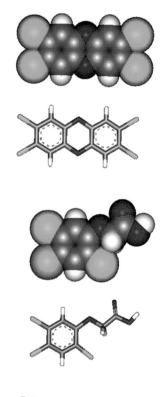

191 TETRACHLORODIBENZO-*para*-DIOXIN $C_{12}H_4O_2Cl_4$

192 2,4,5-TRICHLOROPHENOXYACETIC ACID $C_8H_5O_3Cl_3$

193 PENTACHLOROPHENOL C_6HOCl_5

The compound known formally as tetrachlorodibenzo-*para*-dioxin is commonly called 'dioxin', but it is just one of a family of a similar compounds with different numbers of chlorine atoms and with chlorine atoms in different locations on the benzene rings. One source of dioxin is a side reaction that takes place during the production of 2,4,5-T (more formally, 2,4,5-trichlorophenoxyacetic acid, 192), a herbicide and a component (with the analogous 2,4-D) of the defoliant 'Agent Orange'. The side reaction can be minimized by controlling the concentration and temperature, but the procedure remains hazardous. Less easy to control is the combustion of wood that has been treated with chlorinated phenols, such as pentachlorophenol (PCP, 193), which as it burns produces a variety of dioxins. A mixture of pentachlorophenol and its analogue 2,3,4,6-tetrachlorophenol, in which the chlorine atom at eight o'clock on the ring is absent, is a common herbicide, insecticide, fungicide, and mollucicide (an agent for killing snails). The formation of dioxins can be expected to occur whenever chlorinated organic compounds, such as PCBs (194), are burned, except when combustion takes place at very high temperatures (typically, at 1200 °C).

Dioxins are readily soluble in hydrocarbons, so they dissolve in fats and enter the food chain, but almost always at very low levels. The most toxic are those with chlorine atoms on the edges of the rings (as in 191; technically, the β positions) rather than in the 'top' and 'bottom' positions (technically, the α positions).

The effect of dioxins on human health is highly controversial, and our knowledge of it comes largely from studies of populations exposed to accidental release of the compounds, such as the incident at Seveso in Italy in 1976, which resulted from leaving a reaction running unattended over a weekend. Relatively high concentrations of dioxin do induce cancers, but low concentrations do not seem to be toxic, although there is considerable concern about the ability of dioxins to accumulate in living tissues.

194 2,3′,4′,5′-TETRACHLOROBIPHENYL C$_{12}$H$_6$Cl$_4$

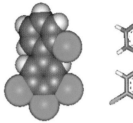

This molecule is an example of a *polychlorinated biphenyl* (PCB). This family of compounds is huge, with over 200 members, with a variety of numbers of chlorine atoms attached in a variety of locations to the two benzene rings. The chlorine atoms in the molecules, with their numerous electrons, result in the molecules having strong attractions for each other, so the compounds are not very volatile. The compounds are stable, and have fire-retarding properties, in part because on combustion they release chlorine atoms that interfere with the radical chain reactions that take place during combustion. They have been used as coolant fluids in power transformers and as plasticizers (52). The compounds are no longer manufactured, but they still remain in place in many installations as past practice has resulted in their widespread occurrence throughout the world. Combustion can result in the formation of dioxins (191); high-temperature incineration avoids the formation of dioxins but still releases chlorine (as chlorine molecules and hydrogen chloride) into the environment.

As for dioxins (191), the effects of PCBs on human health are highly controversial. Some well-publicized incidents, such as the 'Yusho', or 'oil-disease' incident in Western Japan in 1968 and the similar 'Yu-Cheng' incident in Taiwan, were attributed to the consumption of rice-bran oil contaminated with PCBs, but subsequent analysis revealed the presence of toxic thermal-degradation products in the oil.

7

Life

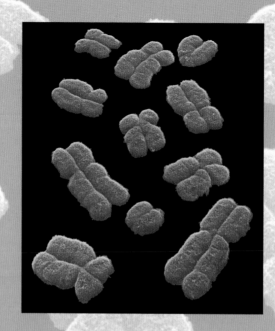

After the horrific activities of the molecules encountered in the last part of the previous section, it is a relief to come to molecules with a more agreeable – indeed glorious – purpose: life. We have already encountered many molecules that contribute to the tapestry of chemistry we recognize as being alive: these include the fats of Chapter 2 and the proteins and carbohydrates of Chapter 3. However, there are also many foot soldiers that contribute to the sustenance and propagation of organisms, such as the pheromones of Chapter 4 and the neurotransmitters of Chapter 6. Here, we concentrate on a couple of hormones that are responsible for the secondary sexual characteristics of humans, and an aid to their propagation.

One of the most stunning achievements of modern chemistry has been to discover the molecular mechanism of inheritance and evolution and to interpret the genetic code. We end this survey of molecules with this astonishing and hugely important achievement.

SEX

Here we meet a couple of molecules that determine how men and women differ, at their molecular roots.

195 TESTOSTERONE C$_{19}$H$_{28}$O$_2$

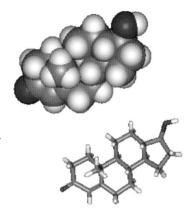

The testosterone molecule is a shortened version of the cholesterol molecule (38): the latter's ring structure is preserved, but the hydrocarbon tail has been lost. In addition, the —OH group of the cholesterol molecule has been replaced by a doubly bonded oxygen atom, so testosterone is a ketone; hence the ending -one.

Testosterone is the male sex hormone. Its secretion from the *cells of Leydig* in the testes is initiated at puberty and controls the development of the secondary sexual characteristics, including differences in the skeleton, the voice, the pattern of body hair, patterns of behaviour, and the organs of reproduction themselves. Testosterone molecules also induce the retention of nitrogen and hence encourage protein formation (*anabolism*), which leads to an enhancement of musculature.

Testosterone is a member of the class of compounds called *steroids*, of which cholesterol is both a member and a metabolic precursor. Related anabolic steroids have been used to encourage the growth of muscles in athletes, but some have the awkward side effect – the term seems inappropriate – of causing a persistent erection.

196 OESTRADIOL C$_{18}$H$_{24}$O$_2$

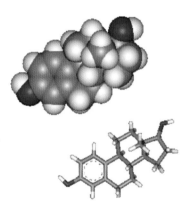

The oestradiol (estradiol in the USA) molecule has the same ring structure as testosterone but carries an —OH group on the benzene-like ring and lacks a —CH$_3$ group at the junction of that ring and its neighbour.

It is extraordinary what difference an extra —CH$_3$ group and the slight rearrangement of a ring of atoms can make: oestradiol is one of the principal female sex hormones. It is released at puberty, maintains the secondary sexual characteristics, and then decreases in abundance at menopause. The tissues on which oestradiol acts bind it strongly; it stimulates the synthesis of the RNA molecules that mediate the interpretation of DNA (203), and growth occurs accordingly.

Oral contraceptives, which are arguably one of chemistry's great contributions to liberty, are often a combination of an *oestrogen* and a *progestogen*. They act by maintaining the high levels of hormones that are characteristic of pregnancy, during which ovulation is suppressed. *Ethynodiol diacetate* (*ethyne* is the modern name for *acetylene*, HC≡CH, and the molecule contains this group) is an example of a progestogen. The ethynodiol diacetate molecule differs from the

In the first edition, this image was a reproduction of *Venus and Mars*, painted in the fifteenth century by Alessandro di Mariano Filipepi (better known as Sandro Botticelli). Human nature – that extraordinarily complex manifestation of the molecules within us – has changed little since then.

oestradiol molecule in being an ester; in it, each of the two —OH groups of the hormone molecule has been combined with an acetic acid molecule (31). In addition, it carries a triply bonded pair of carbon atoms. Progestogen-only preparations are available as 'mini-pills', so called because they are used in smaller doses. They act by thickening the mucus at the neck of the womb, so that it forms a barrier against sperm, and they modify the lining of the womb so that it is unfavourable to the growth of an egg.

197 SILDENAFIL $C_{22}H_{30}O_4N_6S$

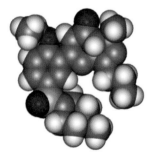

Sildenafil, as its citrate salt, is the active component of the drug sold as *Viagra* for the treatment of male erectile dysfunction. Its full name is 1-[{3-(6,7-dihydro-1-methyl-7-oxo-3-propyl-1H-pyrazolo[4,3-d]pyrimidin-5yl)-4-ethoxyphenyl}sulfonyl]-4-methylpiperazine citrate, but that was perhaps thought likely to prove a mouthful in the heat of the moment. A tablet of Viagra also contains a number of inactive ingredients, including microcrystalline cellulose, calcium

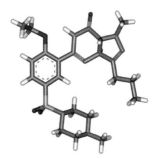

phosphate, magnesium stearate, lactose, and titanium dioxide; it is coloured with Blue No. 2 aluminium lake. Apart from its principal function, of which more below, Viagra temporarily affects discrimination between blue and green and causes a slight drop in blood pressure.

To understand the mode of action of Viagra, we need to understand the mechanics of erection. In brief, a penis contains two expandable sacs, each known as a *corpus cavernosa*, with inner walls composed of smooth muscle tissue. Normally, the smooth muscle is contracted and blood flows unhindered through the *corpus cavernosa*, but, on arousal, the smooth muscle relaxes, the veins draining the blood are squeezed shut, and the blood is trapped, so stiffening the penis. More specifically, signals from the autonomic nervous system result in the release of the neurotransmitter acetylcholine, which binds to the endothelial cells of the arteries feeding the *corpus cavernosa* and causes the synthesis and release of nitric oxide (NO, 9), which relaxes the smooth muscle through a cascade of biochemical reactions that involve the messenger guanosine-3′,5′-cyclicmonophosphate (cGMP). The precise mechanism remains unknown but might involve inhibition of the rise in concentration of calcium ions needed to achieve muscle contraction or the opening of various ion channels. The central point, though, is that, so long as cGMP remains present in the smooth muscle, the muscle is unable to contract. When the cGMP is destroyed, the muscle contracts, blood is released, and the penis wilts.

Now for the role of the active component of Viagra. Sildenafil acts by inhibiting the enzyme responsible for the degradation of cGMP. It follows that the compound does not induce an erection directly: the drug goes into action well down the hierarchy of processes that are initiated by the sexual stimulation itself. Thus, Viagra does not increase desire, but enhances the immediate consequences of desire.

Two enzymes closely related to the cGMP-destroying enzyme are also found in the heart and the retina of the eye: that sildenafil affects these enzymes weakly as well as its main target accounts for its effects on blood pressure and colour discrimination mentioned earlier.

REPRODUCTION

An austere view of the ultimate purpose of sex is that it is a way of ensuring that a particular fragile pattern of hydrogen bonds survives in the world. By this I have in mind the DNA molecule that inhabits the nucleus of every cell and propagates genetic information from generation to generation. That molecule, the famous *double helix*, is held in shape largely by the hydrogen bonds between its components, and its replication depends on the sequence of components acting as a template for the construction of a copy. It acts as a template through the hydrogen bonds its components can form, so reproduction is a replication of hydrogen-bonding patterns. From this point of view, evolution by natural selection is the consequence of the competition among patterns of hydrogen bonds.

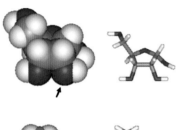

198 RIBOSE $C_5H_{10}O_5$

199 ADENINE $C_5H_5N_5$

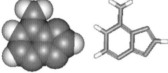

200 GUANINE $C_5H_5ON_5$

201 CYTOSINE $C_4H_5ON_3$

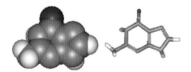

202 THYMINE $C_5H_6O_2N_2$

203 DEOXYRIBONUCLEIC ACID, DNA

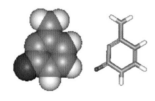

The most famous and possibly most important molecule in the world, deoxyribonucleic acid (DNA), carries genetic information from one generation to the next and controls the function of every cell in our body. The determination of its detailed structure in the 1950s transformed our understanding of inheritance and the genetic modification we call, depending on the outcome and context, either disease or evolution.

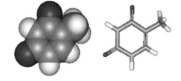

The molecule is a long thread, with units attached regularly along its length. The thread itself is built from alternating sugar molecules and phosphate groups. The sugar molecule is ribose (198), a close relative of glucose (87), from which the oxygen atom marked with an arrow has been removed (hence the 'deoxy' and the 'ribo' parts of the name). Attached to each deoxyribose ring is a *nucleotide base*. Only four nucleotide bases occur in DNA, namely adenine (commonly denoted A, 199), guanine (G, 200), cytosine (C, 201), and thymine (T, 202). These four bases fall into two pairs: adenine and guanine, with two rings of carbon and nitrogen atoms stuck together, are examples of *purines*. Cytosine and thymine have only one ring and are examples of *pyrimidines*.

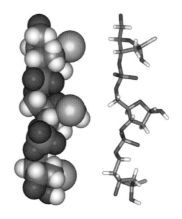

The structure of DNA itself is a right-handed *double helix*, in which one long strand of nucleic acid (203, the yellow spheres denote the bases) is wrapped round another to form an entwined pair. The key feature is that the nucleotide bases of one strand match the nucleotide species of the other in the sense that adenine is always matched with thymine (which we denote A···T) and guanine is always matched with cytosine (G···C). Note that a relatively small purine (adenine and guanine) is always matched with a more bulky pyrimidine (thymine and cytosine), for in that way the double helix is uniform: two big purines would have resulted in a bulge and two small pyrimidines in a pinch. Thymine and adenine have just the right shape and arrangement of nitrogen, oxygen, and hydrogen atoms to form two hydrogen bonds very snuggly. Similarly, cytosine and guanine also fit together snuggly, but by forming three hydrogen bonds. There are about three billion base pairs in human DNA.

If the human DNA in one set of 23 chromosomes (with one DNA molecule in each chromosome) is stretched out and joined together, then it would be about 1 metre long, and all that stuff has to be confined to the tiny cell nucleus. To achieve this wonderful feat of packing, the double helix is wrapped round clusters of protein molecules called histones, which act like spindles. Then these spindles are wrapped round each other. That coil is itself coiled – it is *supercoiled* – around itself, and the tightness of the coil determines whether the

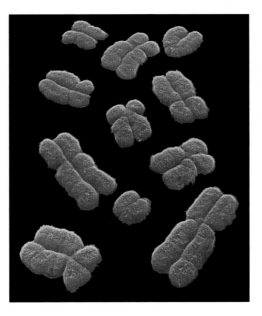

A selection of human chromosomes. Each of our somatic cells contains twenty-three chromosomes, each one of which contains its complement of tightly coiled, and supercoiled, DNA, the repository of information passed down the generations.

chromosomes are bundled up, as they are during the process of replication called *mitosis*, or spread out through the nucleus, as they are for the rest of the cell's life.

The key idea of reproduction is that each strand of the double helix acts as a template for the other. Thus, suppose that the two strands have their nucleotide bases in the following sequence:

... ACCAGTAGGTCA ...
... TGGTCATCCAGT ...

with the first A in the upper strand linked by hydrogen bonds to the first T in the lower strand, C linked likewise to G, and so on. When the strands separate into

... ACCAGTAGGTCA ...   and   ... TGGTCATCCAGT ...

the nucleotide bases in the cell attach to the separated strands, each of which forms a template for the creation of a new strand, and form

... ACCAGTAGGTCA ...   ... TGGTCATCCAGT ...
                      and
... TGGTCATCCAGT ...   ... ACCAGTAGGTCA ...

Now we have two identical double helices where originally we had one.

Each three-letter sequence of DNA is a *codon*, a group of bases that specifies a particular amino acid (80) used to construct the polypeptides we know as proteins (Chapter 3). The genetic code is expressed with U, for uracil (204), in place of thymine, for reasons that will become clear in the following entry. The genetic code is highly redundant, with up to six codons referring to the same amino acid and three meaning stop. This redundancy makes 'mistakes' in replication less likely to have fatal consequences. For instance, CCT, CCC, CCA, and CCG all code for proline, so a mistake in copying the last letter is unimportant. Even where a single-letter change is significant, the outcome is often the replacement of one amino acid by a similar one. For instance, the change from TTT to TAT results in the replacement of phenylalanine by its cousin tyrosine. The code is almost optimal in this respect. The redundancy of the code also leaves open the possibility of its future development, as redundant codons acquire new meaning.

204 URACIL C$_4$H$_4$O$_2$N$_2$

205 RIBONUCLEIC ACID, RNA

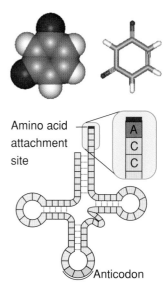

A DNA molecule is too big and too fragile to escape from the cell nucleus and participate directly in reproduction. Instead, the information contained in DNA is carried out of the cell by ribonucleic acid (RNA), a more primitive version of DNA. Ribonucleic acids have the same general structure as DNA, consisting of a sugar–phosphate backbone with nucleotide bases hanging from it. However, the sugar is ribose (198) rather than deoxyribose (hence the R of RNA in place of the D of DNA). Secondly, in place of thymine (202) RNA has the subtly different but very similar pyrimidine uracil (U, 204). It is not entirely clear why U occurs rather than T or why ribose rather than deoxyribose is used in the spine: it is probably due to the slight difference in strengths of the hydrogen bonds that the molecule can form. One major difference from DNA is that RNA consists of a single strand. It is presumed that RNA was the original encoding substance but its function was taken over by the more stable DNA at an early stage in evolution. Some support for this view comes from the observation that RNA can also act as an enzyme. That function resolves one problem about the origin of life: which came first, the chicken (the enzymes needed to make use of genetic material) or the egg (the genetic material needed to specify enzymes).

There are two main types of RNA, namely *messenger-RNA* (mRNA) and *transfer-RNA* (tRNA). We consider first mRNA, the molecule that carries the information encoded in DNA out into the cytoplasm. To pick up the message, mRNA is synthesized rather like DNA replicates, with one strand of DNA exposing itself and enzymes using that strand as a template to produce mRNA. Vertebrate DNA enzymes duplicate about thirty bases a second, and it takes about seven hours to replicate a cell's full complement of DNA; proof-reading enzymes keep a watchful eye and most errors are corrected. When copying reaches a 'stop' codon, mRNA stops forming and is transported away from the DNA and out through pores in the nuclear membrane into the cytoplasm where it encounters the ribosomes. Ribosomes are aggregations of protein and RNA that lurk as two separate blobs, then combine into a single functional unit when they attach to the mRNA that emerges from the cell nucleus.

The component of the cytoplasm that we need to notice at this stage is *transfer-RNA* (tRNA, 205; each square denotes a nucleotide

A part of the sequence of steps by which a protein is constructed by interpreting the encoded message from DNA.

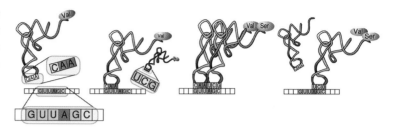

base), the nucleic acid that does the actual construction of the protein. The *anticodon* loop is the part of the molecule that recognizes the codon in mRNA. For instance, if the codon is CGU, coding for arginine, then the anticodon will be the complementary sequence GCA, which can find the codon CGU by a matching of hydrogen bonds and sticks to it like Velcro. The *attachment site* at the end of the nucleic-acid chain is another Velcro-like part of the molecule, and has a sequence of nucleotides that can stick to one and only one amino acid, in this case arginine.

The illustration shows what happens in the cytoplasm once a ribosome has clamped itself round a piece of mRNA. The ribosome pauses over the first codon, various tRNA molecules try their luck, but have the wrong anticodon to stick. Then along comes a tRNA molecule with an anticodon for GUU with valine carried on its attachment site. This molecule sticks, and in doing so allows the ribosome to ratchet along to the next codon, which may be AGC. In due course, a tRNA with an anticodon for AGC comes along, carrying the serine molecule which it has bumped into and captured elsewhere in the cytoplasm. The anticodon attaches to the codon, bringing its serine molecule close to the valine molecule; an enzyme shifts the valine off its tRNA and attaches it to the serine molecule, so forming the dipeptide valine–serine, and the discharged original tRNA ambles off into the solution in unconscious search for another valine. Now the ribosome ratchets along to the next codon, and the process is repeated. Gradually, the protein chain grows and the information originally in the nuclear DNA is converted into functional protein.

Glossary

acid   A molecule or ion that can donate a hydrogen ion (a proton) to another molecule or ion. *Examples*: sulfuric acid (8), nitric acid (10).

aerosol   A cloud of tiny droplets in air.

alcohol   An organic compound containing the group –OH attached to a carbon atom that is not part of an aromatic ring or carries another oxygen atom. *Example*: ethanol (26).

aldehyde   An organic compound containing the group $-C\!\!{\overset{\displaystyle O}{\underset{\displaystyle H}{\big\langle}}}$, which is normally abbreviated to –CHO. *Example*: acetaldehyde (29).

alkali   A water-soluble base. *Examples*: sodium hydroxide (NaOH), ammonia (6).

alkaloid   A naturally occurring organic compound containing nitrogen that acts as a base. Many are poisonous. *Example*: caffeine (179).

amide group   The group $-C\!\!{\overset{\displaystyle O}{\underset{\displaystyle NH_2}{\big\langle}}}$. See *peptide*.

amino group   The group $-NH_2$. Substances that contain this group (including those in which the hydrogen atoms have been replaced by other hydrocarbon groups) are called *amines*. *Examples*: hexamethylenediamine (78), trimethylamine (152).

amino acid   A compound that contains both an amino group and a carboxyl group. *Example*: glycine (80).

anion   A negatively charged ion. *Example*: nitrate ion, NO_3^-.

antioxidant   A substance that decreases the rate at which another substance is oxidized. *Example*: butylmethoxyphenol (42).

aromatic compound   A substance containing one or more benzene rings. Aromatic compounds, even though they are unsaturated, do not undergo the normal reactions of unsaturated compounds. Their special stability is related to the ability of the bonding electrons to spread around the planar ring. *Example*: benzene (22).

atactic   Lacking stereochemical regularity. A polymer is atactic if the groups attached to the backbone are not arranged in any regular geometrical pattern. *Example*: polypropylene (50).

atom   The smallest particle of an element that can exist. An atom consists of a minute positively charged nucleus surrounded by a cloud of electrons.

base   A molecule or ion that can accept a hydrogen ion from an acid. *Example*: ammonia (6), which accepts H^+ to form the ammonium ion NH_4^+.

bond   A chemical link between two atoms. In an *ionic* bond, the attraction is between the opposite charges of neighbouring ions. In a *covalent* bond, the two atoms share a pair of electrons that lie between them. A *single bond* is a shared pair, a *double bond* is two shared pairs, and a *triple bond* is three shared pairs of electrons.

carbonyl group   The group $\diagdown C = O$. When a hydrogen atom is attached to the carbon atom, the resulting compound is called an aldehyde. When only carbon atoms are attached, the resulting compound is called a ketone. See also *carboxyl group* and *amide group*.

carboxyl group   The group $-C \diagup\!\!\overset{O}{\diagdown}_{OH}$. The carboxyl group is normally abbreviated to –COOH. See *carboxylic acid*.

carboxylic acid   An organic compound containing the carboxyl group. The hydroxylic hydrogen can be lost as a hydrogen ion, so substances with this group are acids. *Example*: formic acid (30).

catalyst   A substance that facilitates a reaction without itself being consumed. (The Chinese term for catalyst, 'marriage broker', conveys the sense.) See *enzyme*.

cation   A positively charged ion. *Example*: calcium ion, Ca^{2+}.

chain reaction   A reaction in which the product of one step is a reactant in a later step, which produces a reactant for another step, and so on.

chiral   Handed. A chiral molecule is one that is distinguishable from its mirror image. *Example*: lactic acid (32).

***cis* isomer**   An arrangement in which two groups are on the same side of a double bond, as in $\overset{x}{\diagdown}=\overset{x}{\diagup}$. See *trans isomer*.

cortex   The outer, most highly developed, and evolutionarily most recent part of the brain.

crystal   A solid in which the atoms, ions, or molecules lie in an orderly and virtually endlessly repeated arrangement.

distillation   Separation of the components of a mixture on the basis of their different volatilities.

electron   An elementary particle with a negative charge. In an atom, the electrons are arranged in shells around the nucleus, and only those in the outermost shell take part in the formation of chemical bonds.

electron pair   Two shared electrons, one of each spin, responsible for a chemical bond. See *lone pair*.

electronegative atom   An atom that draws electrons towards itself in a molecule. The most electronegative atoms are fluorine, chlorine, oxygen, and nitrogen.

emulsion   A dispersal of one liquid as minute particles (each one consisting of many molecules) in another liquid. *Example*: milk (39).

enzyme   A substance that facilitates a biochemical reaction; a biological catalyst.

epithelium   A type of tissue consisting of closely packed cells that covers a body or lines a cavity or tube.

ester   The compound RCOOR′, the outcome of the reaction between a carboxylic acid RCOOH and an alcohol R′OH. *Example*: tristearin (35).

fatty acid   A carboxylic acid, especially one with a long hydrocarbon chain. *Example*: stearic acid (34).

fluorescence   The emission of longer-wavelength light immediately following the absorption of shorter-wavelength radiation. *Phosphorescence* is similar, but may persist after the stimulating radiation has been extinguished.

ganglion   An aggregation of nerve-cell bodies outside the central nervous system.

glucoside   A glycoside in which the sugar is glucose (87).

glycoside   A sugar molecule in which the hydrogen atom of an –OH group is replaced by another group.

heterocyclic molecule   A molecule containing at least one ring of carbon atoms, with at least one other type of atom (normally nitrogen or oxygen) in the ring. *Example*: 2,6-dimethylpyrazine (140).

hormone   An organic compound that regulates biochemical reactions.

hydrocarbon   A compound of carbon and hydrogen alone. *Examples*: methane (15), carotene (158).

hydrogen bond   A link formed by a hydrogen atom between two electronegative atoms, and denoted A · · · H–B.

hydrogen ion   The ion left when a hydrogen atom loses its only electron; a bare hydrogen nucleus, a proton. In water, a hydrogen ion is attached to a water molecule to form the *hydronium ion*, H_3O^+.

incandescence   The emission of light by a hot substance. All wavelengths of radiation are present but the greatest intensity shifts from red to blue as the temperature is raised.

infrared radiation   Electromagnetic radiation with a wavelength that is longer than that of visible red light. Infrared radiation is responsible for the transmission of radiant heat.

ion   An electrically charged atom or group of atoms. See *cation* and *anion*.

isomer   A molecule built from the same atoms as another molecule but in a different arrangement. *Examples*: ethanol (26), CH_3CH_2OH, and dimethyl ether, CH_3OCH_3; D- and L-lactic acid (32).

isotactic   Geometrically regular. A polymer is isotactic if all the groups attached to the backbone are arranged in the same geometrical pattern. See *atactic*.

ketone   An organic compound containing the carbonyl group $\overset{\displaystyle |}{\underset{\displaystyle |}{C}}{=}O$, to which other carbon atoms are attached. *Example*: testosterone (195).

limbic system   A network of neurons encircling the brain stem, thought to control the emotions and translate them into actions. (*Limbus* is Latin for 'hem'.)

lipid   A naturally occurring substance that is soluble in organic solvents but not in water.

lone pair   A pair of electrons not involved in bond formation.

luminescence   The emission of light, specifically as a result of a chemical reaction.

magnetism   The tendency of a substance to move into a magnetic field. This property is more strictly called *paramagnetism*. It arises from the presence of one or more electrons that are not taking part in bonding pairs or lone pairs. Each electron behaves as a tiny magnet with a magnetic field that is not cancelled out by the other (missing) member of a pair. *Example*: oxygen (1).

monomer   A unit from which a polymer is built. *Example*: ethylene (47).

osmotic pressure   The pressure needed to prevent the flow of a solvent through a semipermeable membrane (one permeable to solvent molecules but not to solute molecules). A solvent tends to flow through such a membrane from a less to a more concentrated solution. This tendency can be opposed by exerting a pressure on the more concentrated solution.

oxidation   Reaction with oxygen, as in combustion; in general: electron loss. The former definition is the original meaning of the term, and the one used in this book.

peptide group   The group ─C$\underset{HN─}{\overset{\overset{O}{\parallel}}{}}$, denoted ─CONH─. Molecules containing this group are *peptides* and, on a larger scale, the polypeptides. *Example*: insulin (82).

phenol   A substance in which a hydroxyl group is attached directly to a benzene ring. *Example*: vanillin (139).

photochemical reaction   A chemical reaction induced by light.

photon   A packet of electromagnetic energy.

polarized light   Electromagnetic radiation in which the electric field oscillates in a single plane.

polarized-light microscopy   The study of transparent substances, including crystals, with a microscope in which the sample lies between two polarizing filters.

polymer   A molecule formed by connecting smaller molecules together to form a string or network. *Example*: polyethylene (48).

prokaryote   A cell without a nucleus.

protein   A molecule formed by stringing together amino-acid molecules. Proteins, which include enzymes, structural materials, and transport agents, are polypeptides. *Example*: haemoglobin (83).

radical   A fragment of a molecule containing at least one unpaired electron (a lone electron not present as part of a bonding pair or a lone pair).

reduction   The addition of hydrogen and the removal of oxygen; in general, the gain of electrons. Reduction is the opposite of oxidation.

salt   The ionic product of a reaction between an acid and a base.

saturated compound   An organic compound that does not contain carbon–carbon multiple bonds. *Example*: ethane (17), as distinct from ethylene (47).

spin   The intrinsic rotational motion of an electron. The spin of an electron occurs at a fixed rate, and may be either clockwise or counterclockwise.

stereochemistry   The spatial arrangement of atoms in a molecule.

surfactant   A surface-active agent; that is, one that accumulates at the interface between two liquids and modifies their surface properties. *Example*: the stearate ion in sodium stearate (43).

synapse   The junction between two neurons.

thalamus   A part of the vertebrate brain just behind the cerebrum; an important entry point into the brain from the rest of the nervous system.

***trans* isomer**   An arrangement in which two groups are on opposite sides of a double bond, as in $\overset{X}{\diagup}\!\!=\!\!\diagdown_{X}$. See *cis isomer*.

transmutation   The conversion of one element into another by a process taking place in the nucleus.

ultraviolet radiation   Electromagnetic radiation with a wavelength that is shorter than that of visible violet light.

unsaturated compound   An organic compound that contains carbon–carbon multiple bonds. *Example*: ethylene (47), as distinct from ethane (17).

van der Waals interactions   The weak interactions between molecules that are responsible for their adhering together.

vitamin   Organic compounds needed in small quantities in the diet in order to maintain the metabolic processes of the body.

wavelength   The distance between neighbouring peaks of a wave of electromagnetic radiation (or any other periodic wave).

Acknowledgements

Oranges, ©David Waterman/Alamy; **Cornflower,** ©ImageState/Alamy; **Red poppies,** ©Chris Rose/Alamy; **Volcano,** ©Robert Harding Picture Library Ltd/Alamy; **Pigs,** ©Geoffrey Morgan/Alamy; **Shells,** ©Pictures Colour Library/Alamy; **Blue-ice tunnel,** ©Malie Rich-Griffith/Alamy; **Sea slug,** ©Maximilian Weinzierl/Alamy; **Soap bubbles,** ©Robert Harding Picture Library Ltd/Alamy; **Gas flame** ©ImageState/Alamy; **Smoked Salmon,** ©Foodfolio/Alamy; **Nettle stem,** ©Dr Jeremy Burgess/Science Photo Library; **Ant,** ©Holt Studios International Ltd/Alamy; **Cocoa pod,** ©Holt Studios International Ltd/Alamy; **Oil seed rape,** ©David Noton Photography/Alamy; **Thyme,** ©The Garden Picture Library/Alamy; **LM polyethylene,** ©John Durham/Science Photo Library; **Rubber tree,** ©Stock Connection, Inc./Alamy; **SEM nylon,** ©Eye of Science/Science Photo Library; **SEM velcro,** ©Eye of Science/Science Photo Library; **SEM wool,** ©Science Photo Library; **SEM silk,** ©Andrew Syred/Science Photo Library; **Cobweb,** ©The Garden Picture Library/Alamy; **Sugar cane,** ©Photo Resource Hawaii/Alamy; *E. coli,* ©Science Photo Library; **Sea squirts,** ©SNAP/Alamy; **Cicada insect wing,** ©Gusto/Science Photo Library; **Hops,** ©TH Foto/Alamy; **Chilli pepper,** ©Ian Davidson Photographic/Alamy; **Mushrooms,** ©ephotocorp/Alamy; **Raspberry fruit on canes,** ©Holt Studios International Ltd/Alamy; **Coffee beans,** ©Wayne Hacker/Alamy; **Garlic,** ©The Garden Picture Library/Alamy; **Nutmeg,** ©The Garden Picture Library/Alamy; **Chocolate and cocoa products,** ©Tropicalstock.net/Alamy; **Civet Cat,** ©John Foxx/Alamy; **Red-crested turaco,** ©Kevin Schafer/Corbis; **Persimmons,** ©TH Foto-Werbung/Science Photo Library; **Sweetcorn,** ©imagestopshop/Alamy; **Prawn** ©ImageState/Alamy; **Autumn leaves,** ©ImageState/Alamy; **octopus,** ©Up the Resolution/Alamy; **Ripe coffee** ©Chris Fredriksson/Alamy; *Papaver orientalis* **poppy seedhead,** ©David Hughes/Alamy; **Cannabis plant,** ©The Garden Picture Library/Alamy; **Naked couple,** ©Oscar Burriel/Science Photo Library; **Chromosomes,** ©Andrew Syred/Science Photo Library.

Index

TEACHING AND LEARNING PRIMARY SCIENCE WITH ICT

Edited by

**Paul Warwick, Elaine Wilson and
Mark Winterbottom**

Open University Press

Open University Press
McGraw-Hill Education
McGraw-Hill House
Shoppenhangers Road
Maidenhead
Berkshire
England
SL6 2QL

email: enquiries@openup.co.uk
world wide web: www.openup.co.uk

and Two Penn Plaza, New York, NY 10121-2289, USA

First Published 2006

A catalogue record of this book is available from the British Library.

ISBN-10: 0335 21894 6 (pb)   0335 21895 4 (hb)
ISBN-13: 978 0335 21894 3 (pb)   978 0335 21895 0 (hb)

Library of Congress Cataloging-in-Publication Data
CIP data applied for

Typeset by RefineCatch Limited, Bungay, Suffolk
Printed in Poland by OZ Graf. S.A.
www.polskabook.pl

CONTENTS

vi CONTENTS

LIST OF CONTRIBUTORS

Paul Warwick is a lecturer at the Faculty of Education University of Cambridge (UK) and is engaged in a range of research and teaching activities that link directly with his interests in primary science education and to the professional development of trainee and beginning teachers. Previously he was a primary school deputy head teacher and an adviser for science for a local education authority. He is a member of the editorial board responsible for developing web-based materials associated with *Reflective Teaching* (Pollard *et al.* 2005). His recent publications include work on procedural understanding, data interpretation and the scaffolding of speech and writing in primary classrooms. His interest in the learning affordances of new technologies for primary science can be traced to his work in the 1990s on data logging in the primary classroom and has been stimulated by recent work on a Gatsby-funded research project with a primary school cluster in Cambridgeshire.

Mark Winterbottom taught science in upper schools in England for five years. During that time he was head of biology, ITT mentor, newly-qualified teacher mentor and lead-teacher responsible for developing interactive learning activities across the school using ICT. Mark has written a variety of textbooks and a handbook for newly qualified teachers of biology. He joined the Faculty of Education at the University of Cambridge (UK) in 2002 and teaches on the Science/Biology secondary

PGCE course and on the BA in educational studies. His research interests are in ICT and the psychology of education.

Elaine Wilson has taught secondary science in a range of schools. She is now course leader for secondary science at the Faculty of Education University of Cambridge and is involved in initial teacher education and early careers professional development. She has published work on activity theory and on classroom-based action research projects. Her current research involves working with new teachers in their early years of teaching. Elaine is a Higher Education Academy National Teaching Fellow and is using her prize money to develop a science education website.

Colette Murphy is Head of Learning and Teaching (pre-service) at the Graduate School of Education, Queen's University Belfast (UK). Before coming to Queen's in 1999 she was a primary science teacher educator (principal lecturer) at St Mary's University College, where she was also director of the in-service programme. She has been involved in several short and long term science development projects in primary schools in Northern Ireland and has been commissioned to deliver programmes for primary teachers organised by the Education and Library Boards. She has delivered practical science programmes for primary school teachers all over Northern Ireland and the Republic of Ireland, including the ASE Certificate in Primary Science Teaching, DASE modules in primary science and the Certificate in Professional Development for Teachers (Primary Science). Her principal research area is in primary science teaching and the use of ICT to enhance it. She directed the AZSTT-funded Science Students in Primary School (SSIPS) project and is currently directing the Science in the New Curriculum (SiNC) project, also funded by the AZSTT. Colette is a reviewer for three international education journals and is a member of the editorial board of the *International Journal of Science Education*.

John Williams has worked in primary education for over 30 years as class teacher, head teacher, adviser for science, governor and as a senior lecturer in higher education. As a head teacher in Kent, his school was one of the first in the country to use computers for teaching primary age children. He has written over 20 classroom books on primary science and design and technology. He has recently retired from full-time teaching at Anglia Polytechnic University, where he was the admissions tutor and senior lecturer responsible for the teaching of design and technology to trainee teachers. By dividing his time between Italy and the UK he remains involved with primary education, lecturing, supervising students in schools and, of course, writing. His other interests include research into various aspects of the history of science.

Nick Easingwood is a senior lecturer for ICT in education and acts as the ICT coordinator for the School of Education, Anglia Polytechnic University in Chelmsford, UK. He leads a post-graduate certificate of education secondary ICT course and contributes to primary and other secondary initial teacher education courses as well as in-service Bachelors and Masters degree courses. Having himself spent 11 years as an Essex primary school teacher, he maintains regular contact with schools through visiting students on their school experiences. His research interests include the development of online resources for student teachers and their ICT capabilities on entering higher education. His publications include *ICT and Literacy* (2001) which he jointly edited, a contribution to the second edition of *Beginning Teaching, Beginning Learning* (2002), *ICT and Primary Science* (2003), and *ICT and Primary Mathematics* (2004).

Derek Bell is Chief Executive of the Association for Science Education and has extensive experience not only of teaching and learning in science but also of the wider range of education issues including teacher education, higher education, subject leadership in schools, research, project management and network development. He has taught in schools and higher education institutions and been involved in science education research and development including the coordination of a major curriculum project. In addition he has undertaken a wide range of consultancies in the UK and overseas and is a member of several advisory/expert panels including the National Co-ordinators Group for the National Network of Science Learning Centres, the WISE National Co-ordinating Committee and the Astra-Zeneca Science Education Forum. He is a member of the Board of the Science Council, the Engineering Technology Board (ETB) and SETNET. From 2002 to 2004 Derek was Chair of the Wellcome Trust Society Awards Panel. Throughout his career Derek has maintained a strong and active interest in the enhancement of teaching and learning in science and is keen to strengthen the links between science, technology, engineering and mathematics through new and existing partnerships across the education, industrial and business sectors.

Adrian Fenton works as the curriculum support manager for the Association for Science Education (ASE). He has been the project officer leading Inclusive Science and Special Educational Needs developments in collaboration with NASEN, with outcomes including the Inclusive Science CD-ROM and website www.issen.org.uk. This has led to a continuing involvement with schools and organisations, writing articles for various journals and delivering professional development training. He has contributed to ICT in Science developments in collaboration with Becta, is part of the development team for Science UPD8 www.upd8.org.uk and has a broad understanding of curriculum developments and assessment, providing support for current science teachers. As a qualified science teacher (with a physics specialism) he has a broad experience of teaching in differing environments (1993–2001) including mainstream science teaching, teacher training with VSO in Tanzania and previous work with deaf students in higher/further education.

Ben Williamson is a learning researcher at NESTA Futurelab (UK), a not-for-profit initiative established to investigate the potential of new technology in learning. His research interests are learning outside school, literacy and ICT, the potential of videogames in education and participative approaches to new technology design. Previously, he trained as an English teacher at Bristol University and taught at a local secondary school. Ben is also currently a doctoral research student at the University of the West of England and a visiting fellow at Bristol University.

Patrick Carmichael is a lecturer in the Faculty of Education, University of Cambridge (UK) where he principally teaches on MEd programmes. He was previously Director of the MA in IT and Education at the University of Reading and has taught science and ICT in secondary, primary and special schools, mainly in London. He is also a member of 'Learning how to Learn' an ESRC-funded project exploring the development of formative assessment practice in classrooms, schools and networks.

Ruth Kershner is a lecturer at the Faculty of Education University of Cambridge (UK), with teaching and research interests in the psychology of learning and intelligence, the primary classroom as an environment for learning and special educational needs and inclusion. She has previously worked as a lecturer at Homerton College, Cambridge, and before that as a learning support teacher, primary teacher and childcare worker. Her current research is in the areas of the developing uses of ICT in the primary classroom environment, gender and special education, and teaching strategies for children with learning difficulties.

John Siraj-Blatchford is a lecturer at the Faculty of Education University of Cambridge (UK) and Associate Director of the ESRC Teaching and Learning Research Programme. His experience of teaching and researching in early years classrooms has been considerable. His publications include *Supporting Science, Design and Technology in the Early Years* (1999, with Ian MacLeod-Brudenell), *Supporting Information and Communications Technology in the Early Years* (2003, with David Whitebread), *101 Things to do with a Buzz Box: A Comprehensive Guide to Basic Electricity Education* (2001), *Developing New Technologies for Young Children* (2004) and an edited account of the 11 m. € Experimental School Environments (ESE) initiative from the European Commission Intelligent Information Interfaces (i3) programme. His latest book is a curriculum development guide to *ICT in Early Childhood Education* (2004, with Iram Siraj-Blatchford).

Helena Gillespie is a tutor in the School of Education and Lifelong Learning at the University of East Anglia (UK). She specialised in working with children with special educational needs in mainstream settings and has taught for ten years across the primary age range. She has published in a range of journals and books on primary education and in the field of information technology. Her research interests centre around learning environment and virtual learning environments.

Angela McFarlane is a professor at the Graduate School of Education, University of Bristol (UK). She is a director of the TEEM project on evaluation of digital content in the classroom, and is on the steering committee of the NESTA Futurelab project. Angela has a PhD in biological sciences and has taught science at school level for five years. She ran a software research and development unit at Homerton College, Cambridge and has experience of educational software development in a range of subjects including science. In addition, Angela has designed and directed national research and evaluation projects on ICT and learning, and was part of the team that designed the longitudinal study of the impact of networked technologies on home and school learning – Impact2. She was a member of the OECD expert group on quality in educational software and the first Evidence and Practice Director at Becta, the UK government agency for ICT. Her current research includes the role of e-learning in professional development, ICT in science education and computer games in learning.

Reference

Pollard, A., Collins, J., Simco, N., Swaffield, S., Warin, J. and Warwick, P. (2005) *Reflective Teaching* (2nd edn). London: Continuum.

ACKNOWLEDGEMENTS

We are grateful to the following companies, organisations and people for granting permission to use screenshots and other materials in this volume:

- Data Harvest © for screenshots of 'Sensing Science' in Chapter 3.
- Black Cat © for screenshots of 'Information Workshop' and 'Number Box' (Granada Learning Ltd, Television Centre, Quay Street, Manchester M60 9EA) in Chapter 3.
- Soda Creative/NESTA Futurelab (© 2004) for screenshots of 'Moovl' in Chapter 5.
- Kidspiration © for the mind maps created on 'Kidspiration' software that feature in Chapter 7.
- Professor G.K. Salmon, University of Leicester, for the five-stage model of e-moderating used in Chapter 9, http://atimod.com/e-moderating/5stage.shtml (Salmon, G. (2004) *E-moderating: the key to teaching and learning online*, 2nd edn. London: Kogan Page).
- Netmedia Education © for screenshots of Netmedia VLE images in Chapter 9.

The chapter authors would like to express their heart-felt thanks to the many primary teachers and children around the UK who participated in the projects that made the writing of this book possible.

1

CONSIDERING THE PLACE OF ICT IN DEVELOPING GOOD PRACTICE IN PRIMARY SCIENCE

Paul Warwick, Elaine Wilson and Mark Winterbottom

Good practice?

Let us think about a science lesson for primary pupils. In our fictional classroom the pupils are aged between 9 and 10 and the lesson has two central foci – understanding the insulating properties of materials and understanding the importance of fair testing procedures in experimentation. The teacher ensures a 'real-world' context for the practical investigation, linking it to ongoing design technology work on creating insulated cups to keep tea warmer for longer. Initially the teacher talks about fair testing, drawing on the children's understandings from recent work and thus setting the lesson within a framework of children's prior learning. Having set the procedural framework of the investigation, she encourages the children to express some of their initial ideas about why some materials may be better insulators than others. She demonstrates part of the task, deliberately ignoring some of the fair testing principles that have been discussed and inviting comment on her procedures. In doing so she uses the interactive whiteboard (IWB) connected to data-logging equipment and reminds the children of some of the functions of the data logger. Throughout this introduction the nature of group discussion around procedural and conceptual aspects of the task is modelled in the nature of the exchanges between the teacher and the children and in the guided inter-actions between the children themselves. In setting groups to work, the teacher emphasises that she is looking for a group consensus on what to say about fair testing in this experimental circumstance and clear evidence for asserting that one material is a better insulator than another. She provides a writing frame as a guide to thinking and discussion in the groups and, whilst most groups use more conventional equipment, one group is set to work using the IWB and data-logging hardware. During group work the teacher circulates, trying to ensure that group members have the opportunity to have their voices heard and challenging groups to justify their approach and findings. In drawing the lesson to a close the work of the group using the IWB is used as the 'springboard' for discussion of the results from all groups and differences between results are considered, primarily in terms of how investigational procedures might influence the nature of the evidence collected.

We have chosen this approach and context because it is familiar and also because it illustrates how ICT has become embedded in primary science teaching. Importantly, we also believe that this Year 5 classroom illustrates how social constructivism is informing pedagogical approaches. Social constructivism suggests that learning is a 'transactional' process which takes place through a complex interweaving of language, social interaction and cognition (Vygotsky 1978; Bruner 1985). These theories propose that the learner must be encouraged to make sense of newly developing knowledge within an already established personal knowledge framework. In our Y5 classroom pupils are encouraged to engage in actively constructing their own meaning through orientation, elicitation/structuring, intervention/re-structuring, review and application (Ollerenshaw and Ritchie 1993; Howe *et al.* 2005).

Wells (1999) goes further and suggests that pupils also need to have the opportunity to talk through their ideas and be allowed time for conjecture and argument. He believes that students need to be able to articulate reasons for supporting a particular concept and provide justification to their peers. The desired co-construction of knowledge takes place during this group interaction. Cooper and McIntyre (1996) describe the teacher's role in this transaction as providing the 'grammar and scripts' needed to set up the circumstances that will enable the learner to integrate this understanding through the process of 'scaffolding'. Cooper and McIntyre's definition of 'grammar' is the way the pupils behave in the learning situation and 'scripts' are the specialised language being introduced in the classroom. Science teaching can therefore be conceptualised in terms of introducing the learner to one form of the social language of science, namely school science. The teacher has a key role to play in mediating this language for students. Bruner (1985) draws attention to this central role of the teacher:

... [the] world is a symbolic world in the sense that it consists of conceptually organised, rule bound belief systems about what is to be valued. There is no way, none, in which a human being could master that world without the aid and assistance of others; in fact, that world is others'.

(Bruner 1985: 32)

Alexander argues that it is the interactions which take place in the classroom which have the biggest impact on learning, and that 'classroom discourse gives purchase, provides a balance and exercises power and control over the teaching and learning' (Alexander 2004: 424).

Mortimer and Scott (2003) draw too on the Vygotskian constructs of internalisation of concepts where the learner makes personal sense of the new social language with the active support of the teacher through the Zone of Proximal Development (Daniels 2000) from assisted to unassisted competence. They argue that it is through this interaction for 'meaning making' that learning takes place, and define the challenge in terms of the need to 'engage students in the patterns of talk, almost of argumentation, that are characteristic of science.'

This last notion is important when we consider what we are trying to achieve in science education more broadly. In the debate that surrounded the publication of *Beyond 2000* (Millar and Osborne 1998), it became clear that 'educating for scientific literacy' must necessarily include a focus on scientific understanding, not merely of content but of the nature of science. From this perspective consideration of the 'how and why' of scientific approaches to enquiry, together with the development of an understanding of science as a social process, can be seen as fundamental to science education (Driver *et al.* 1996). In trying to define just which 'ideas about science' may be central to a science curriculum for 5- to 16-year-olds, the Evidence-based Practice in Science Education (EPSE) project has identified the following elements: science and certainty; historical development of scientific knowledge; scientific methods and critical testing; analysis and interpretation of data; hypothesis and prediction; diversity of scientific thinking; creativity; science and questioning; and cooperation and collaboration in the development of scientific knowledge (Osborne *et al.* 2001). It will be clear that the approach taken in the classroom is fundamental to the development of such components of scientific understanding and that the approach taken is also likely to have a substantial effect on the attitudes of children towards science.

Placing this discussion in the reality of a 'target-oriented' education system Murphy *et al.* (2001), in looking at effective science teaching in Year 6 classrooms, confirm that *the* most commonly accepted measure of effectiveness used by schools, local education authorities (LEAs) and government is the end of Key Stage test results. They note that in the quest for this 'holy grail' two effective teaching models can be defined. The first is a teacher who might be described as a social constructivist (Light and

Littleton 1999), seeing the relationship between members of the class, including the teacher, as collaborative. Here, even though the curriculum may be subject structured, subject boundaries are often crossed by the teacher's approach as s/he looks at ways of making learning meaningful to the children by connecting knowledge that is presented in meaningful contexts. Concern about children *understanding* is of paramount importance. The best of such teachers get high end of Key Stage test results.

The second teacher type identified by Murphy *et al.* is one who represents science only as knowledge to be acquired. The subject is presented to the children as disconnected ideas and learned as disconnected ideas, which are re-inforced through revision testing. The teacher is in authority and the children tend to lack autonomy. Essentially the teacher inputs and the children output in the form of responses to the end of Key Stage tests. Interestingly, this is also a very effective model for achieving high test results. In the light of all that we have said so far, however, whether the children in the classes of such teachers are actually receiving a science education – let alone a good science education – must be strongly open to question.

In other words, what teachers of primary science do in the classroom, and how they do it, matters. In presenting our Year 5 class vignette we are arguing that good practice in primary science envisages learning as an active process of genuine engagement with the world, involves the teacher in scaffolding learning (Wood *et al.* 1976) through acting sometimes as an instructor and sometimes as a guide and facilitator, emphasises social negotiation and mediation between the children and the teacher and also between the children in group settings, helps children to become self-aware with respect to the intellectual processes that led them to specific conclusions, and encourages them to articulate ideas, explain, postulate and argue because the idea that 'scientific reasoning is a linguistic process' (Wollman-Bonilla 2000: 37) is taken seriously.

Where does ICT 'fit'?

If this is our view, in what ways might ICT be seen to 'fit' within this framework, and perhaps develop through providing new perspectives on pedagogy in primary science? Harrison *et al.* (2002) have shown how difficult it is to arrive at clear evidence of ICT directly enhancing teaching and learning, but there does now seem to be a gathering body of work suggesting that 'when teachers use their knowledge of both the subject and the way pupils understand the subject, their use of ICT has a more direct effect on pupils attainment. The effect on attainment is greatest when pupils are challenged to think and to question their own understanding . . .' (Cox *et al.* 2003: 3). Clearly, the use of ICT employed in our Year 5 lesson was influenced partly – as ever – by resource constraints, but most importantly it was influenced by the subject understanding and pedagogical intentions of the class teacher, with a clear emphasis on encouraging the pupils to challenge and develop their science understandings through collaboration and talk. The teacher demonstrated an understanding of the relationship between the ICT resources being used and the science concepts and processes that were the basis for the lesson objectives. She appreciated that the presentation of information using ICT can have an impact on the pupils' ability to engage with the subject matter of the lesson. And she recognised that the ways in which the pupils might be organised to collaborate in response to the lesson activity was, in part, influenced by the ways in which she wanted the ICT to be used.

This vignette therefore reflects some of the pedagogical principles for science education within ICT classrooms that are identified by Linn and Hsi (2000):

• that teachers should scaffold science activities so that pupils participate in the enquiry process;
• that pupils should be encouraged to listen and learn from each other;
• that teachers should engage pupils in reflecting on their scientific ideas and on their own progress in understanding science.

Recent developments in our understanding of cognition and metacognition seem to add weight to the prevailing assumption that the use of ICT is changing the pedagogical role of teachers and that it may have the potential to act as a catalyst in 'transforming' learning. One aspect of this may be in the development of the teacher's ability to develop powerful explanations (Moseley *et al.* 1999). In the classes researched by Moseley, primary teachers employed examples and counter-examples and involved pupils in explaining and modelling to the class as part of an emphasis on collaborative learning, enquiry and decision making by pupils. Even earlier projects (Somekh and Davies 1991) identified some transformational possibilities in the use of ICT, emphasising teaching and learning as complementary activities, with 'communicative learning' seen as central, and with technology seen as just part of the complex of interactions that takes place with learners.

Within our Year 5 class the use of data logging *might* be seen as simply replacing the existing technology of thermometers and removing the necessity of recording readings manually. Yet McFarlane *et al.* (1995) found that, with the use of data-logging hardware and software, the direct physical connection drawn between environmental changes (such as heat) and their immediate representation on the screen in the form of a line graph had a profound effect on pupils' later ability to note significant features on such graphs. Further, the pupils seemed more prepared to behave in a genuinely investigative manner, changing independent variables and analysing the effect of these changes, because they knew that it was comparatively easy to do so. Consider this, combined with the potential of the interactive whiteboard for opening up findings immediately to the whole class for discussion, and it begins to become clear that the nature of ICT tools themselves may have much to do with the quality of learning that may take place.

Yet it is clear that the tools themselves are only part of the picture. All that we have said so far illustrates that it is the mediation of those tools by the teacher, and the pedagogical practices adopted by the teacher in relation to those tools, that is likely to determine the extent to which the use of ICT in primary science is ever likely to transform learning. For example, Jarvis *et al.* (1997) evaluated the effect of collaboration by e-mail on the quality of 10- to 11-year-old pupils' investigative skills in science, in six rural primary schools. They found that the influence of the teacher was the crucial element in whether learning is enhanced.

Several themes thus seem evident when considering teachers' pedagogies and the use of ICT in science. A focus on developing a student-centred environment (Boyd 2002) seems to be connected with effective use of ICT. Linked to this, the development of new behaviours (Cox *et al.* 2003) to support collaborative learning in the classroom may be necessary in maximising learning potential for pupils, and this may include developing strategies to give pupils space to work with one another without the constant presence of the teacher. Indeed, Hennessey *et al.* (2005) found that the increased use of ICT did provide scope for such practices, with the teacher becoming more involved in supporting learning, keeping the pupils focused on the subject matter of the lesson and encouraging analysis. Some demonstrable cognitive benefits do seem to be associated with pupils sharing perspectives and understandings in collaborative learning supported by ICT. Fundamentally, the development of a conversational framework (Laurillard *et al.* 2000) seems central to developing learning with ICT in primary science. Different software (for example, for manipulating data, modelling or simulating activity) and different hardware (IWBs, video connected to the computer, data loggers) all require a different, carefully planned approach. Yet it is our view that the story of developing science ideas needs to be articulated in the classroom. As Wegerif and Dawes (2004: 86) suggest, 'children need to be given the opportunity to make their ideas public, that is, to participate in extended stretches of dialogue, during which concepts are shared and vocabulary put to use to create meaning.' It seems to us that developing and extending learning through using ICT in the primary science classroom has to be linked to this central pedagogical intention.

Contributing to emerging understandings

In this book we try to illustrate where we are with respect to the uses of ICT in primary science and to indicate, through chapters with specific foci, where we may be going. We have done this by combining chapters that are more practical and more theoretical in nature yet which are all founded in some way on existing practice and which broadly reflect the perspectives expressed thus far in this chapter. The book can thus be taken 'as a piece' and read in order, or individual chapters can be read in isolation depending on the reader's interests. To aid readers in their endeavour, the following outline of the book's chapters may prove helpful.

Chapters 2 and 3 are primarily about 'where we are now' and provide readers with a clear introduction to the uses of ICT in primary science education. In Chapter 2, Colette Murphy draws upon her recent work for NESTA Futurelab (Murphy 2003) in evaluating how ICT is currently being used to support primary science and in doing so provides another perspective on the 'good practice' themes of this chapter. She focuses primarily on how ICT might aid the development of children's skills, concepts and attitudes in science and on the development of primary teachers' confidence and skills in science teaching. She calls for specific and systematic research into various applications and their potential for enhancing children's learning in primary science. Finally, she suggests implications – drawn from her experience, research and reading – for software and hardware developers who are seeking to serve the needs of the learner and teacher in the primary science classroom.

In Chapter 3, John Williamson and Nick Easingwood argue passionately for the need for work in primary science to have a substantive practical base and illustrate how various forms of ICT use can be central to such work. They thus re-emphasise the need for active learning advocated by Murphy and then set about illustrating how the teacher's planning and classroom interactions are essential in promoting effective science learning in contexts where ICT is used. They give a clear overview of the numerous ways in which ICT might be used to enhance primary science learning, with their illustrations focusing primarily on applications that are concerned with the storage and use of data – databases, spreadsheets and data logging. In a piece that has practical examples of existing good practice at its heart, this chapter thus links with the previous one in providing a clear outline of practice and associated issues for those who may be relatively new to this area.

The central issue of inclusion is tackled by Derek Bell and Adrian Fenton in Chapter 4, where they argue that the key to genuinely inclusive science lies in the extension of existing good practice. They focus particularly on catering for the needs of those children with learning difficulties, demonstrating that science education can be enhanced where the development of pupil choice, self-advocacy, confidence and autonomy are promoted. Emphasising the central role of the teacher, they provide vignettes that illustrate how ICT might be used to boost physical accessibility to tools and ideas, to increase engagement with lessons and with ideas, to extend and develop the teaching dialogue, to aid the recording and reporting of work and, for the teacher, to extend the professional community's consideration of inclusion in science. In so doing they make a strong plea for appropriate and extended continuing professional development for teachers.

Chapter 5 introduces four chapters based upon the authors' personal research. In it, Ben Williamson considers how ICT might facilitate children's ability to create and manipulate visual illustrations and drawings of science concepts. In reporting on exploratory work with 'Moovl' software created by NESTA Futurelab, he aligns creative thinking with both conceptual and procedural thinking in science, showing how young children can be encouraged to engage in collaborative enterprise around drawing tasks where physical conditions such as air resistance can be duplicated. In the 'Moovl classroom' ideas are provisional and the iterative process of resolving approaches to a problem can provide real insights into children's scientific thinking. Links between notions of creativity and ICT are considered, as are the pros and cons of simulation software, before the potential of the Moovl software is explored. The multi-modal nature of representation and of thought is central to this chapter, as is the alignment of creativity with the development of understanding in science. Those interested in how a greater emphasis on children's creativity and visual literacy in the classroom can impact on their ability to become scientifically inquisitive will find this chapter provocative.

Continuing and developing the discussion about simulations and modelling in another context, Patrick Carmichael uses Chapter 6 to analyse the value of computer models to develop aspects of the 'characteristic and authentic' activity that scientists engage in when addressing positive, negative and neutral analogies in developing theory. Whilst noting the potential problems associated with the almost inevitable over-simplification of computer models, this chapter re-inforces the notion that ICT can prove to be a powerful tool in enabling young children genuinely to 'think like scientists'. The nature and role of analogical modelling in science is considered in a piece that is illustrated with excerpts from transcripts of interviews and conversations with young children who used a variety of computer programmes designed to represent individual animals, communities and whole ecosystems. The surprising sophistication of children's thinking in these circumstances, including an awareness of the computer itself as an 'active agent' in control of the simulation, reveals the value that such engagement might have in developing the capacity for a 'meta-level' of learning about the significance of modelling itself.

In Chapter 7 Paul Warwick and Ruth Kershner examine work carried out with pupils who used 'mind mapping' software to negotiate procedural and conceptual understandings in science lessons. This chapter looks in particular at the types of interaction and collaboration associated with the uses of such tools as interactive whiteboards and laptops within science activity in primary classrooms. An examination of the differences between pupil groups working independently at the IWB and on laptop computers allows a consideration of the affordances and constraints of differing hardware. In considering the affordances of the software, the focus on the use of mind mapping software enables a discussion of specific aspects of communication – such as questioning, explaining and pointing – together with an examination of how teachers might use a range of classroom tools to mediate learning. Interestingly, in examining the differences between the uses of the IWB with younger and older primary children the chapter emphasises this mediating role.

In asking for contributions to this volume it seemed essential to provide a contemporary account of the development of science and ICT in the early years. Though a volume in this series deals exclusively with ICT in the early years, we wanted to include a perspective related explicitly to science to give readers a sense of how 'pre-primary' learning in science can be seen as allied to, yet different from, later school experiences. In Chapter 8, therefore, John Siraj-Blatchford provides an analysis of the conjunction of ICT and 'emergent science' in the early years. He notes the problems of defining science learning for this age group but then takes the reader on a journey of discovery, pointing to the kinds of early years experiences that are likely to promote in children a strong orientation towards later scientific endeavour in their primary schooling. The strong link between learning in science and the 'playful curriculum', the centrality of the role of supporting adults and the importance of educators appreciating children's personal frameworks of understanding are all highlighted in a piece that then illustrates the relevance of ICT use from these perspectives. This chapter reveals the incredible capabilities of very young children and indicates that many of the uses of ICT suggested in other chapters for primary pupils are likely to be well within their capabilities.

In Chapter 9, Nick Easingwood and John Williams get 'a second bite of the cherry' within this volume. Here, they consider the development of science education outside the context of the school – specifically, they look at the science learning opportunities provided by museums and review the possibilities for enhancing this learning through the use of still and video digital technology. If science education is indeed a 'journey of enquiry', then this chapter reflects upon how that journey can be guided, recorded and developed through the dialogue that surrounds the planning, filming and presentation of experiences and activities with real objects in museums.

Chapters 10 and 11 look firmly ahead to possible futures. In Chapter 10, Helena Gillespie explores the possibilities inherent in the use of virtual learning environments (VLEs) for extending and developing children's work in science in the primary school. She considers how such environments might be used in compiling and cross-referencing ICT tools into subject-based or cross-curricular learning tools and notes the clear potential that this may have to develop Laurillard's (2002) intriguing notion of technology-based 'conversational framework(s)' that support learning. The key word in this chapter is surely 'imagine' – imagine how the capacity of such environments, now used increasingly with older pupils and adults, might impact on the primary science classroom in the future.

Finally, Chapter 11 gives the thoughts of someone with an international reputation in the world of education and ICT. In it, Angela McFarlane argues that the conceptualisation of the science curriculum has to change if we are to fully exploit the benefits of development in ICT in schools. She argues that if we are in the 'knowledge age' then education systems must change to prepare learners for this age. She notes that science is central to understanding the work of developing societies and yet the established science curriculum seems to do little to encourage reasoned, evidence-based discussion of science and science-related issues. The potential of ICT for developing webs of communication to support such discussion seems only to be partially exploited and one reason may be a somewhat slavish adherence to work based upon the three traditional school science disciplines. Angela thus provides a highly individual and thought-provoking piece to end this volume.

We hope that all readers, of whatever background, will find something of interest and value in this book and that it fulfils our intention of raising issues, debates and even arguments about the ways in which ICT can and should be used to enhance science learning in schools.

References

Alexander, R. (2004) *Towards Dialogic Teaching: Rethinking Classroom Talk.* Cambridge: Dialogos.

Boyd, S. (2002) *Literature Review for the Evaluation of the Digital Opportunities Project.* Wellington: New Zealand Council for Educational Research.

Bruner, J. (1985) 'Vygotsky: an historical and conceptual perspective', in Wertsch, J. (ed.) *Culture, Communication and Cognition: Vygotskian Perspectives.* Cambridge MA: Cambridge University Press.

Collins, S., Osborne, J., Ratcliffe, M., Millar, R. and Duschl, R. (2001) 'What "ideas about science" should be taught in school science? A Delphi study of the expert community'. Paper presented at the Annual Conference of the American Educational Research Association, Seattle WA.

Cooper, P. and McIntyre, D. (1996) *Effective Teaching and Learning: Teachers' and Students' Perspectives.* Milton Keynes UK: Open University Press.

Cox, M., Webb, M., Abbott, C., Blakeley, B., Beauchamp, T. and Rhodes, V. (2003) *ICT and Pedagogy: A Review of the Research Literature.* Norwich: HMSO.

Daniels, H. (2000) *Vygotsky and Pedagogy.* London: Routledge Falmer.

Driver, R., Leach, J., Millar, R. and Scott, P. (1996) *Young People's Images of Science.* Buckingham: Open University Press.

Harrison, C., Comber, C., Fisher, T. *et al.* (2002) *ImpaCT2: The Impact of Information and Communication on Pupil Learning and Attainment. Strand 1 Report.* London: DfES.

Hennessey, S., Deaney, R. and Ruthven, K. (2005) 'Emerging teacher strategies for mediating "Technology-integrated Instructional Conversations": a socio-cultural perspective', *Curriculum Journal*, 16 (3).

Howe, A., Davies, D., McMahon, K., Towler, L. and Scott, T. (2005) *Science 5–11: A Guide for Teachers.* London: Fulton.

Jarvis, T., Hargreaves, L. and Comber, C. (1997) 'An evaluation of the role of email in promoting science investigative skills in primary rural schools in England,' *Research in Science Education*, 27 (1): 223–236.

Laurillard, D. (2002) *Rethinking University Teaching: A Framework for the Effective Use of Learning Technologies.* London: Routledge Falmer.

Laurillard, D., Stratfold, M., Luckin, R., Plowman, L. and Taylor, J. (2000) 'Affordances for learning in a non-linear narrative medium,' *Journal of Interactive Media in Education*, 62.

Light, P. and Littleton, K. (1999) *Social Processes in Children's Learning.* Cambridge: Cambridge University Press.

Linn, M.C. and Hsi, S. (2000) *Computers, Teachers, Peers: Science Learning Partners.* London: Erlbaum.

McFarlane, A., Friedler, Y., Warwick, P. and Chaplain, C. (1995) 'Developing an understanding of the meaning of line graphs in primary science investigations, using portable computers and data logging software', *Journal of Computers in Mathematics and Science Teaching*, 14 (4): 461–480.

Millar, R. and Osborne, J. (1998) *Beyond 2000, Science Education for the Future.* London: Nuffield Foundation.

Mortimer, E.F. and Scott, P.H. (2003) *Meaning Making in Secondary Science Classrooms.* Maidenhead: Open University Press.

Moseley, D., Higgins, S., Bramald, R. *et al.* (1999) *Effective Pedagogy Using ICT for Literacy and Numeracy in Primary Schools.* Newcastle: University of Newcastle.

Murphy, C. (2003) *Literature Review in Primary Science and ICT.* Bristol: NESTA Futurelab.

Murphy, P., Davidson, M., Qualter, A., Simon, S. and Watt, D. (2001) 'Effective practice in primary science: a report of an exploratory study funded by the Nuffield Curriculum Projects Centre'. Unpublished. Available from Patricia Murphy at The Open University, Milton Keynes.

Ollerenshaw, C. and Ritchie, R. (1993) *Primary Science: Making it Work.* London: David Fulton.

Osborne, J., Ratcliffe, M., Collins, S., Millar, R. and Duschl, R. (2001) *What Should We Teach about Science? A Delphi Study.* London: King's College.

Somekh, B. and Davies, R. (1991) 'Towards a pedagogy for information technology', *The Curriculum Journal*, 2 (2): 153–170.

Vygotsky, L.S. (1978) *Mind in Society: The Development of Higher Psychological Processes.* Cambridge, MA: Harvard University Press.

Wegerif, R. and Dawes, L. (2004) *Thinking and Learning with ICT.* London: Routledge Falmer.

Wells, G. (1999) *Dialogic Inquiry: Towards a Sociocultural Practice and Theory of Education.* Cambridge: Cambridge University Press.

Wollman-Bonilla, J.E. (2000) 'Teaching science writing to first graders: genre learning and recontextualisation', *Research in the Teaching of English*, 35 (1): 35–65.

Wood, D., Bruner, J. and Ross, G. (1976) The role of tutoring in problem-solving, in *Journal of Child Psychology and Child Psychiatry*, 17: 89–100.

2

THE IMPACT OF ICT ON PRIMARY SCIENCE

Colette Murphy

Introduction

This chapter reviews ways in which ICT currently impacts on the learning and teaching of science in primary schools. It evaluates how ICT is currently being used to support primary science in terms of how effectively it promotes 'good' science in relation to children's skill, concept and attitude development and to the development of primary teachers' confidence and skills in science teaching. In doing so it focuses primarily on 'where we are now' and reflects some of the broad perspectives on science and learning that are given consideration in Chapter 1. It also seeks to highlight the relative lack of research into how, when, how much and how often ICT can be used to enhance the development of children's science skills, concepts and attitudes. It calls for specific and systematic research into various applications and their potential for enhancing children's learning in primary science. Finally it suggests implications for software and hardware developers which are aimed at enhancing children's learning experience in primary science.

ICT and the improvement of children's scientific skills, concepts and attitudes in science

Primary science is centrally concerned with developing a beginning understanding of physical phenomena, materials and living things, laying the foundations for an understanding of physics, chemistry and biology respectively. Whilst these are the broad areas of study, primary science is not just concerned with knowledge but particularly with scientific ways of working and the ways in which these link to the development of both procedural and conceptual understanding. It is therefore 'child active', developing both manipulative and mental activity; and it is child focused, concentrating on the world as experienced by the child. Further, primary science education intends to develop an array of learning attitudes, some of which are shared with attitudinal learning intentions in other curriculum areas.

So, primary science has three central aims: to develop scientific process skills, to foster the acquisition of concepts and to develop particular attitudes:

Skills

The process skills are:

1. Observation – a fundamental skill in which children select out information using all our five senses.
2. Communication – the ability to say clearly through many media – e.g. written, verbal, diagrammatic, presentation software – what one has discovered or observed.
3. Measurement – concerned with comparisons of size, time taken, areas, speeds, weights, temperatures and volumes. Comparison is the basis of all measurement.
4. Experimentation – children often experiment in a trial and error way. To experiment means to test usually by practical investigation in a careful, controlled fashion.
5. Space–time relationships – ideas of time and space have to be developed. Children have to learn to judge the time that events take and the volume or area objects or shapes occupy.
6. Classification – children need to recognise, sort and arrange objects according to their similarities and differences.
7. The interpretation of data – the ability for children to understand and interpret the information they collect.
8. Hypothesising – a hypothesis is a reasonable 'guess' to explain a particular event or observation – it is not a statement of a fact.
9. Inference – based on the information gathered, a child, following

careful study, would draw a conclusion which fits all the observations he or she has made.

10. Prediction – to foretell the result of an investigation on the basis of consistent, regular information from observations and measurements.
11. The control and manipulation of variables – the careful control of conditions in testing which may provide a fair test and give valid results.

Whilst it is desirable that children acquire these skills it must be said that it is unlikely any of these skills can be taught or acquired in isolation but are involved and developed in many, if not all, science activities.

Concepts

Examples of concepts fostered by primary science learning are:

Time	Life cycles
Weight	Interdependence of living things
Length	Change
Volume	Adaptation
Energy	Properties of materials

Children will gradually acquire the above concepts, primarily through practical, scientific activities.

Attitudes

Science can also develop important learning attitudes and, some would argue, a child's character. Specific attitudes which are highly treasured by teachers and society and which can be achieved through hands on, enquiry-based investigations are noted in Figure 2.1.

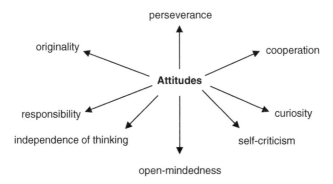

Figure 2.1   Attitudes that might be developed through science education.

The National Curriculum documentation for primary science in England and Wales interprets these skills, concepts and attitudes in the four sections of the Programme of Study as Scientific Enquiry (Sc1), Life Processes and Living Things (Sc2), Materials and their Properties (Sc3) and Physical Processes (Sc4) (DfEE/QCA 1999). The skill areas are identified as planning experimental work, obtaining evidence and considering evidence. There are variations in the Programmes of Study for Northern Ireland and Scotland. The Northern Ireland Programme of Study for primary science has reduced to two attainment targets: Exploring and Investigating in Science and Technology (AT1) and Knowledge and Understanding (AT2) (DENI 1996). The skills areas are planning, carrying out, making and interpreting, and evaluating. In Scotland, science is a component of the national guidelines for Environmental Studies (SOED 1993). Here the skills are categorised as planning, collecting evidence, recording and presenting, interpreting and evaluating, and developing informed attitudes.

How do children acquire these skills, concepts and attitudes?

If the development of these skills, concepts and attitudes are central intentions of primary science education, what kinds of activity are children likely to engage in to acquire them? The following would seem to be fundamental:

1. Observing – looking, listening, touching, testing, smelling.
2. Asking the kind of question which can be answered by observation and fair tests.
3. Predicting what they think will happen from what they already know about things.
4. Planning fair tests to collect evidence.
5. Collecting evidence by observing and measuring.
6. Recording evidence in various forms – drawings, models, tables, charts, graphs, tape recordings, data logging.
7. Sorting observations and measurements.
8. Talking and writing in their own words about their experiences and ideas.
9. Looking for patterns in their observations and measurements.
10. Trying to explain the patterns they find in the evidence they collect.

The teacher's role

Chapter 1 notes the centrality of the teacher's role in facilitating the emerging understanding of science primary pupils. In practical terms this means helping pupils to raise questions and suggest hypotheses, encouraging children to predict and say what they think will happen and encouraging closer and more careful observation. It also often means helping children to see ways in which their tests are not fair and ways to make them fairer, encouraging pupils to measure. For many practical enquiries pupils need help in finding the most useful ways of recording evidence so that they can see patterns in their observations. This may lead to the need for help in seeing the uses they can make of their findings. Central to all of this is the teacher's role in encouraging children to think about their experiences, to talk together, and to describe and explain their findings and thoughts to others.

The teacher's role in facilitating children's learning in science is explored more deeply in the next section which reviews the research into various aspects of children's science learning, particularly those linked to neuroscience.

The role of ICT in enhancing children's science learning

Recent studies of the brain, such as reported by Greenfield (2000), have led to 'network' models of learning. Such models consider ways in which computers appear to 'think' and 'learn' in relation to problem solving. They describe the brain behaving like a computer, forging links between neurons to increase the number of pathways along which electric signals can travel. As we think, patterns of electrical activity move in complex routes around the cerebral cortex, using connections we have made previously via our learning. The ability to make connections between apparently unrelated ideas (for instance the motion of the planets and the falling of an apple) lies at the heart of early scientific learning in terms of both creativity and understanding. As children explore materials and physical and biological phenomena, physical changes are taking place in their brains. These physical changes help to explain Ausubel's assertion over 35 years ago that 'the most important single factor influencing learning is what the learner already knows' (Ausubel 1968).

In the context of this discussion recent work in this area carried out by Goswami (2004) dispels three 'neuromyths'. The first is that the two hemispheres of the brain work independently – neuro-imaging has shown that there are massive cross-hemisphere connections in the normal brain and that both work together in every cognitive task so far explored. Secondly, it is not the case that education in certain tasks must happen at 'critical' times to be effective. Although there are sensitive periods for learning language, for example, this does not prevent adults from acquiring competence in a foreign language later in life. Thirdly, new neural connections can be made at any time if there is specific environmental stimulation.

The 'network' model of learning predicts that the active learning promoted by constructivist teaching approaches – in which children are actively engaged in knowledge construction – enables more pervasive neural connectivity and hence enhanced science learning. Of course, constructivist approaches present many challenges – the unique ideas and experiences 30 individuals bring to each new science topic; the challenge to these ideas that is presented by established scientific understandings of phenomena; the level of involvement in scientific enquiry required for each child; and the role of collaboration in group work that may be conducted with limited resources. So what role can ICT play in helping the learner to develop the skills, knowledge and attitudes associated with science and in helping the teacher to develop a constructivist approach to learning in her classroom?

McFarlane (2000a) illustrated the relationship between the use of ICT and the development of children's science skills (see Figure 2.2). Technology moves quickly, and, although McFarlane's scheme is still valid for mapping the process skills enhanced through using ICT during practical

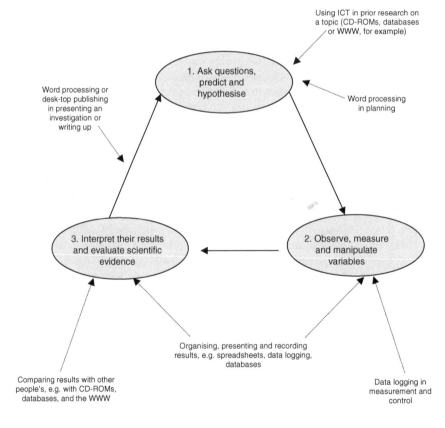

Figure 2.2   The relationship between the use of ICT and the development of children's science skills (McFarlane 2000a).

work, it might be usefully updated by including the recent increased use of multimedia in the writing up phase of science practical work and the extensive use of presentational technology, such as interactive white-boards or active slate systems. Perhaps the approach towards integration of ICT into primary science should focus more on functionality rather than specific ICT applications, for example: content versus content-free software, data logging, information handling and control technology. Which types of application, therefore, are best suited towards the development of the range of skills, concepts and attitudes outlined above?

O'Connor (2003) describes a methodology for implementing ICT into the primary science classroom which is rooted in constructivist pedagogy, 'where the children are agents of their own development'. She describes how multimedia is most effectively used as a tool 'to construct knowledge *with*', as opposed to learning *from*. She argues that the effective use of content-free software enables children to assume control of their own learning and illustrates this with a description of 10- to 11-year-old children creating PowerPoint presentations to demonstrate and communicate their understanding of electric circuits.

ICT can support both the investigative (skills and attitudes) and more knowledge-based aspects (concepts) of primary science. Recent approaches to science learning, particularly social constructivist methodologies, highlight the importance of verbal as well as written communication as being vital for children to construct meaning. ICT use can greatly enhance the opportunities for children to engage in effective communication at several levels. Communication, however, is only one use for ICT in the primary science classroom. Ball (2003) categorises four ways in which ICT is used in primary science: as a tool, as a reference source, as a means of communication and as a means for exploration. There is little *systematic* research on the use of ICT in primary science teaching, other than reports of how it has been used to support specific projects – for example, those included in the ICT-themed issue of the *Primary Science Review* in Jan/Feb 2003. Despite this, Chapter 1 has already made clear that the effective use of ICT in primary science is strongly linked to an understanding of effective pedagogical practice and in the following examples it is hoped that this link will become clearer.

The following section comprises an account of instances of practice derived from different sources in which usage of ICT in various primary science contexts has been reported (Murphy 2003). The author provides commentary on these from two standpoints, first, from working with students and teachers from a range of science backgrounds in the role of primary and secondary teacher educator, and, second, from directing a research project funded by the AstraZeneca Science Teaching Trust (AZSTT). In the project, science specialist student teachers co-planned, co-taught and co-evaluated science lessons with primary classroom teachers, using ICT to promote collaboration between students, teachers, teacher educators and subject matter experts (including curriculum developers, advisors and ICT specialists). The data from confidence audits carried out by students and teachers at the start and at the end of the project indicated a highly significant increase in students' confidence in ICT use during the project but less so for the teachers (Murphy, Beggs and Carlisle 2005). The ICT used for this work comprised a virtual learning environment (called *Blackboard*) which facilitated communication and document sharing between all participants, and training in software (called *Black Cat*) which could be specialised for primary science. When they were in college the science students had further opportunities to develop their ICT skills outside the classroom science context.

ICT as a tool

Spreadsheets

Spreadsheets are mainly used in primary science for data entry, tabulation and graph production, and form an essential element of fair testing and seeking patterns. Children at primary level are expected to use spreadsheets but not to create them for themselves, enabling concentration on the science aspects (Ball 2003). Poole (2000), however, warns that primary children have sometimes used spreadsheets without going through all the preliminary stages such as selecting axis scales and deciding on the best type of graph to explore patterns in the data. He suggests that the key issue is the pupil's ability to handle and interpret the data, so that the use of ICT for graphing needs to be part of a well-coordinated programme for teaching graphical skills. When the use of spreadsheets is considered in terms of the skills, concepts and attitudes summarised earlier it might be argued that the only added value of using a spreadsheet in terms of primary science is the speed with which the data can be presented graphically. This could indeed prove to be problematic because, if the children are not drawing the graphs for themselves, they may experience a 'conceptual gap' between measurements and their graphical representation. McFarlane and Sakellariou (2002), however, argue that using the graphing applications of spreadsheets can allow the teacher or pupil the choice of data handling to focus on presentation and interpretation rather than simple construction. The issue could be analogous to that of children using calculators routinely instead of mental arithmetic.

Databases

Ball (2003) is fairly dismissive of the value of databases in primary science, especially in relation to the fact that data or samples collected by the children are not often suitable for effective interrogation of the database. Feasey and Gallear (2001), however, provide some guidelines for using databases in primary science and illustrate two examples. In the first, 10-year-olds were building up a database about flowers. For some, much of the data collected may seem inappropriate for children of this age (length of anther, length of filament, length of carpel), raising questions as to the benefits of such an exercise in terms of scientific understanding or indeed for the development of ICT skills for children in primary school. However, in terms of understanding how measurement skills might lead to a greater understanding of variability in plant populations – raising questions as to why this variability might exist – this work might be seen as perfectly reasonable for such pupils and may well lead to the kind of discussions characteristic of the 'constructivist classroom'. The second example was a similar activity for infants who were creating a database of their class. This exercise might be viewed as more immediately relevant for pupils of this age. It enables children to produce bar charts and histograms for interpretation more quickly than by hand and, once again, it is the discussion around the meaning of these that leads to the development of scientific understanding.

Perhaps the most exciting use of a database with young children (6- and 7-year-olds) that I observed was an instance in which children were able to interrogate a prepared database of dinosaurs, whilst working with a science specialist BEd student. The children were fascinated to discover that some of these huge dinosaurs were vegetarian! They were stimulated to ask questions and wanted to find out more. In this context the children were using a database as a means of exploration. In addition to developing science knowledge it is clear that working with databases can directly enhance children's classification skills and, indirectly, could develop their powers of inference (Murphy 2003).

Data logging

Data logging is a highly versatile ICT tool for use in experimental science at any level. Higginbotham (2003) describes 6- to 7-year-old children 'playing' with a temperature sensor and discovering that they could find out whether it was in hot or cold water by watching the screen – they were effectively interpreting graphical data. McFarlane and Sakellarious' (2002) work presented supporting evidence of actual transferable learning take place. Ball (2003), however, argues that many primary teachers are not confident enough to use data loggers effectively in their science lessons. From my own experience of facilitating data-logging sessions with student teachers, I would add that many sensors are not sufficiently robust for use in the 'normal' classroom. Sensors that seem to 'work' perfectly well in one session may prove entirely useless in the next. That apart, the potential value of using sensors in primary science is considerable in terms of the development of the skills of observation, measurement, experimentation, space–time relationships, interpretation of data, inference, prediction and the control and manipulation of variables. The concepts of time and change can also be developed via the process of data logging, as can the attitude of curiosity and, if working in groups, children can learn to be cooperative in their approach.

ICT as a reference source

CD-ROMs

Amongst the most common ICT reference sources used in primary science classrooms are CD-ROMs. These range from encyclopaedic resources, such as *Encarta*, to the *ASE Science Year* CD-ROM, which contains a wealth of science-related activities. CD-ROMs are relatively permanent, physical entities which can be catalogued and stored like books. As such, schools and other institutions have 'banks' of CD-ROMs available for use. Undergraduate student teachers in a Northern Ireland University College who were science specialists preparing to teach in primary schools evaluated several of the most popular CD-ROMs which were used in primary schools. Their comments were most interesting since they were asked to evaluate in terms of their own enjoyment as well as from a teacher's perspective (Murphy 2003). Table 2.1 summarises some of the student

Table 2.1 BEd science students' comments on primary science CD-ROMs

Name	Positive	Negative	Suggestions
Light and Sound	Diagrams and animations	Not very exciting start Written explanations complex	More interaction Integrate assessment of pupil learning
Mad about Science – matter	Good graphics Games and rewards Flash questions – would keep children's interest	'Upper class' English accent Children would need relatively good knowledge of materials to benefit	Voice-over to read questions Use for only short time periods – games become repetitive
Mad about Science – 2	Voice-overs Good explanation of terms	No differentiation for different ability levels No instructions No 'second chance' to answer questions in games	Different levels 'Second chance' option for questions
I love science	Interactive diagrams Safety messages Reward system	Too difficult for 7–11 age range 'Word attack' confusing	

My first amazing science explorer (5–9)	Incentives and rewards Personal record and progress chart Varying difficulty levels Clues given to help answer questions Fill explanation of correct answers	Some parts too advanced for age range Could not find purpose for the worksheets	Programme adapted to take account of pupil's understanding before awarding 'badges'
My amazing human body	'Secret file' section	Too advanced for 6–10-year-olds – some questions difficult for a BEd science student!	
Magic School Bus	Entertaining and enjoyable Links body organs	Too much clicking of icons required Difficult navigation	
Science explorer 2	3D graphics Animations Virtual labs – book facility Website Safety warnings	Difficult animation; lack of instructions Boring voice-over Some investigations too complicated	Include an interactive 'character' as guide to involve children More colour, excitement and interaction Use only with small group of children
Science: forces, magnetism and electricity	Graphics and music	No explanation of experimental results Little variety	Only use for five minutes or so – becomes boring

views which could be useful for both developers and teachers when designing and using CD-ROMs.

The students' comments highlight the pedagogical issues surrounding the use of different CD-ROMs as reference sources. In terms of the skills, concepts and attitudes primary science aims to develop in children, the use of CD-ROMs has the potential both to enhance and to inhibit children's learning. The developers have a vital role in this regard to ensure that they provide a learning experience which ensures that children are highly motivated by the software to enable the development of specific skills, concepts and attitudes. For example, difficult navigation and lack of clear instructions are immediate 'turn-offs' for both teachers and children.

All software development should include several phases of formative evaluation by the target audiences. From my own experience of developing courseware, I can state that packages look, sound and run completely differently in the absence of input from the children at whom they are aimed.

The Internet

The Internet is used in primary science both as a reference source and as a means of communication. Problems of lack of access to the Internet in primary classrooms restrict its use in lessons, though this is changing dramatically with the 'revolution' in large group access being brought about by the introduction of interactive whiteboards (IWBs) in many primary classrooms. Even in the absence of such hardware, teachers are able to download and use many excellent resources with the children. It is also common for children to use the Internet as a reference source at home. Indeed it appears that children use the Internet more than teachers. A survey of more than 1500 primary children and over 100 primary teachers (November 2001) reported a highly significant different mean response ($p<0.001$), with 23 per cent of the children claiming to use the Internet often, compared with only 13 per cent of the teachers. There was no significant difference, however, between those reporting no use of the Internet – 54 per cent children and 55 per cent teachers. In the same study, 13 per cent of primary children responded that they often used a computer for homework (Murphy and Beggs 2003).

The Internet provides a wealth of resources for primary science learning and teaching. Cockerham (2001) has produced a resource called *Internet Science*, which details a series of activities aimed at 7- to 11-year-old children. These activities largely comprise comprehension questions based on children's navigation and interpretation of relevant websites. Such activities might aid children's concept development in specific content areas and have the potential to arouse curiosity and, depending on connectivity and the availability of specific URLs, might have a strong effect on developing the attitude of perseverance! More recently Becta (2002) produced guidance for using web-based resources in primary science.

An example of Internet use for a primary science investigation involving hundreds of schools took place in Northern Ireland in March 2002. Over 5,000 children took part in a Science Year project in which they used the Internet to enter and analyse their data. Children (or the teacher) entered either 'R' or 'L' into a prepared database to indicate which hand they used for the following tasks: writing their name and throwing a tennis ball into a box (for 'handedness'); kicking a tennis ball and hopping on one leg (for 'footedness'); identifying a quiet sound in a box ('earedness') and looking at a friend through a cardboard tube ('eyedness'). Children could obtain immediate feedback as to how their data fed into the total set and an update on the analysis. The study concluded that 'handedness' did not relate directly to 'footedness', 'earedness' or 'eyedness' (Greenwood, Beggs and Murphy 2002). As an exercise in understanding the importance of the collaborative, cooperative nature of scientific endeavour (Osborne *et al.* 2001) this could hardly be more striking.

ICT as a means of communication

E-mail and online discussion

The use of e-mail in primary science learning and teaching is restricted because not all classrooms are online. However, this is changing and the potential for children to exchange a wide variety of experiences and information with those from other schools, both locally and globally, via e-mail is huge, particularly for environmental projects. A current difficulty with teaching about global environmental issues is that children feel powerless to do anything about them and consequently do not change their behaviour in ways which could alleviate problems (Murphy 2001). Greater communication with children from other areas of the world would enable pupils to empathise more and consider the wider implications of their actions in an environmental context.

Using e-mail has the potential for enhancing children's communication skills in primary science, particularly as it enables children to communicate about science directly and informally with their peers. There is much progress to be completed in terms of connectivity in primary classrooms before this facility can be exploited on a wide scale, but the progress is encouraging.

Digital camera, PowerPoint and the interactive whiteboard

Apart from the more obvious e-mail and Internet applications, the digital camera, PowerPoint and interactive whiteboards have proved to be highly versatile in helping children develop a range of communication and other skills. Lias and Thomas (2003) described their use of digital photography in children's meta-learning. A class of 8- and 9-year-old children used photographs of themselves carrying out science activities to describe what they had been doing, their reasons for doing it, what they had found and why. The children's responses to the photographs (displayed on an interactive whiteboard) generated far more confident and fluent descriptions which needed a lot less prompting and support than had ever been observed previously. In addition, their responses were more detailed and complete. When tested several months later, the children's recall of the activity and their understanding of the associated scientific concepts were significantly improved when they were shown the photographs. Lias and Thomas (2003) aim to extend this work by using digital photography to help children to critically evaluate their own progress, identify ways to improve what they have done and to recognise the usefulness of what they have learned.

Presentation tools such as PowerPoint – and interactive whiteboards used for this purpose – provide excellent opportunities for children to consolidate knowledge, assume responsibility for and ownership of their learning, engage in high-level critical thinking and communicate their learning to peers, teachers and wider audiences. O'Connor (2003) illustrates slides developed by children as part of a presentation on electricity which she describes as an example of how ICT and primary science can be integrated and linked successfully.

In terms of skills, concepts and attitudes, presentation tools have enormous potential for enhancing children's learning in primary science. By preparing a presentation, children could be involved in communicating all aspects of planning and carrying out experiments, rehearsing hypotheses, describing methods and discussing their recording procedures. They might then be involved with data interpretation, inference and drawing conclusions, which would be required for them to 'tell the story' of their work to their peers. The attitudes of cooperation, perseverance, originality, responsibility, independence of thinking, self-criticism and open-mindedness can all be fostered. Having to communicate their understanding of scientific concepts, and perhaps answer questions based on that understanding from less informed peers, enables constructivist learning (Vygotsky 1978). I would argue that it is in the area of presenting scientific information, as reported by O'Connor (2003), that children's learning in primary science might benefit most by their classroom use of ICT.

ICT as a means for exploring

Control technology

ICT can be used in an experimental and exploratory manner allowing children a safe and supportive context in which to work (Dorman 1999), though the connections to mathematics, to design and technology and to the development of collaborative learning are, as yet, more obvious than are specific links to science education.

Simulators and virtual reality

Probably the least exploited use of ICT in primary science classrooms currently is exploration using simulators and virtual reality (Murphy 2003), though later chapters in this volume make it clear that this picture is changing. An example of simulator use is illustrated in the Teacher Training Agency (TTA, now the Teacher Development Agency –

TDA) guidelines for using ICT in primary science (TTA 2003). The teacher used a program that simulated the speed of fall of different sizes of parachutes. She scheduled groups to use the program on the classroom computers over a week. She emphasised that they were to predict the results of their virtual experiments before carrying them out and asked each group to write a brief collaborative report on what they had learned from using the program. The teacher did not intend the 'virtual lab' work to replace the practical activities, but felt that carrying out experiments on the computer was a good way to enable the children to predict and hypothesise using their knowledge of air resistance. They would get instant feedback to reinforce their learning of how air resistance operates.

Case study of integrating ICT into primary science

The Teacher Training Agency produced explicit guidelines and exemplification materials for using ICT in primary science aimed at mentors and initial teacher training institutions working with primary student teachers (TTA 2003). They illustrated their guidance with reference to three case studies in the areas of

- grouping and changing materials (6/7-year-olds);
- the environment and invertebrate animals in their school grounds (8/9-year-olds);
- forces (10/11-year-olds).

Each of the case studies indicates links to the curriculum documents and gives background information and notes about the context and computer resources. The case studies follow the investigations step by step, indicating teacher decisions about what, how and when to use different ICT applications, for example:

> The teacher found that the Internet and CD-ROMs did not provide as much useful information as the book sources she used. In addition, the books were portable and she was able to use them outside.
> The teacher knew that temperature and light levels could be measured using simple devices such as a thermometer or a light meter, but she wanted pupils to appreciate the way in which each habitat changed over a longer period. This was most easily done using a data logger. The teacher used a data logger, which did not need to be connected to a computer, to take readings of light, temperature and moisture over a 24-hour period.
> She decided to allow the use of the digital camera to take photographs of each animal because she realised that pupils would enjoy having photographs for use later in their work. She restricted each child to a

single image to supplement their hand-drawn pictures. She felt that printing out each image 32 times (one for each child) would take too much time, be expensive and have little or no educational value. In retrospect, she felt that even this limited use of the digital images had little educational benefit especially since the quality of the close-ups was not good.

She decided not to let pupils word process their writing this time, since she only had two computers available for this work and realised that it would take too long for each child to write his or her account using a computer. In any case, the two classroom computers were being used for searching for information and printing the images. The teacher wanted pupils to use the information from books and CD-ROMs selectively so she showed pupils how to make brief notes rather than indiscriminately using a whole entry.

(TTA 2003)

Although clearly idealised and extensive, these case studies do provide a useful source of information about ways to use ICT in primary science. Comments relating to children's responses and classroom restrictions could provide valuable insights for software developers in the design of courseware for primary science.

Specific research areas to explore how ICT use can enhance primary science learning

Some questions raised in this chapter point towards gaps in the research into primary science and ICT. In relation to the role of ICT enhancing children's science learning, the question is raised about how ICT use can aid the constructivist approach to science teaching. More particularly, there is a dearth of research into which types of application might enhance different aspects of science learning. Is content-free software most useful in helping children to 'construct' and communicate ideas? If so, which applications are best suited, and how, for the construction of ideas and which for communication? Or is it the case that presentation software, for example, can enhance both processes?

When ICT as a tool is considered, are the use of spreadsheets and databases creating conceptual gaps in children's development of graphing and key construction skills? Indeed, do we need to acquire such skills in order to interpret, interrogate and manipulate data successfully? McFarlane *et al.* (1995) have suggested that the use of data logging with 'live graphing' can enhance an understanding of the meaning of graphs without the need for the mechanical skills of graph drawing. This raises a debate similar to that which raged with the introduction of calculators in schools. If graph-drawing skills are found not to be required for successful graphical interpretation, then ICT use can substitute for the less exciting aspects of scientific investigation such as the manual plotting of data. If not, then the two must be used in tandem, so that children can conceptualise how the data record was produced.

When exploring the use of ICT as a reference source, Table 2.1 presents reactions of student teacher users of a variety of CD-ROMs. A more systematic survey of attitudes of teacher and child users towards CD-ROMs might lead to the incorporation of particular generic features which should be included in all such packages to facilitate the 'uptake' of information from a computer screen.

Implications for software and hardware designers

In the light of this chapter there are several messages for software and hardware designers. Software designers need to work much more closely with their target audiences of both children and teachers, at least in the formative evaluation phase. It would be even more beneficial to involve teachers at earlier stages, say in the specification and design phases of courseware production. Some groups, such as NESTA Futurelab, are at the forefront of such activity (Murphy 2003).

The pedagogical element of much software designed for use in primary science is frequently lacking. In Table 2.1, an evaluation of several published primary science CD-ROMs by student teachers indicated problems such as:

- content too difficult for the target age group;
- no differentiation for different ability levels;
- not enough pupil interaction possible;
- poor assessment elements, for example no 'second chance' facility;
- no explanation of experimental results.

These problems can be addressed by more consultation with pedagogical experts in the area and more evaluation by the target groups at each stage in the production. The author of this chapter suggests a set of generic pedagogical issues which developers, in consultation with subject matter experts, should address in all courseware:

1. Is the software (e.g. a CD-ROM) an appropriate delivery medium for the particular content or skill area being addressed?
2. Is the pedagogical approach (e.g. branched tutorial) the most appropriate to enhance learning of the material?
3. Has the navigation been fully piloted and evaluated by the target group?
4. Is the terminology appropriate for the target group – is there a hyperscript facility and is it sufficient?

5. Has the material been checked for bias towards any particular group of users?
6. If the package is intended for class use, has differentiation in pupil ability levels been addressed?
7. Have the developers made provision for pupils with special educational needs?
8. Are there measurable learning outcomes (if appropriate)?
9. Have the developers taken expert advice about an appropriate assessment strategy for the target group?
10. Are learners sufficiently motivated by this package?
11. Is there a voice-over? How does it contribute to the learning intentions? Might the accent distract learners?
12. Are interactions fairly frequent and meaningful? Do longer periods of working with this package render the interactions repetitive and menial?
13. Are the graphics pleasing?
14. Do the graphics distract the user in any way?
15. Are there directions and are they clear?
16. Is the lesson length satisfactory?
17. Does the pupil fully determine the pace of learning?
18. Is there inclusion of a book marking facility (where appropriate)?

In the case of software designed specifically for primary science, developers should also ensure that courseware design addresses the aims of primary science.

The implications for hardware developers highlighted in this chapter are many. Though there are good examples, in general data loggers must be far more robust for use in both primary and post-primary schools. Remote data loggers would be ideal, particularly if they could be reliable in providing replicable data. Too often the present generation of data loggers, in the experience of this author, have been found wanting in this regard. The digital microscope has been a welcome and potentially valuable tool for use in the primary classroom. Unfortunately, whilst the technical aspects were very carefully addressed in its development, the pedagogical issues associated with how teachers and pupils can maximise its potential for use in primary science were neglected. Consequently, it is this author's experience that there is widespread under-use of this equipment in primary schools (Murphy 2003).

In an ideal world I would also like to see custom-made computer hardware in primary classrooms. I am sure that there is a huge market for lighter, more mobile machines with infra-red connections which are designed for use specifically by children in classrooms. Current machines are, by and large, designed for adults who work in offices. I would also advocate that developers of such machines lobby for 'school' as opposed to 'office' software to be installed. Children's books, desks and microscopes, are specifically designed to enhance their learning environment – why not computers? The situation is slowly moving forward, but it is as yet far from ideal.

References

Ausubel, D.P. (1968) *Educational Psychology: A Cognitive View*. New York: Holt, Reinhart and Winston.

Ball, S. (2003) 'ICT that works', *Primary Science Review*, 76: 11–13.

Becta (2002) *Using Web-Based Sources in Primary Science*. Coventry: Becta.

Cockerham, S. (ed.) (2001) *Internet Science*. Hemel Hempstead: KCP Publications.

Department for Education and Employment (DfEE)/Qualifications and Curriculum Authority (QCA) (1999) *Science – The National Curriculum for England: Key Stages 1–4*. Norwich: HMSO.

Department for Education in Northern Ireland (1996) *The Northern Ireland Curriculum*. Belfast: HMSO.

Dorman, P. (1999) 'Information technology: issues of control' in David, T. (ed.) *Teaching Young Children*. London: Paul Chapman Publishing.

Feasey, R. and Gallear, B. (2001) *Primary Science and Information Communication Technology*. Hatfield, Herts: ASE Publications.

Goswami, U. (2004) Neuroscience and education, *British Journal of Educational Psychology*, 74: 1–14.

Greenfield, S. (2000) *Brain Story*. London: BBC Books.

Greenwood, J., Beggs, J. and Murphy, C. (2002) *Hands up for Science! Report for Science Year. Northern Ireland*.

Higginbotham, B. (2003) Getting to grips with datalogging, *Primary Science Review*, 76: 9–10.

Lias, S. and Thomas, C. (2003) 'Using digital photographs to improve learning in science', *Primary Science Review*, 76: 17–19.

McFarlane, A. (2000a) The impact of education technology, in Warwick, P. and Sparks Linfield, R. (eds) *Science 3–13: The Past, The Present and Possible Futures*. London: Routledge Falmer.

McFarlane, A. (2000b) *Information Technology and Authentic Learning: Realising the Potential of Computers in the Primary Classroom*. London: Routledge Falmer.

McFarlane, A., Friedler, Y., Warwick, P. and Chaplain, C. (1995) 'Developing an understanding of the meaning of line graphs in primary science investigations, using portable computers and data logging software', *Journal of Computers in Mathematics and Science Teaching*, 14 (4): 461–480.

McFarlane, A. and Sakellariou, S. (2002) 'The role of ICT in science education', *Cambridge Journal of Education*, 32 (2): 219–232.

Murphy, C. (2001) 'Environmental education 2020', in Gardner, J. and Leitch, R. (eds) *Education 2020: A Millenium Vision*. Belfast: Blackstaff Press.

Murphy, C. (2003) 'Literature review in ICT and primary science. A report for NESTA Futurelab'. Bristol: NESTA Futurelab Series.

Murphy, C. and Beggs, J. (2003) 'Primary pupils' and teachers' use of computers at home and school', *British Journal of Educational Technology*, 34 (1): 79–83.

Murphy, C., Beggs, J. and Carlisle, K. (2005) 'Computer conferencing in coteaching; paper presented at the August 2005 European Science Education Research Conference in Barcelona, Spain.

O'Connor, L. (2003) 'ICT and primary science: learning "with" or learning "from"?' *Primary Science Review*, 76: 14–17.

Osborne, J., Ratcliffe, M., Collins, S., Millar, R. and Duschl, R. (2001) *What Should We Teach about Science? A Delphi Study*. London: King's College.

Poole, P. (2000) 'Information and communications technology in science education: a long gestation', in Sears, J. and Sorenson, P. (eds) *Issues in Science Teaching*. London: Routledge Falmer.

Scottish Office Education Department (1993) *Environmental Studies 5–14 National Guidelines*. Edinburgh: SOED.

Teacher Training Agency (2003) *ITT Exemplification Materials: Using ICT in Primary Science. Using Information and Communications Technology to Meet Teaching Objectives in Primary Science*. London: TTA.

Vygotsky, L. (1978) *Mind in Society*. Cambridge, MA: Harvard University Press.

—— **3** ——

POSSIBILITIES AND PRACTICALITIES: PLANNING, TEACHING, AND LEARNING SCIENCE WITH ICT

John Williams and Nick Easingwood

Introduction

In 2004 an article in *Biobits* (the newsletter of the Institute of Biology, Issue 3) suggested that there had been a decline in the amount of practical science taught in secondary schools. Whilst the evidence was mostly circumstantial, two reasons given for this apparent decline were health and safety regulations and a shortage of equipment. In responding, Dr Ian Gibson, the Chair of the House of Commons Select Committee for Science and Technology, questioned whether such a decline was actually taking place and, if it was, whether these were indeed the reasons. He wondered if the problem was perceptual rather than actual.

Although Dr Gibson was referring to secondary school science, we have found that in many English primary schools the amount of practical science taught does seem to have decreased in recent years and, in consequence, the use of much relevant ICT. These are personal observations on our part, and it is difficult to suggest reasons for this without more systematic evidence. Safety issues may play a part, in that even a small amount of disruptive behaviour during practical lessons (and if we are honest we have all suffered that at some time in our teaching career) can be a problem. However, there is no need to use dangerous chemicals or equipment in primary science. Indeed, the authors have found that practical science will interest, motivate and engage primary school children, including those with special needs. Our experience suggests that the factors in this decline in practical work are more likely to be the demands of the literacy and numeracy strategies, the all-pervading demands of formal assessments and the amount of planning that is now required from all teachers. Nevertheless, there are schools and individual teachers within schools who manage to include practical science in their curriculum. We have even visited schools in England that have decided to dispense with such things as the QCA Schemes of Work and return to a limited topic-based curriculum that includes extensive, practically based scientific enquiry.

In this chapter we suggest why it is important for primary science to provide a practical basis for classroom discussion, collaboration and learning, and we consider how and when ICT can be used both to enhance its content and to record its findings. We include examples of what we think are appropriate science topics so as to establish what skills are needed, by the teacher as well as the children. We will constantly have in mind that it is science that is being taught and that ICT, whilst being a vital element of the lesson, should be used to support the science and not dilute it or take its place altogether. We will include some practical suggestions as to how this can be accomplished within the school and classroom.

Practical science and ICT in the classroom

In our book *ICT and Primary Science* (Williams and Easingwood 2003) we have suggested two main reasons why science in the primary school should almost always be taught with reference to practical experiences:

1. **Science in the wider world is essentially practical**. It is carried out in laboratories, workshops, observatories and even in the 'field', which can be any part of the natural world from the arid desert to the depths of the oceans. Whatever the science involved it requires a set of skills that can only be learnt through practice and experience and which in turn will illuminate the body of knowledge which we call 'science'. Of course there are examples of pure theory which appear to contradict this. Yet, however imaginative these theories may be they are usually based not only on sound scientific principles but also on practical scientific activity. Darwin, for example, based his theory of evolution on his scientific observations. He was in fact a very practical scientist. We should allow children as far as possible to learn these skills, observation being perhaps the first and most basic. They are not only important in themselves, but through their use children will be more able to develop a better understanding of the essential scientific

concepts. For example, children may learn something about a simple animal such as a woodlouse by copying a picture from a book. However, they are more likely to have an understanding of how it lives, what kind of animal it is, and its place in the animal kingdom if they study it in its natural environment. If the children then collect some of these creatures to make a series of careful choice chamber tests back in the classroom, then they will be able to check the observations that they first made in the field. They may even go further and formulate and test their own hypotheses, all fully supported by various aspects of ICT which will be described later. This kind of learning simply cannot be done only with reference to the secondary material, although a CD-ROM might help in some cases! Without these practical applications teaching science would be akin to teaching art without ever touching a paint-brush, or learning music without handling an instrument or even being allowed to sing. It would also be very dull and the enthusiasm of the children would be lost, which brings us to our second point.

2. **In our experience young children are highly motivated by practical work of any kind, and even the most reluctant learners seem to enjoy it.** In the primary school, by practical science we do not only mean practical experiments, but also role play, drama, some technology, investigation and observation (as described above with the woodlice) as well as the appropriate use of ICT. When young children are first introduced to science, they are faced with abstract concepts such as life processes, electricity and forces. Surely the only way they can hope to understand such things is by practical application and study? They may well have an instinctive idea of, for example, the nature of electricity, which can be surprisingly sophisticated, but the best way for them to explore these ideas is to work with real bulbs and batteries. Life processes must surely include actually growing some-thing, but could also involve both drama and role play when it comes to learning about such things as bacteria and their effect on the body. If it required Galileo to carry out numerous experiments whilst investigat-ing forces (perhaps the most abstract of the three), then we would argue that children should carry out their own practical investigations of this highly abstract area of science. What we are suggesting is, of course, the notion that before an abstract idea can be fully understood it must first go through a concrete stage that can form the basis of thought, discussion and the comparing of ideas.

Science, ICT and the National Curriculum

Despite the fact that the issues that we discuss in this chapter hold true in a 'National Curriculum-free' world, it is in the context of the English National Curriculum that most of our work with teachers and children has taken place. We would therefore like to make a few key points about it here, and we will link some of our later comments to the specifics of National Curriculum documentation only where that seems appropriate.

By using National Curriculum (DfEE/QCA 1999) documentation creatively and imaginatively those teachers who believe that science is a practical subject will be able to teach it in that way. Within the National Curriculum ICT has its own section. However, there is also a separate statement at the beginning of the document emphasising its use across the whole curriculum. In the teaching requirements there is an emphasis on the links with other subjects, and in the programmes of study there are explicit guidelines which state that the computer should play an integral part in other areas of the curriculum, and particularly in science. Many schools seem to have a slot for both ICT and science in their timetables and clearly there are ICT skills that need to be learnt. However, once these are understood they should be an integral part of any science investigation. We believe strongly that it essential to integrate the ICT with the science, thus allowing the teacher to find the time for the practical science work described above.

As we have stressed, ICT should not be an 'add on' part of any subject, but should be a carefully planned and integrated area of the curriculum. In this way it will not only allow more time for the practical work, but will enhance and stimulate students' learning. ICT capability is a key feature of teaching and learning where knowledge, skills and understanding are developed in a practical and meaningful context. It is also important to remember that the practical investigation and the collection of data is as much an integral part of the ICT component as using the hardware and software.

So how should the class teacher go about integrating ICT into science?

Planning the lesson

When making any decision as to whether or not ICT should be used to support the teaching and learning of science, the teacher needs to be completely clear as to why it is being used, as well as being convinced that its use will actively enhance the teaching and learning experience. There is little point in using ICT if the intended teaching and learning outcomes could be more easily and efficiently achieved by not using the computer. We have already stated that the use of ICT should not replace the practical experience of handling scientific equipment and engaging in genuine scientific investigation and discovery. There is also a 'value added' component to this aspect – there is little point in using £1,000 worth of ICT hardware and software to make a teaching point that could just as easily be achieved with a set of plastic beakers costing 10 pence.

So what clear advantages can the use of ICT bring to the teaching and learning of primary science? First, quite apart from being a great motivator, there is the ability of ICT to act as a means to encourage and facilitate collaborative and active learning. A very powerful feature of the computer is that it can act as a focus for group work, where raw data is transformed into information through the use of graphs, tables and charts, or where the reports of scientific investigation can be created and presented, perhaps by the use of presentation software incorporating still and video digital imaging, text, sound and animations. The capacity, speed, range and automatic function of a computer enables large amounts of different kinds of data and/or information to be handled quickly, automatically and in an integrated way. It enables the removal of the 'manual' element from work, so pupils can access higher levels of intellectual engagement and learning. For example, rather than spending a significant proportion of the lesson drawing and colouring in histograms, the computer can produce these charts for the pupils, thus enabling them to spend the time saved in engaging in the higher-order scientific thinking skills of reflection and analysis. It is the computer's ability to act as a word processor, desk-top publisher, database or spreadsheet, as well as acting as a vehicle for the Internet and e-mail, that gives it its pedagogical power and potential. Additionally, when used for more specialised scientific applications, such as data logging and control technology, it can truly provide opportunities for primary age pupils that would have hitherto been impossible. And where laptops, personal digital assistants (PDAs) or handheld data loggers are used, ICT is no longer confined to the classroom or the specialist science lab. The pupils can take these 'real' pieces of hardware outside into the real world and can use them in a realistic and meaningful context. Indeed, within the school grounds, they may be able to access the Internet through the use of a radio network, giving them ready and immediate access to a whole range of online opportunities, from researching key scientific concepts, ideas and knowledge, to e-mailing experts in universities or museums 'on the spot'.

So, bearing the above in mind, what, as teachers, do we need to consider when we plan any science lesson where ICT is to be used? What are the key features of an effective lesson?

Above all, the lesson should be interactive. Active learning is a crucial part of any lesson, but particularly in ICT and science. The pupils must interact with the computer in that they should not be passive recipients of the data or information on the screen. They must be in control of the computer, not the other way round. Additionally, it is critically important that the teacher should interact with the pupils and the computer. It is when the teacher intervenes by asking key questions that pupil learning is greatly extended. The questions asked need to be sufficiently focused to ensure that the pupil thinks carefully about the concepts being taught, but also sufficiently open-ended to ensure that considerably more than a simple yes/no answer is required. This will invariably mean questions of a 'what if?' 'why?' 'how?' nature. Example questions might include:

- What would happen if the variables in this spreadsheet were changed?
- Why do you think that the crosses on the scattergram are clustered together? What is this telling you?
- How might the variables in the spreadsheet be changed?

This demonstrates clearly that computers can never replace teachers! In fact, their role becomes absolutely crucial to the success of the lesson. It is teacher's ability to provide the detailed subject and pedagogical knowledge and the ability to ask the right question at an appropriate moment that makes the use of ICT such a powerful tool for the teaching and learning of science. It is these abilities that the teacher needs to utilise in order to ensure that the pupils are interested and on task in such a way that both their ICT and scientific capabilities will fully develop.

Quite apart from extending pupil learning, this kind of questioning is an extremely powerful means of assessing pupils through formative assessment, or assessment for learning (AfL). This provides the opportunity to assess pupil progress in line with the stated objectives for that particular lesson. Clearly, assessment must be planned for at the appropriate stage and must reflect the intended learning outcomes of the lesson. In this way, pupil progress can be ascertained qualitatively. Indeed, through this constructivist, questioning method, many of the pupils, even the youngest ones, will be able to engage in self-assessment – and record their thoughts and findings.

This may prove to be very useful, as assessing ICT is a potentially difficult area. What exactly is it being assessed – the use of the technology, the technology itself, or the context in which it is being used? We have already seen that as far as primary school ICT is concerned, it is in fact all three, as we are seeking to develop ICT capability – the knowledge, skills and understanding underpinning its use. However, in the primary school, if ICT is invariably and quite rightly taught through the context of another subject – in this case science – then the question of 'appropriate' assessment is brought into sharp focus. Thus the key objectives for the lesson will be scientific ones, with perhaps a secondary objective concerning the use of ICT. Indeed, the National Curriculum document for ICT makes it quite clear that there is an expectation that ICT capability will be developed in this way.

Although there are many programs available to children, we should take care not to assume that by their use alone children are learning science. Some databases will allow children to produce graphs, pie charts and tables from an already selected range of data. Whilst these are fine for children to use to familiarise themselves with the programs and the computer, once this has been accomplished, surely it is better for them to use their own data collected during one of their own science investigations? If they use their own statistics this may also provide more time for further work. When planning what kind of program to use, teachers need also to be aware of exactly what they want the children to learn. The authors recall that when ICT was first introduced into schools, there were many programs (and they still exist) that showed various circuits with bulbs and batteries, which required the users (the children) to decide whether the bulb would light or not. Whilst this might perhaps be a useful reinforcement exercise, we do not think it can take the place of actual 'hands on' experience. Children will enjoy manipulating the wire and looking closely at a bulb to see what is inside it, and we have found that even teachers can discover what happens when a 6-volt battery is used together with bulbs which are labelled 1.5 volts!

CD-ROMs are another aspect of ICT which need to be used with care. Obviously when studying certain areas of science it is just not possible to learn in an entirely practical way. For example, there are parts of work on life processes and living things that require children to learn about the human blood system. Elements of work on the Earth and Beyond also reveal some constraints on the use of practical enquiry! A user-friendly, interactive animated CD-ROM can be invaluable for these, although in their planning teachers need to make sure – perhaps by producing clear and simple guidance sheets – that their pupils know what to look for when using a CD-ROM. These can include specific questions for the children to answer, so that the teacher will know that they have understood what they have been watching. At another level it will help the children not be distracted by other parts of the program. We well remember watching children using this kind of CD-ROM, who without these guides had become more interested in the reproductive system rather than the skeleton which is what they had been asked to study. We did not want to stifle their curiosity, but it just was not part of the day's project!

Although there are other aspects of ICT that should be used whenever possible, such as the digital microscope (a 'must' and not just for life processes), the digital camera and of course simply using the computer as a word processor, it is in the storage and utilisation of scientific data that the computer comes into its own. A computer is able to store vast amounts of data, in many different forms, and at great speed. For this the children will use database, spreadsheet and data-logging programs. These can come in various forms and it will be helpful if we remind ourselves just what they are designed to do, and how they can be used in the classroom.

Database programs

A database program allows for the storage of considerable amounts of information, which can subsequently be retrieved, sorted and researched, and be produced at a later stage in a variety of graphical or tabular forms. We think that there are three basic kinds of database that can be used in the primary school.

Free text database

This is used to search for information on the World Wide Web or on a CD-ROM. The children will simply use the 'search' function of the web page or the appropriate software to find specific information. This could be anything from the habits of a particular kind of animal to the details of a painting in an art gallery.

Branching database

This asks the children to describe an object so that it can be identified by answering a sequence of simple questions. The database may already be set up for the pupil to follow and may be specially written to identify any number of different things from insects to rock types. Of more interest, however, are the blank databases. The structure of the database is provided for the children so that all they need to do is to fill in the necessary questions that will eventually lead to the identification of their chosen object. It is important that these questions must allow for a simple 'yes' or 'no' answer. This, at first, may be quite difficult for the children to manage. They will, we hope, be used to answering questions of the 'what if? or 'why do you think?' type. To actually be required to formulate questions of their own which must only have a binary answer, particularly when they might equate 'no' with 'wrong', is quite a challenge. However, it is an imaginative one, and a considerable learning process in itself. Once completed, this database can form a part of the school's science resources.

Random access database

Arguably the most useful of the three; this will also be the most familiar to teachers, although it may only have been used for very simple topics such as those based on hair and eye colours. Although there are several database packages available for primary schools, most use the same basic structure. A whole topic, such as 'birds', will form a file and is saved in the same way as any other program application file. An individual object within the file is referred to as a 'record', and will contain specific information about that object – in this example it will be details about the bird. Each item of information on the record is contained in a field, a further category of information under which the original birds can be sorted, which might be by type of nest or their special habitat.

If the children are to use their own information gathered during their science work then they will obviously need practice with this kind of

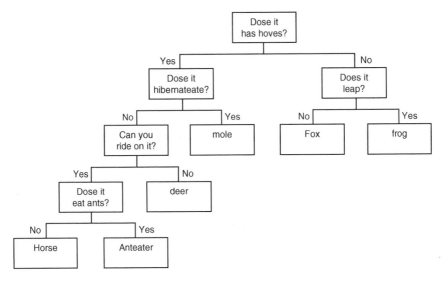

Figure 3.1   An example of a branching database, taken from a case study used in *ICT and Primary Science* (Williams and Easingwood 2003).

database. There are several that have been designed specially for primary schools such as 'First Workshop' and 'Information Workshop' (produced by Granada Learning) that can be used with a minimum amount of adult help. These programs contain 'ready made' databases at three different levels of difficulty, which can enable the children to find out how they work, and what they do. Nevertheless, before the children start using their own information the teacher will need to see that the information gathered during the practical work will be appropriate to the database, so that the children can become familiar with and understand the meaning of the various entries, i.e. 'file name', 'field' and 'record'. It is often a good idea for the children to keep a record of their discoveries, not as long handwritten texts but in short note format and under headings such as 'habitat' or 'feeding habits' (we are still using our example of the bird topic). These will become the fields or records, and the data can be entered directly into the computer.

Constructing the database, as important as this may be, is but half the picture. Once the database is complete the children can ask it to list the animals under different headings depending on the fields used. Birds could be listed under habitats or geographical areas, under their feeding habits or whether they are predators, carnivores or herbivores. Moreover, where relevant, this information can be displayed and printed out in a graphical or tabulated format. This complete activity allows the children to use the higher-order scientific skills of data collection, preparing and entering the data into the program, together with the subsequent computer activities of sorting, searching and retrieval.

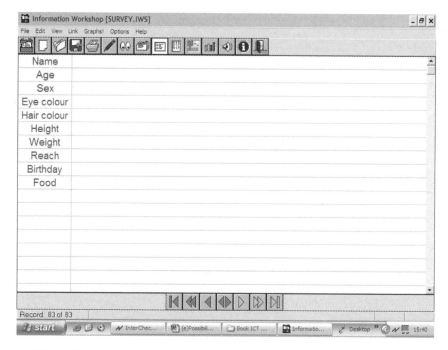

Figure 3.2   A blank database showing the fields in Information Workshop.

Spreadsheets

At first sight it might be difficult to decide when it would be more appropriate to use a spreadsheet rather than a database. As we have seen, the latter are good for collecting and manipulating data, and to a certain extent a spreadsheet can also do this. However, it will also allow the user to change the data, to make calculations with it – such as finding an average – and can utilise the given data to produce further information, such as a trend or an estimated outcome (Feasey and Gallear 2001).

To look at a blank spreadsheet on the screen is to see a simple blank table, the kind that children have often produced on paper to be filled in later with the results of their experiments. Indeed this blank spreadsheet can be used for such a purpose. However, if an average of these results needs to be entered in a subsequent column, then using the appropriate formula the spreadsheet will do this for you. (No doubt some will argue that this does not 'teach' averages. We agree that it does not teach children how to calculate them, but it does help to show children what they are and therefore what they mean by providing a real context for their use.)

As an introduction to the spreadsheet, teachers could use some of the ready-prepared ones that are available, for example those in the Black Cat Suite Number Box. Although some of these pre-prepared spreadsheets are more strictly mathematical, some are based on science topics such as 'Pulse Rates' or 'Growing a Plant'. These could be used for practice, but, as we have suggested, they would have more meaning for children if the information came from their own projects.

Once the children have become conversant with these the teacher can introduce them to a whole-screen blank spreadsheet. The children would fill in each space or cell with names, measurements or numbers, depending on what is needed for the topic. As we have already suggested, one of these columns could be the average of several measurements. This can be obtained by highlighting the name and average columns from the original spreadsheet, which can be done by holding down the 'CTRL' key and clicking and dragging in the usual way.

We have often found that as children use these sheets for recording their findings they soon learn how to fill the 'cells' and how to alter their size and format. As they become more confident in their use the children are able to use the information displayed before them as a starting point for discussion and even take note of any trends and relationships that appear in the data itself.

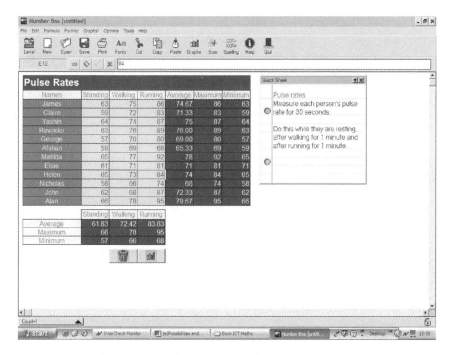

Figure 3.3   Pulse rates spreadsheet in Number Box.

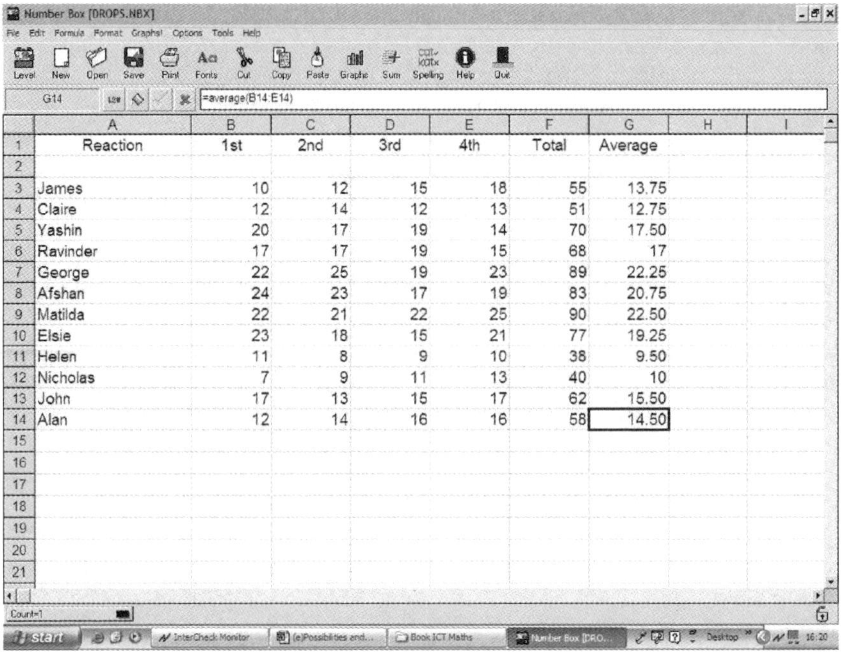

Figure 3.4   Number Box spreadsheet used to record dropping the ruler as part of a topic on 'Reaction Time'.

Data logging

As we have seen in the previous chapters this allows for the collection, collation and displaying of data, gathered directly from the environment or workbench, through the help of special electronic sensors attached to the computer. Although there are several different sensors available, those which log temperature, light and sound are the most commonly used in the primary school. There are several advantages in using a program of this sort for most primary science projects. One obvious advantage is that it saves that most valuable commodity – time! For long-term investigations such measuring light, temperature and sound changes in the classroom or an outside habitat over a period of a day or more, once the sensors are in place the children do not have to stand and watch. They can just go away whilst the sensors and computers do this work for them. The children can carry on with something else, ideally related to their science investigation. Of course they will eventually need to study the results and recordings, but that is when the science learning takes place. It is this reflection on and the analysis of the data displayed that is of fundamental importance.

Another advantage of using data logging in primary science is that it provides an instant but long-term visual representation of 'what is going on' in either a macro or micro environment (Porter and Harwood 2000). The children can use more than one sensor at a time, so if they needed to find out temperatures, light intensities or sound levels all at once this could be done and the information displayed for further investigation.

It is not difficult to imagine the many science projects in which data-logging equipment could be used. One such piece of equipment is the Ecolog system, produced by Data Harvest. This consists of a small interface box (to connect with the computer), leads and the software. There is also a manual which suggests many possible uses for the equipment. Data Harvest and other manufacturers are continually updating their products, both hardware and software.

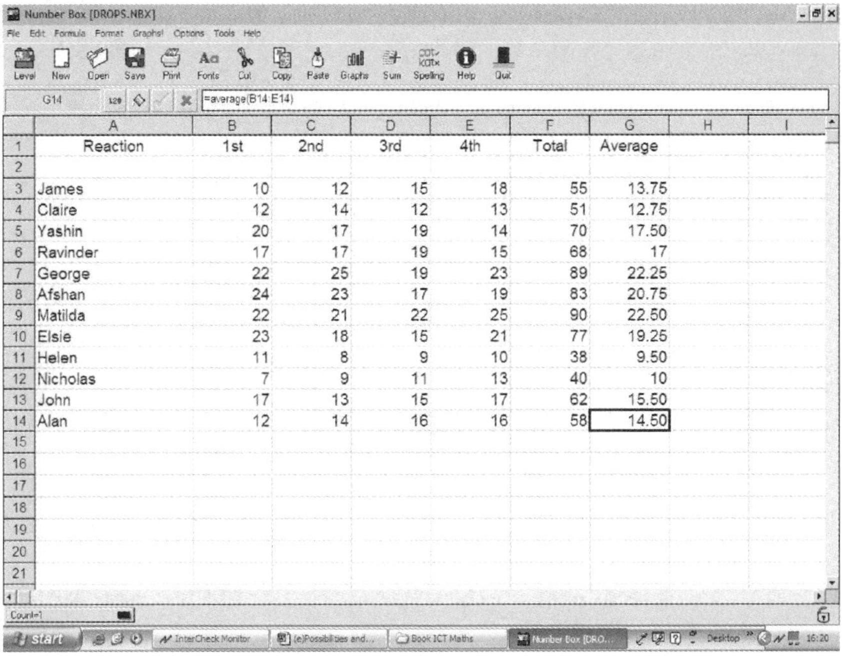

	A	B	C	D	E	F	G	H	I
1	Reaction	1st	2nd	3rd	4th	Total	Average		
2									
3	James	10	12	15	18	55	13.75		
4	Claire	12	14	12	13	51	12.75		
5	Yashin	20	17	19	14	70	17.50		
6	Ravinder	17	17	19	15	68	17		
7	George	22	25	19	23	89	22.25		
8	Afshan	24	23	17	19	83	20.75		
9	Matilda	22	21	22	25	90	22.50		
10	Elsie	23	18	15	21	77	19.25		
11	Helen	11	8	9	10	38	9.50		
12	Nicholas	7	9	11	13	40	10		
13	John	17	13	15	17	62	15.50		
14	Alan	12	14	16	16	58	14.50		

Figure 3.5 A typical sample of the graphical information obtained from sensors during a data-logging project (reproduced by kind permission of Data Harvest Limited).

Widening the use of applications for primary science

Obviously not all science – primary or otherwise – needs to have a computer input. There was much good science done in primary schools before computers became widely available. However, we hope that we have made clear the advantages of some specific uses of ICT. These can be for logistical purposes (that is, saving time by removing many of the time-consuming repetitive tasks) or as an essential part of the recording of science, or even as an aid to the understanding of the science learning process itself. Once the teacher has decided on and planned a topic then it should become clear which of the types of programs will be needed.

If, for example a topic such as 'mini-beasts' is to be studied then databases will be invaluable. It is hoped that the children will be studying a variety of habitats so that not only will they be able to observe and identify different types of animals, they will also be able to make comparisons between where and how these animals live, what food they might require and what environmental conditions they prefer. All these observations can be entered into a suitable database. When the information has been collated, analysed and displayed in a relevant format then the resulting discussions might lead the children to answer the important 'why' questions. These should, where possible, always follow the 'what is there, what can you see, what have you found?' type of enquiry so that evidence forms the basis of discussion.

So far most of our examples have shown the use of ICT in biological science projects. Indeed, we have used such examples in our book *ICT and Primary Science* (Williams and Easingwood 2003) to show in detail how all these programs can be used in one single ecological project carried out by Year 6 children. This is not to suggest that other areas cannot benefit. Databases and data logging may appear to lend themselves specifically to the natural history aspect of the curriculum but this need not always be the case. Any information gathered during a project on materials, particularly if the work is related to the grouping of materials, can easily be collected and used on a database, whilst the sensors of a data logging program immediately lend themselves to light and sound or to a topic involving traffic surveys and noise pollution. Spreadsheets, as our pictures show, are very definitely structured so that they can be invaluable for any of the sciences.

An example of how all these programs could be used for one topic

For this section we will use as an example a topic on 'Forces'. At Key Stage 2 this involves motion, types of force, friction and the measurement of force. In our view all this might be taught as one topic, which is perhaps better than only considering one example, such as only teaching about gravity. Without 'forces' we could not move, aeroplanes could not fly, ships would not float, indeed it is 'forces' that keep our solar system in place. Surely this is what we need to teach? Perhaps not all at once, but within such an all-embracing topic several conceptually linked activities can provide the focus for the work. We could use the children's own footwear to show the importance of friction when walking as well as to measure friction by pulling the shoes over different surfaces with a spring balance. By utilising the children's first-hand experiences we can find out which of the shoes best resists the pull; is it the one with the most tread? We can also experiment with a simple paper dart to show the four forces involved in flight – thrust, drag, lift and gravity. We can return to the experiments on floating and sinking first carried out at Key Stage 1 to show that given the right surface area anything will float if the force of gravity can be overcome by the upward 'thrust' of the water. By allowing children to attach magnets to model cars they will soon discover that magnets push as well as pull and need not even touch each other. Finally, children can replicate Galileo's experiments with gravity and motion and learn some history of science as part of the process towards the ultimate goal of scientific literacy.

Can all the computer programs we have described be utilised for this topic? As we shall see, databases and spreadsheets certainly can, but data logging – at first sight at least – may seem to be out of place here. We certainly do not advocate any contrived situation, invented just as an excuse to introduce various aspects of ICT for their own sakes. However, there are aspects of data logging that are not only appropriate but will add a completely new dimension to this experimental work. We are helped here by the work of Galileo and Newton who first gave us a consistent view of the dynamics of movement, both for the universe as a whole and here on the Earth. Galileo's experiments are often to be found in the primary classroom, although usually in a different context. We have seen children running model cars down an incline fitted with different surfaces to discover if a rougher surface affects the distance travelled by the car. This gives a good indication of the force of friction, although there is seldom any indication as to how this might connect with other forces or to the science of motion, or indeed even to Galileo. The children will often measure just how far the cars travel and note how friction can affect this distance. This is good, but for this work teachers need to understand that when Galileo studied what he called 'local motion' he was studying the way different objects moved through the force of gravity. He at first studied them as they fell freely to Earth, but because this was unsatisfactory from an experimental point of view (it was too quick) he later constructed an incline plane. The original of this is in the History of Science Museum in Florence. The most striking thing about this construction, apart from the fact that Galileo actually used it, are the little bells placed at intervals down the ramp. A ball was rolled down and in order to make careful measurements of its speed, the ball rang the bells on its way down. By noting the time between each strike, Galileo was able to describe mathematically the way objects behave under freefall, their accelerated motion as well as their constant speed, or, as it is now called, the inertial motion.

We are not for one moment suggesting that primary children need study this in any detail, but as they do use ramps and toy cars as just described then why not give them some of the background and actually make some accurate measurements? Several manufacturers produce 'light gates' (a series of light-sensitive cells) for use with data loggers and these can be used as accurate timers. There is a light gate at the top of a ramp which is activated by the model car to start the timing and one at the bottom to stop it. The slope of the ramp can be altered and times compared. This seems to be an ideal way of utilising data-logging technology and using ICT to record the results and to look for patterns.

Any enquiry to show the effect of magnetic force, such as the strength of a magnet using measurements of the distance of the attractive force, can be entered onto a spreadsheet, as can work on the measuring of friction, such as those mentioned earlier using the children's own shoes.

As we have seen, databases (or at least random access databases) tend to lend themselves to such things as population studies, be they biological or geographical. However, why not, in the database suggested for the materials project, include a field for magnetic attraction? There could also be one for electrical conductivity, for as we know certain metals – whilst they are good conductors of electricity – are not magnetic. As with all these examples, when making this entry into a database we should bear in mind the requirement to develop children's understanding of scientific enquiry. For example, they will be able to use their collected data to assess evidence, search for patterns within that evidence, and compare and contrast it with previous information. Finally, they will be able to communicate their findings clearly and simply.

There is no reason why, in any science topic, we should not take advantage of other aspects of ICT. Digital video, or at least still digital images of the children's experiments, could be a part of their record of work. Word processing, desk-top publishing or presentation software could be used so that the children can explain, when necessary, what they have done. This could incorporate still or moving images, sound and text, and these in turn could be subsequently displayed in a variety of forms, including on the World Wide Beb. Simple design programs, or even simple pictures from various programs now in common use, can help to explain various ideas and concepts such as those forces described earlier.

There may also be some particular aspects of science at Key Stages 1 and 2 that for the moment do not lend themselves to practical work, although we do not think there are many. These could include the so-called 'minds on' activities (Watt 1999) as distinct from the 'hands on' activities of practical science, which lend themselves to the appropriate use of ICT. We have, in another context, suggested that constructing food chains could be such an activity, where a simple design program would help to guide and later illustrate the children's thinking.

Some final thoughts

We hope that we have shown here, even putting aside the requirements of the National Curriculum, that science in the primary school should be largely practically based and that ICT must be an integral part of the work. ICT can be used at different times during a scientific enquiry – it can be used for research, collecting data, analysing information, recording findings and displaying and presenting the results of the scientific investigation. Time can be found for both within even today's crowded syllabus. The new National Primary Strategy (DfES 2003) gives us the opportunity, for within it you will find the phrase 'empowering primary schools to take control of their curriculum and to be more innovative and to develop their own character'.

References

Department for Education and Employment (DfEE)/Qualifications and Curriculum Authority (QCA) (1999) *Science – The National Curriculum for England: Key Stages 1–4*. Norwich: HMSO.

Department for Education and Skills (2003) *Excellence and Enjoyment: A Strategy for Primary Schools*. London: HMSO.

Feasey, R. and Gallear, B. (2001) *Primary Science and Information and Communication Technology*. Hatfield: The Association for Science Education.

Institute of Biology (2004) *Biobits*, 3. London: Institute of Biology.

Porter, J. and Harwood, P. (2000) *Data Logging and Data Handling in Primary Science: MAPE Focus on Science*. MAPE, Newman College.

Watt, D. (1999) Science: learning to explain how the world works, in Riley, J. and Prentice, R. (eds) *The Curriculum for 7–11 Year Olds*. London: Paul Chapman.

Williams, J. and Easingwood, N. (2003) *ICT and Primary Science*. London: Routledge Falmer.

—— 4 ——

MAKING SCIENCE INCLUSIVE: EXTENDING THE BOUNDARIES THROUGH ICT

Derek Bell and Adrian Fenton

In writing this chapter we are faced with a dilemma. On the one hand there should be no need for a separate chapter on making science inclusive through the use of ICT. This is because, almost by definition, the whole of this book is about making science more accessible for all the pupils we teach. Each chapter demonstrates ways in which ICT offers all of us established and new opportunities for extending the boundaries of our teaching and of our pupils' learning.

On the other hand, we know that all too often, when faced with particular types of pupils, we find our teaching has to go beyond our 'natural boundaries' in order to engage them. In other words, we are challenged to put our understanding of teaching and learning to the test in our efforts to help all pupils make progress. However, despite the wealth of research and curriculum resources that are available to support science, ICT and inclusion, these three issues are all too often dealt with in isolation. Murphy (2003), for example, highlights the separation of these areas and the comparative lack of research into how, when, how much and how often ICT can be used to enhance the understanding of science held by specific groups of children. Similarly, in their review of science education and ICT (which has a more secondary focus) Osborne and Hennessy (2003) only make passing reference to the benefit to 'low ability' pupils when they argue:

the use of ICT changes the relative emphasis of scientific skills and thinking: for example, by diminishing the mechanical aspects of collecting data and plotting graphs – particularly beneficial for low ability pupils – while enhancing the use of graphs for interpreting data, spending more time on observation and focused discussion, and developing investigative and analytical skills.

(Osborne and Henessey 2003: 23)

Both of these excellent reviews simply reflect that there is very little consideration of how science, ICT and inclusion can be brought together in order to enhance our practice. In this chapter we aim to redress the balance.

ICT and science can be a very powerful combination in supporting inclusion in the primary classroom if we bring them together in appropriate ways. Although Wall (2001) provides effective insights, with specific examples, of how this might be achieved for pupils from 5 to 16, we can envisage a more general model of the situation in terms of a dynamic Venn diagram, in which the extent of overlap of the circles reflects the degree to which the three elements combine to support each other. Figure 4.1 attempts to illustrate what we mean, by showing just three of an almost

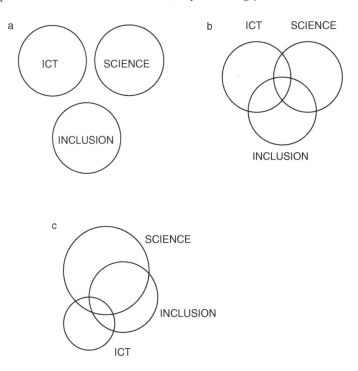

Figure 4.1 Primary science, inclusion and ICT: a model of interactions.

infinite number of possible combinations. The first (4.1a) shows the extreme situation in which ICT, science and inclusion are considered virtually in isolation, with any overlap being purely coincidental. The second (4.1b) illustrates a planned and balanced approach, bringing the three elements together in order to capitalise on ways in which they can be mutually supportive. The third (4.1c) illustration indicates one way in which, in a specific situation, there might be a particular emphasis that has been planned to meet the needs of either an individual or group of pupils. In this last example, the focus would be on the science, hence the larger circle, but with ICT supporting an aspect of the work in order to make the activity more inclusive.

The important point behind the model is that inclusion is a central principle in all our teaching. Science is a discipline which makes a major contribution to the education of all our pupils and, in this situation, the use of ICT is a vehicle for extending the learning opportunities for pupils. We should remember that, while inclusion is a principle which should underpin virtually everything we do, there will be occasions when teaching science does not require the use of ICT and conversely times when we will use ICT for other purposes. Hence it is unlikely that the three circles in Figure 4.1 will ever fully overlap.

In what follows, we will outline what we mean by inclusion in the context of this chapter, what science has to offer and how the use of ICT might be exploited to enhance the learning experiences of different pupils. We will go on to explore in more detail some of the ways in which we can develop our own teaching approaches to maximise the benefits of the opportunities made possible through using ICT. Vignettes of actual classroom activities will be used to illustrate how the principles we highlight can be put into practice. More specific examples and case studies of the ways in which ICT has been used to support science and inclusion can be found on a wide range of websites, some of which are listed at the end of this chapter.[1] We hope that, after reading this chapter, you will be able to recognise and justify ways in which the array of ideas discussed elsewhere in this book can contribute to making science even more inclusive by extending the boundaries of teaching and learning through the use of ICT.

What do we mean by inclusion?

The meaning of the term 'inclusion' is not a simple matter of providing a definition. Rather, as discussed by many authors (Farrell 2001; Lindsay 2003; Wedell 2005), it is a complex set of ideas. At the heart of the concept is the aim to ensure that all pupils, regardless of their background, culture, ethnic origin, gender, physical abilities or learning capabilities have the opportunity to engage proactively in their education. Inclusion thus involves social, political and cultural issues, as well as matters relating to teaching and learning. In writing this chapter we have focused very deliberately on the latter, with a particular emphasis on exploring ways in which ICT can be used to support children with learning difficulties and how it can overcome some of the barriers to their learning. We would agree with Rose (2002) in arguing that 'there is a need to move the inclusion debate forward through a consideration of classroom practice to address the needs of all pupils including those identified as having special educational needs'.

Furthermore, in taking this stance, we feel strongly that, as discussed elsewhere (i.e. Bell 2002, 2003; Davies and Florian 2004), making our teaching more inclusive is fundamentally an extension of our own good practice. As Davies and Florian (2004) concluded, 'questions about whether there is a separate special education pedagogy are unhelpful . . . The more important agenda is about how to develop a pedagogy that is inclusive of all learners.'

A key assumption in our approach to making science inclusive is that 'children will learn with appropriate teaching' (Solity 1995) and that 'effective teaching for those with special needs has direct relevance to effective teaching in general . . . [and] . . . a key element in teaching and learning approaches is the recognition of the learner as an active rather than a passive participant' (Wedell 2005).

Effective science teaching for children with special educational needs can take place in mainstream settings, special schools or in specific learning environments (such as a hospital school or at home). However, we should also take note of the 2002 joint statement on inclusive science by the Association for Science Education (ASE) and the National Association for Special Educational Needs (NASEN):

Both nationally and internationally, there is a trend towards inclusion for children with special educational needs. This has been interpreted as attendance at a mainstream school for learners with special educational needs. Our view is that inclusion is not simply about placement but related to the quality of the educational experience.

The current context provides challenges and opportunities to educators. Those working in a mainstream environment are engaging with a wider range of students and need appropriate support and guidance on effective inclusion and provision for the students. Some special schools are faced with the new challenge of providing an appropriate science curriculum. There exists a need for the sharing of good practice between those with different expertise.

Inclusive science involves issues of access, quality, relevance and purpose. This joint statement encompasses the notion that all students with special educational needs are entitled to access high quality science education that recognises and responds to diverse learning needs.

(ASE and NASEN 2002)

Although, in this chapter, we have used the phrase 'special educational needs' in line with the ASE/NASEN Statement and most of the existing literature, it is worth noting that other terminology is being introduced. For example in Scotland, the phrase 'additional support for learning' has been adopted and enshrined in legislation through the introduction of the Education (Additional Support for Learning) (Scotland) Act 2004 (HMSO 2004). Our use of the terms 'special educational needs' and 'inclusion' encompasses the ethos of providing 'additional support for learning' through the use of ICT. Examples of how ICT can help to break down barriers to learning and enrich learning have been included but other aspects of inclusion such as gender, ethnicity and social or cultural backgrounds have not been dealt with explicitly.

Our emphasis is on children with learning difficulties but we also recognise the potential for supporting pupils identified as gifted and talented, particularly in science. Work with 'gifted and talented' and 'more able' pupils is an area of inclusion that has recently been given higher recognition in the education community, encouraging schools to identify and develop their support for such students. As stated by the DfES-supported National Centre for Technology in Education (NCTE 2005), ICT plays an important role in providing opportunities for gifted students to progress at a rate that is appropriate to their abilities, accommodating their individual learning styles, whilst developing and practising higher-level thinking skills. Networks and website support for working with more able pupils have continued to develop; for example see Becta's web publication *How to Use ICT to Support Gifted and Talented Children* (Becta 2002), or the *London Gifted and Talented* web pages. ICT can be used as a vehicle to further gifted pupils' understanding through the additional supplementary activities and extension materials that are available in different software packages or web-based resources. Furthermore, 'many ICT tasks do not require the use of a specific classroom or laboratory. They can, therefore, extend learning beyond the teaching space and class contact time' (University of York Science Education Group 2002).

To further illustrate the complexity of making science inclusive, there are some students who might be gifted and talented but have other special educational needs. In these circumstances, it is necessary to explore ways in which the barriers to learning can be effectively overcome in order to engage the talents of the individual. Montgomery (2003) has considered this issue, which is referred to as 'double exceptionality', in more detail. However, as with much of this field, it is the teacher who has to tailor the learning situation to meet the needs of the pupil. Once again we are reminded of the importance of the teacher, as Osborne and Hennessy (2003) stated in their extensive review, 'we need to acknowledge the critical role played by the teacher, in creating the conditions for ICT-supported learning through selecting and evaluating appropriate technological resources, and designing, structuring and sequencing a set of learning activities.'

What does science offer?

As argued in more detail elsewhere (i.e. Bell 1999, 2002) it is widely acknowledged that learning in science provides opportunities for children with learning difficulties to develop a better understanding of the world around them, with all the possibilities and challenges that it brings. More specifically, science allows such children to (QCA 2001):

- develop an awareness of, and interest in, themselves and their immediate surroundings and environment;
- join in practical activities that link to ideas, for example, doing and thinking;
- use their senses to explore and investigate;
- develop an understanding of cause and effect.

Although written to support the National Curriculum in England, the publication *Planning, Teaching and Assessing the Curriculum for Pupils with Learning Difficulties* (QCA 2001) provides helpful material for planning learning opportunities and activities in science for pupils from 5 to 16 and includes a set of 'performance descriptions' (the p-scales) which describe stages of achievement in the early learning of children with a wide range of learning difficulties.

The contribution of science to the education of children with learning difficulties, however, goes beyond the scientific concepts and skills that might be acquired. Science also provides opportunities for children to develop self-advocacy (Mittler 1996) through, amongst other things, an understanding of choice, the development of skills and competencies, confidence in taking risks and feelings of being regarded, encouraged and supported as they develop their confidence and autonomy.

Whilst there is a wealth of material available relating to special educational needs generally and children with learning difficulties more specifically, there is little available which examines teaching and learning of children with learning difficulties in particular subject domains, and science is no exception. Yet there is some evidence which suggests that children's perceptions of their academic abilities are specific to different content areas (Carlisle 1996) and that it is not unreasonable to suggest that they may be more able to succeed academically in some content areas than in others. Thus it is important that, as teachers, we are sensitive to the response of individual pupils to science as a subject, as well as the opportunities it provides.

What can ICT in science add?

Before considering in more detail the potential of ICT to extend the boundaries of teaching and learning, we should remind ourselves that our understanding of teaching and learning should underpin the way in which we use ICT. Chapter 1 has already emphasised some of the perspectives that are important here. Harlen (2005), for example, provides support for these in her excellent account of teaching, learning and assessment in science for pupils aged between 5 and 12 years. In particular she emphasises the value of children's ideas, the importance of asking questions and dialogue, the need for developing process skills to underpin conceptual understanding and the major contribution of positive attitudes and values to learning. These underlying principles apply in all situations but, as Bell (2002) has argued, in supporting inclusion particular attention must be paid to:

- the value of being able to understand, recognise and, most importantly, make explicit the incremental steps that are required to help children develop their use of process skills and early understanding of concepts and;
- the need to adapt and modify our teaching strategies, often in small but significant ways, in order to meet the learning needs of individual and groups of children, paying particular attention to the use of language, questions and dialogue, the relevance of activities to the children and the selection of resources.

Appropriate, and we would stress *appropriate*, use of ICT in all its forms has the potential to enhance teaching and learning in science for children with special educational needs in much the same way as for other pupils. ICT can add to their motivation, develop their social interactions and improve their confidence in their work. More specifically, in relation to science it can, amongst other things, extend and enhance observations, provide records of events, improve presentation and communication of findings and support dialogue in reaching conclusions. Furthermore, we would suggest that ICT can:

- make science activities more physically accessible;
- increase the levels of engagement between pupils, teachers and the topics being studied;
- extend and develop the teaching and learning dialogue;
- facilitate the recording of evidence, reinforcement of experiences, ideas, evidence and concepts, and reporting of achievements and progress;
- develop an extended learning community through dissemination and networking.

Making science more physically accessible

For many children, ICT enables them to do things they could not otherwise achieve. Appropriate modification of ICT equipment allows pupils with physical disabilities to, for example, prepare reports to a high presentational standard, construct diagrams, make selections from option lists and use simulation software to test their ideas. This can be done through a range of devices including roller balls, joysticks, sticky keys, concept keyboards and touch screens. Specific modifications can be made for particular disabilities.

Pupils with hearing difficulties benefit enormously from high-quality electronic equipment which picks up sounds, and which can in turn be amplified through the use of appropriate software. The ease of using more visual material further increases the potential for such pupils to gain insights into the ideas being explored and for them to receive feedback of a more detailed nature.

Similarly, pupils with visual impairment are able to benefit through, for example, the use of increased font sizes and variations in the colour balance of texts made available electronically. Such control regarding the presentation of text can also greatly assist students with dyslexia (Becta 2003a). Pupils with very little or no sight can benefit enormously from the use of 'text to voice' software, and programmes which provide commentaries of events.

The range of possibilities for providing access to learning for pupils with physical disabilities is expanding all the time and is probably one of the best documented areas related to the use of ICT. Reports such as *Tools for Inclusion: Science and SEN* (Wall 2001) provide specific advice and further ideas can be obtained from organisations catering for specific disabilities. We should, however, remember the importance of matching the solution to the needs of individual pupils; this may involve trying out several possible devices or software packages. The children concerned should be engaged in this process because not only will it result in a more appropriate solution, but it is also part of the wider learning arena of developing self-worth and key skills such as negotiation, decision making and communication.

Gaining physical access through the use of ICT is, of course, not restricted to students who are in the classroom. Individuals who have a long-term illness, for example, and need to work from home or hospital, are also able to gain much benefit. The use of the Internet and appropriate software packages stimulate interest and spark students' imagination regardless of their schooling environment. Projects such as satellite schools provide education and lessons across most subjects via e-mail and the Internet.

Vignette 1: Making science more physically accessible

In some schools, ICT has become pervasive (but not intrusive) in the classroom. At the Fleming Fulton School in Belfast, for example, the whole science classroom has been redesigned so that ICT is an integral element, providing pupils with full access to the curriculum (Fleming Fulton School 2005). Clearly, this may not be possible in all situations, but there are many resources that can be introduced into virtually every context.

A talking thermometer is one such resource appropriate for visually impaired students, but which can also be useful with other groups of students. The durable probe can be placed in any liquid and, when the button is pushed, an audible spoken temperature reading is given. Other students who are less confident with using thermometers and reading off scales can use the same piece of equipment to occasionally check their readings. The RNIB have developed a wide range of equipment which supports visually impaired students across the curriculum and which are collated in a freely available catalogue (RNIB website: http://www.rnib.org.uk/xpedio/groups/public/documents/code/InternetHome.hcsp).

DiagramMaker (Wilkinson 2003) is another simple to operate resource that allows students (and teachers) to produce accurate, clear diagrams of experimental set ups. This can be particularly motivating for pupils who might not have the patience or motor control to produce accurate freehand drawings of their equipment. The planning stage of an experiment can be frustrating for some students, since they know they are not very good at 'art', and they just want to get on with the practical. Using such a tool to produce diagrams can encourage them to express their own ideas and engage with planning their experiment.

Increasing engagement

In providing an overview of ways in which science can be made more inclusive for children with learning difficulties, Bell (2002) highlights the value of 'hands on' approaches – in other words, of active learning. Children with learning disabilities are more likely to succeed using these approaches because of the reduced emphasis on the use of texts and abstract textual learning in favour of more concrete experiences and physical interaction with the scientific phenomena. Clearly the use of ICT has a role to play in this context. The use of appropriate material provides striking visual and moving images, interactive exercises and games, and authentic sounds and other facilities, all of which can rapidly secure students' attention. Interactive whiteboards (IWBs) and digital projectors, in particular provide, extensive opportunities for individual, group or class involvement. The physical engagement of students using an interactive whiteboard to choose objects or select answers enables them to make non-verbal choices whilst developing physical coordination.

Involving children in quizzes can be easily facilitated through IWBs in particular, making the learning fun and improving the level of engagement. Though not yet extensively available to schools (largely due to the costs involved) interactive voting systems have some potential with pupils being invited to 'press your buttons now'. The immediate feedback opens up a variety of possibilities for a teacher to assess the understanding of a group. Furthermore, the anonymity of submitting opinions in this way encourages the less confident to participate and enables the introduction of a range of controversial questions linked to science which pupils may usually feel inhibited from contributing to openly, for example, 'Should the school canteen sell healthy salads instead of burgers?'.

We cannot stress too strongly the importance of adaptability and flexibility in using such engaging resources, since they must be tailored to the needs of the group. It can be very frustrating to discover an impressive Internet-based animation, only to find that insufficient thought has been given to the accompanying on-screen text, which is hard for the user to access. In this sense, there is much to be said for self-created, adaptable presentations.

Vignette 2: Increasing engagement

The commercial production of robust, easy-to-use digital microscopes has been a valuable addition to primary science resources in recent years. When used to capture close up images of a variety of materials, they can both stimulate and engage students, whilst contributing to valuable pedagogic advancements for the students. Such images can be viewed 'live' as they happen but can also be stored for use on other occasions.

A particularly useful way to do this and produce tailor-made resources for catching pupils' attention is through the development of digital presentations using appropriate software, of which PowerPoint is only one. One teacher, having captured a variety of images of everyday materials, produced a PowerPoint presentation with brief appropriate prompts and questions. The materials and objects featured included sand, a wood knot, a lightbulb filament, ice and a tooth. When using the PowerPoint images the teacher had as many as possible of the objects available, so that having discussed the students' thoughts, their ideas could be related to the real object, making a cognitive link between the image seen and what the object really was (see ASE 2002a for examples).

This is seen as effective inclusive practice since there is limited written language required, it encourages a different approach to the topic of materials and the quiz presentation is engaging for all. Digital microscopes and cameras can be used by students who might like to choose their own objects to be included in the presentation, encouraging a participatory approach and enabling development by individuals who were enthusiastic to experiment further (ASE 2002b).

Although the use of ICT has enormous potential for gaining students' attention so that they engage with the topic being studied, there are limitations. Indeed, it is important to remember the principles of good practice that apply to all teaching but are particularly critical when working with children with learning difficulties. Three issues in particular seem to be relevant here (see Bell 1999 for a more extensive discussion). The first is the attention span of the children involved, who often find it difficult to focus on a task over a sustained period of time. The second is range of problems that children with learning difficulties can have in recognising the key features that are relevant to the task in hand and their tendency to focus on things that attract their attention, but are not directly relevant to the learning objective. The third is the way in which some children with learning difficulties rely to a significant extent on the external cues they pick up from their surroundings in order to respond to questions. This may involve repeating things said or done by other children, mirroring teacher actions and taking information from pictures and other objects in the room, regardless of their relevance. Thus, the use of presentations and other media has many advantages, but there are dangers too. Over-elaboration may become counterproductive and result in students becoming disengaged rather than involved.

Extending and developing the teaching and learning dialogue

At the heart of most, if not all, learning situations is the interaction which takes place between the pupil, the subject matter being studied and the 'teacher'. Science education has been strongly influenced by constructivist approaches to teaching and learning, in which learners are considered to be actively involved in the construction of meaning and understanding of concepts for themselves (see for example Osborne 1996). There has been increased emphasis on the role of the teacher in helping children construct meanings based on their existing ideas and experiences and on the process of scaffolding in creating opportunities for children to engage with new ideas (Morroco and Zorfuss 1996; Bell 1999). We, like others in this volume and elsewhere (Murphy 2003; Harlen 2005), would argue that these principles are central to good teaching in any situation, but that by using ICT we can endeavour to make science accessible to all students in a manner which, at least in part, overcomes the barriers to learning that they experience.

Many of the ideas outlined in the previous section would, if used slightly differently, be effective in extending the dialogue that is such an integral element of teaching and learning. The interaction that can be developed using an interactive whiteboard can involve pupils in, amongst other things, indicating their ideas, showing examples of their work, watching change sequences and predicting what comes next (see Chapter 7). All these and many other ideas also enable formative assessment to take place naturally as part of the learning process.

In describing the development of science materials to support pupils with special educational needs, Bancroft (2002) highlights the importance of developing a multi-sensory approach, using flexible materials which are relevant and age-appropriate to the children involved. While there is much that can be, and should be, done without its use, ICT allows pupils and their teachers to:

- extend the range of their senses so that it is possible to see and hear things that otherwise would be impossible, for example, 'watch' things grow over long time scales, 'slow' things down so they can be recorded, experience things that are very small and very large, and 'visit' places that we cannot otherwise get to;
- capture and monitor changes using sensors, computers and cameras whilst gathering data from experiments;
- access other materials in much the same way that we would with other children.

One of the big differences, however, is the fact that, having captured the information, data and other forms of evidence, it is possible to review them as often as required, enabling students to recall earlier events without having to rely entirely on memory. Effective use of ICT (for example, by rearranging objects on screen, putting symbols in order or ordering pictures to 'tell a story') also helps in the sequencing of events, which many children with learning difficulties find hard to do.

The ability to revisit activities and lesson materials electronically also makes it easier to adapt to meet the needs of a particular group of students by producing differentiated materials. Clarity of written instructions and use of appropriate diagrams can be reconsidered after initially trialling the materials with one group, without having to start all over again. This is a particularly useful facility when working with more able students who require additional stimuli or more challenging questions.

Vignette 3: Extending the teaching and learning dialogue

The use of 'special effects' simulations provides opportunities to help pupils get below the surface, or in this example 'under the skin', of an object. When a class of students were studying bones, muscles and movement, the running person animation was used as part of the starter activity. This allowed students to see the skeleton inside the body moving as the person either runs or walks, visually reinforcing that the skeleton exists to add structure to the body and that specific joints can move in specific ways. The interactive software was used after the student had already been encouraged to feel their own bones and joints. The animation created a focus for the extended discussion, with the group being able to choose which joint or part of the body they wished to explore. It could, of course, also be used when revising or revisiting the topic at a later stage (Evans 2003).

Language in all its forms is often a major barrier to learning in most subjects. As Wellington and Wellington (2002) explain, in science it can create particular difficulties, partly because of the specific vocabulary that is used, but also because of the need to help children develop and describe abstract ideas. Again ICT has a contribution to make in helping children overcome this barrier. *Writing with symbols*[2] is an ICT tool that produces a symbol to go with every word that is typed onto the computer. The symbols are used in different ways by students who find it hard to read. For example, some students might have to rely on a symbol-supported timetable to give meaning and structure to their school day.[2] Other students may use a symbol-supported topic summary word sheet, with the symbols helping them to find the particular word they need to spell. Websites have now been developed incorporating symbols to explain science-related concepts or to provide general information on a topic. One good example is the *Rainforest with Symbols* website.[2] Students can be invited to submit their own pieces of work (or stories) to the website, recognising and sharing their successful work. This is a developing area and there is further scope to be explored in the use of symbol-supported text in science education.

Reinforcement, recording and reporting

As we have already indicated, the potential for ICT to be used as a means by which children can record events and monitor changes during investigations is almost endless. This is a major step forward in helping to overcome some of the barriers to pupils' learning. By building up a bank of information it is possible to help students look for patterns across a range of items in order to, for example, identify similarities and differences between organisms.

With appropriate support and guidance, pupils can build up their own records and reports of their investigations. Given suitable software, they can prepare good quality work for display because the difficulties of writing and drawing can be reduced. For those who find use of the written word difficult, the production of an audio or visual record is now a relatively easy option.

Vignette 4: Reinforcement, recording and reporting

When studying floating and sinking, a group of pupils were given a collection of objects and their task was to predict which would float or sink. The teacher had prepared an on-screen grid in the software with appropriate key words contained in the grid, and students clicked on words from the grid with a mouse to include them in the word processor part of the package. When the group had made a prediction for a particular object, they used Clicker 3 software (see Becta 2005 and Figure 4.2) to record their ideas.

This use of ICT enables students to produce well presented, high-quality outcomes through the use of Clicker. This shows that the students have the ideas but barriers exist relating to them recording or explaining their predictions. The adaptability of the software means that teachers can use it in a variety of teaching topics across the curriculum.

Importantly, the use of ICT also provides increased opportunities for recording pupils' progress, supported with evidence. In the day-to-day bustle of the classroom, it is all too easy to miss the small steps by which children with learning difficulties progress. By integrating the use of ICT, in its range of forms, as part of teaching and learning, evidence of such improvements can be gathered and, when necessary, reflected upon. For example, when exploring bulbs, wires and batteries for the first time, a group of pupils had to try to get the bulb to light by creating a simple complete circuit (ASE and NASEN 2003). As a student successfully completed this task they demonstrated this to the teacher who took a picture of them and their completed task. For students who were not happy to be included in the picture, the teacher took only an image of their hand pushing the switch to complete the circuit. The pictures were saved for assessment purposes and some were used in a classroom display relating to the circuits work. This recognised the students' achievements and provided a visual reminder of the work that had been completed, which could be referred to later as a reminder when revisiting the subject. Approaches such as

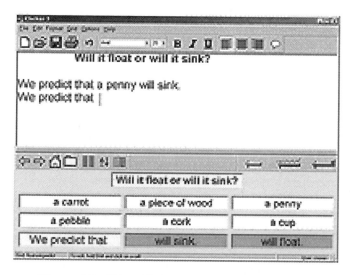

Figure 4.2   Using Clicker to support science writing.

this enable the compilation of electronic portfolios for each child, to which they contribute; these can be particularly valuable when developing and reporting on learning plans for individual children. The use of scanners allows children's handwritten work or drawings to be included as well.

Extending the learning community

A feeling of isolation is quite often felt by those teaching science to children with specific special educational needs. It might be the first time that a mainstream teacher has taught a child with autism, or it might be that the science coordinator in the special school has always struggled to teach a specific topic in science. ICT introduces a means for those in geographically diverse locations to share their experiences, ideas and resources in a virtual environment. This connectivity with others can be very reassuring and has been particularly effective for those working with children with special educational needs. For e-mail forums to be successful the numbers registered must reach a 'critical mass', with subtle prompting or leading from the coordinator of the forum, since many of those registered will at first not feel comfortable sharing their views in what is seen as a public domain. This can be illustrated by looking at findings based on the SENCO Forum, which is a well established e-mail forum that has been monitored and researched (Lewis and Ogilvie 2002) during its development. Other successful SEN e-mail forums are operated by Becta (Becta/Ngfl SEN forums) and the ASE also operates the Inclusive Science e-mail group (ISSEN website: http//www.issen:org.uk/).

It can be hard to find specific resources that are identified as really addressing inclusion and special educational needs in science. In the recent past, some manufacturers may have been timid in promoting a resource as applicable 'for SEN' for fear that it might marginalise the appeal of the resource. However, with the inclusion agenda having become a higher priority, this does not seem so much the case today, with manufacturers beginning to refer to accessibility and special educational needs in their promotional materials. However, it is a small start – Becta (2003b) stated that 'only a small percentage of curriculum materials are currently available in alternative formats accessible to those with special needs.' Resources such as the Ngfl/Becta Inclusion website (Ngfl/Becta) have provided a portal for identifying suitable materials and sharing ideas that address these important considerations.

Vignette 5: Extending the learning community

Several groups have been established to provide a means of linking those working with different groups of children. For example, the ISSEN group was set up with an ethos of bringing together expertise in making science more inclusive, and the Becta SEN forums, including the SENCO forum, provide links for ICT and inclusion.

An example of a more local initiative is the *Science To Raise And Track Achievement* (STRATA) project (Oswald *et al.* 2002; STRATA website: http// www.ase.org.uk/sen/sen/strata-schemes.htm) which brought together teachers in Cambridgeshire special schools to develop topic-based schemes of work for science incorporating the p-scales (QCA 2001) and going up to level 4. The process was not only worthwhile for participating teachers, but the resulting schemes have been further disseminated through the Astra Zeneca Science Teaching Trust website. A Continuing Professional Development unit has also been developed to enable other teachers to gain a better understanding of how to make appropriate use of the p-scales with their students. The schemes, having been adopted and adapted by other teachers working in similar environments with ICT, have been the key to disseminating this good practice beyond the original group, saving others from 'reinventing the wheel'.

Some final thoughts

New forms of technologies are rapidly developing and these will inevitably continue to provide new ways of supporting inclusion in science, particularly regarding accessibility and enabling further independent learning. There has already been increased regulation of website design in the UK (Disability Rights Commission 2004) which will increase the overall accessibility of the Internet. Software tools that automatically convert written electronic text on a website into a symbol-supported text version are already being introduced and recently developed tools to aid those with visual impairment include handheld devices that can scan newspaper text and then be plugged into a television to produce enlarged, legible text.[2]

Recent technology has also supported the development of tactile diagrams that incorporate speech for visually impaired students and others to hear the parts of the diagram named as they touch it. Such tactile diagrams might be a map of the world, part of a car, or the human digestive system, produced in a raised or textured version on a piece of board, so that the diagram can be felt. In the past a support teacher may have explained the diagram as it was being felt, or Braille labels may have been used, but the use of speech synthesis makes them far more engaging and able to be used more independently. The use of electronic notepads that allow students to express their opinions or submit their results to a computer that would collate them centrally would further assist student engagement. Interactive whiteboards have begun to make a large impact in some schools and, as greater numbers are being installed, their use is beginning to be explored by teachers and researchers, reflecting on their potential as a teaching and learning tool, and not only as novel, short-lived practice. This view is further supported by the increasing level of support and guidance that is becoming available through web portals (Becta 2003c; E-learning centre 2005). However, as is concluded by O'Sullivan (2004), who studied the use of interactive whiteboards by students with profound and multiple learning difficulties, being a new, developing technology, there is a need for further detailed research into their impacts and implications for students with special educational needs.

However, as with all areas regarding ICT in science education, there is a need for teachers to receive training relating to its effective use:

> Teachers cite the lack of time, insufficient knowledge of the pedagogical uses of technology, and a lack of information on existing software as three major barriers to integrating technology. Teachers and support staff need ongoing training in order to make informed decisions regarding the technological needs of all students, including those with special needs.
>
> Becta (2003b)

This emphasises that if the use of ICT is to make a real difference for all students then the priority, for the immediate future at least, must be supporting teachers to integrate its use into their everyday practice. Although it is perhaps less exciting than speculating about the potential capability of new ICT hardware and software, it cannot be emphasised too often that the role of the teacher remains key to the effectiveness with which ICT can enhance the learning that takes place. To this end, teachers need to develop confidence in using ICT in combination with subject knowledge and teaching skills. If, by adapting and modifying teaching strategies in small ways, teachers can use ICT to help overcome barriers to learning and make explicit the small incremental steps that are required to help children develop their process skills and understanding of concepts, then we can truly claim we are making science more inclusive.

Notes

1 Further sources of research, policy, pedagogical and curriculum information are available at the following sites:

ACE Centre http://www.ace-north.org.uk/
Astra Zeneca Science Teaching Trust http://www.azteachscience.co.uk/
Dyspraxia Foundation http://www.dyspraxiafoundation.org.uk/
Inclusive Technology http://www.inclusive.co.uk/
Inclusive Science and Special Educational Needs (ISSEN) http://www.issen.org.uk
London Gifted and Talented http://www.londongt.org/homepage/index.php

National Association for Able Children in Education (NACE) http://www.nace.co.uk

National Association for Special Educational Needs (NASEN) http://www.nasen.org.uk

National Grid for Learning/Becta Inclusion website http://inclusion.ngfl.gov.uk/

Royal National Institute for Deaf people (RNID) http://www.rnid.org.uk/

Royal National Institute of the Blind (RNIB) http://www.rnib.org.uk/

The British Dyslexia Association http://www.bda-dyslexia.org.uk

Down's Syndrome Association http://www.downs-syndrome.org.uk/

The National Autistic Society http://www.oneworld.org/autism_uk/

Satellite Schools http://satellitevs.com/ or http://www.satelliteschool.co.uk

STRATA, Astra Zeneca Science Teaching Trust, Cambridgeshire project resources http://www.azteachscience.co.uk/code/development/strata.htm

Symbols World website http://www.symbolworld.org/index.htm

Further contacts relating to special educational needs can be found at http://www.ase.org.uk/sen/

2 Webwide (2005) *Communicate: webwide*, http://www.widgit.com/products/webwide/powerpoint/index.htm (July 2005).

Widget at http://www.widgit.com for *Writing with symbols 2000, Rainforest with symbols, Class timetable produced using Widgit Rebus symbols* and *Guide to symbol-supported timetable* (July 2005).

References

ASE (2002a) *SY Primary CD-ROM*. Hatfield: ASE, http://www.sycd.co.uk/primary (July 2005).

ASE (2002b) 'Digital microscope PowerPoint presentation', in ASE *SY Primary CD-ROM*. Hatfield: ASE, http://www.sycd.co.uk/primary/ict/cameras-and-microscopes.htm (July 2005).

ASE and NASEN (2002) *Inclusive Science and Special Educational Needs Joint Statement by the Association for Science Education and the National Association for Special Educational Needs*, http://www.ase.org.uk/sen/pdf/sen/docs/docguid_stat.pdf (July 2005).

ASE and NASEN (2003) *Case studies*, http://www.ase.org.uk/sen/further/case-studies.htm (July 2005).

Bancroft, J. (2002) 'Developing science materials for pupils with special educational needs for the Key Stage 3 Strategy', *School Science Review*, 83 (305): 19–27.

Becta (2002) *How to Use ICT to Support Gifted and Talented Children*, http://www.ictadvice.org.uk/index.php?section=tl&rid=631&catcode=as_inc_sup_03&pagenum=5&NextStart=1&print=1 (July 2005).

Becta (2003a) *ICT Advice: Dyslexia and ICT*, http://www.ictadvice.org.uk/downloads/guidance_doc/dyslexia_ict.doc (July 2005).

Becta (2003b) *What the Research Says about Supporting Special Educational Needs (SEN) and Inclusion*, http://www.becta.org.uk/page_documents/research/wtrs_ictsupport.pdf (July 2005)

Becta (2003c) *What the Research Says about Interactive Whiteboards*, http://

www.becta.org.uk/page_documents/research/wtrs_whiteboards.pdf   (July 2005).

Becta (2005) *Examples of Practice: The Primary Classroom 2 – Year 4*, http://www.becta.org.uk/teachers/teachers.cfm?section=1_6_1&id=1017 (July 2005).

Becta/Ngfl (2005) *Discussion Forums*, http://www.becta.org.uk/teachers/display.cfm?section=1_1 (July 2005).

Bell, D. (1999) 'Accessing science in the primary school: meeting the challenges of children with learning difficulties'. Paper presented to the Australian Association for Research in Education Conference, Melbourne 1999, http://www.aare.edu.au/99pap/bel99150.htm (July 2005).

Bell, D. (2002) 'Making science inclusive: providing effective learning opportunities for children with learning difficulties', *Support for Learning*, 17 (4): 156–161.

Bell, D. (2003) 'Putting your teaching on the line: making science accessible for all pupils', *Times Educational Supplement Extra Special Education*, December.

Carlisle, J.F. (1996) 'Evaluation of academic capabilities in science by students with and without learning disabilities and their teachers', *The Journal of Special Education*, 30 (1): 18–34.

Davies, P. and Florian, L. (2004) *Teaching Strategies and Approaches for Pupils with Special Educational Needs: A Scoping Study* (Research Report 516). London: HMSO.

Disability Rights Commission (2004) *Formal Investigation Report: Web Accessibility*. http://www.drc-gb.org/publicationsandreports/report.asp (July 2005).

E-learning centre (2005) *Using Interactive Whiteboards*, http://www.e-learningcentre.co.uk//eclipse//Resources//whiteboards.htm (July 2005).

Evans, S. (2003) 'Bones, muscles and movement animated resource', in ASE/NASEN *Inclusive Science and Special Educational Needs CD-ROM*. Hatfield: ASE, http://www.ase.org.uk/sen/focus/muscles.htm (July 2005).

Farrell, P. (2001) 'Special education in the last twenty years: have things really got better?' *British Journal of Special Education*, 28(1): 3–9.

Fleming Fulton School (2005) *Recording and Demonstrating Achievement (UK: Northern Ireland)*, http://131.246.30.23/ita/senistnet/cs8.php (July 2005).

Harlen, W. (2005) *Teaching, Learning and Assessing Science 5–12*. London: Sage Publications.

HMSO (2004) *Education (Additional Support for Learning) (Scotland) Act 2004*. London: The Stationery Office, http://www.opsi.gov.uk/legislation/scotland/acts2004/20040004.htm (July 2005).

Lewis, A. and Ogilvie, M. (2002) *The Impact of Users of the National Grid for Learning SENCO Forum Email List*. Birmingham: University of Birmingham, http://www.becta.org.uk/page_documents/teaching/senco-forum1.pdf (July 2005).

Lindsay, G. (2003) 'Inclusive education: a critical perspective', *British Journal of Special Education*, 30 (1): 3–12.

Mittler, P. (1996) 'Preparing for self-advocacy', in Carpenter, B., Ashdown, R. and Bovair, K. (eds) *Enabling Access: Effective Teaching and Learning for Pupils with Learning Difficulties*. London: David Fulton Publishers.

Montgomery, D. (2003) *Gifted and Talented Children with Special Educational Needs – Double Exceptionality*. London: NACE/David Fulton Publishers.

Morroco, C.C. and Zorfass, J.M. (1996) 'Unpacking scaffolding: supporting students with disabilities in literacy development', in Pugach, M.C. and Warger,

C.L. (eds) *Curriculum Trends, Special Education and Reform: Refocusing the Conversation*. Williston, VT: Teachers' College Press.

Murphy, C. (2003) *Literature Review in Primary Science and ICT*. Bristol: NESTA Futurelab, http://www.nestafuturelab.org/research/lit_reviews.htm.

NCTE (2005) *Exceptionally Able*, http://www.ncte.ie/SpecialNeedsICT/Technology Advice/AdviceSheets/ExceptionallyAble (July 2005).

Osborne, J. (1996) 'Beyond constructivism', *Science Education*, 80 (1): 53–82.

Osborne, J. and Hennessy, S. (2003) *Literature Review in Science Education and the Role of ICT: Promise, Problems and Future Directions*. Bristol: NESTA Futurelab, http://www.nestafuturelab.org/research/lit_reviews.htm.

O'Sullivan, S. (2004) 'The use of interactive white boards and touch screens by pupils who have profound and multiple learning difficulties'. Unpublished undergraduate dissertation, Anglia Polytechnic University, http://www.parkroadict.co.uk/dissertationtext.html.

Oswald, S., Brewer, P., Cornwell, R. and Grant, J. (2002) 'A scheme of work for special school science: the Cambridgeshire story', *School Science Review*, 83 (305): 29–39.

QCA (2001) *Planning, Teaching and Assessing the Curriculum for Pupils with Learning Difficulties: Science*. London: QCA Publications.

Rose, R. (2002) 'Including pupils with special educational needs: beyond rhetoric and towards an understanding of effective classroom practice', *Westminster Studies in Education*, 25 (1): 67–76.

Solity, J. (1995) 'Assessment through teaching and the Code of Practice', *Educational and Child Psychology*, 12 (3): 29–35.

University of York Science Education Group (2002) *ICT in Support of Science Education, A Practical Users Guide*. York: University of York.

Wall, K. (2001) *Tools for Inclusion: Science and SEN*. London: Institute of Education, University of London, http://www.ioe.ac.uk/nof/tfi/science.pdf.

Wedell, K. (2005) 'Dilemmas in the quest for inclusion', *British Journal of Special Education*, 32 (1): 3–11.

Wellington, W. and Wellington, J. (2002) 'Children with communication difficulties in mainstream science classrooms', *School Science Review*, 83 (305): 81–92.

Wilkinson, S. (2003) DiagramMaker software in ASE and NASEN *Inclusive Science and Special Educational Needs CD-ROM*. Hatfield: ASE, http://www.ase.org.uk/sen/focus/muscles.htm (July 2005).

5

ELEPHANTS CAN'T JUMP: CREATIVITY, NEW TECHNOLOGY AND CONCEPT EXPLORATION IN PRIMARY SCIENCE

Ben Williamson

Introduction

This chapter explores the possible implications for primary science education of children using new technology to create and manipulate visual illustrations and drawings of science concepts. In doing so, it addresses three distinct fields of recent analysis. First, it explores how children's creative activities can be promoted by using ICT to enable science learning to become meaningful to them. Secondly, it identifies how work in children's visual literacy from the field of social semiotics impacts on the ICT-enabled science classroom. Finally, it discusses how previous work on using drawing in the science classroom has allowed children to explore and develop their conceptual understandings of science. This three-pronged approach leads into an analysis of a recent prototype development of a computer-mediated drawing tool, Moovl, which allows children to construct and manipulate dynamic drawings. The chapter then discusses how a greater emphasis on children's creativity and visual literacy in the classroom can impact on their ability to become scientifically inquisitive and exploratory when beginning to investigate science concepts.

Creativity

What does 'being creative' mean? More narrowly, what does 'being creative' mean in the context of science? History has revealed many examples of creative scientists whose discoveries have shocked the world, but often the enduring stereotype of these – as in the cases of Stephen Hawking, Albert Einstein or Isaac Newton – is of the solitary genius scientist surrounded by instruments and chemicals or working on equations at a chalkboard (Driver *et al.* 1996; Osborne *et al.* 2002). Certainly the solitary genius does exist, but this popular view is unhelpful if, as many now believe, we wish to encourage children and young people to be creative while learning science in school. It implies that only the most intellectually able can really 'do science' and that the capacity for 'being creative' is something that these people possess as an innate resource (Robinson 2001). Neither creativity nor science should be seen in such narrow terms. With new technology now becoming more widespread in the classroom it is also necessary to conceptualise the relationship between these tools and the creative learning processes they can promote.

The issue of creativity in science was largely sidelined by the introduction of science as a core subject in the National Curriculum in England and Wales in 1989. For many the National Curriculum for Science reconfirmed the science classroom as a preparatory lab for the minority of students who might go on to study science later at university or beyond (Osborne 2002; Osborne and Hennessy 2003). According to some commentators, the attitudes of many school leavers after 12 years of compulsory National Curriculum science are at best ambivalent and at worst entirely negative (Newton and Newton 1992; Jarvis and Rennie 1998), with many of them lacking familiarity with the core scientific ideas that they will meet outside of school (Millar and Osborne 1998) or holding on to misconceptions that have never been challenged (Vosniadou 1997; Murphy 2003). Studies of children's science education in the primary years suggest that many of their misconceptions and attitudes towards science are formed early on as a consequence of their interactions with particular areas of subject matter (Millar and Driver 1987; Kelly and Waters-Adams 2004).

As a consequence, many have comparatively recently come to recognise that if we wish children to find science exciting and stimulating then we need to make its complexities somehow more accessible. The research literature has increasingly emphasised the importance of children's 'quest for *meaning*' in science (Warwick and Stephenson 2002), the development of children's scientific reasoning skills (McFarlane and Sakellariou 2002) and the promotion of 'scientific literacy' (Osborne 2002; Osborne and Hennessy 2003). These approaches, it is argued, will make science more meaningful for pupils. Broadly concerned with how children construct meanings and understandings through science activities, rather than seeing science as content to be practised and remembered, these are views commensurate with the growing literature in creativity.

Creativity, however, is still not well understood. It remains a well-intentioned, but elusive and ill-defined concept, often used as an umbrella term for disparate activities, skills and processes (Harlen 2004). The case for recognising its value is often made in general terms that simply assert it is a good thing for all individuals, or that define it narrowly in instrumental terms linked to the economy (Prentice 2000). Further, it is important to recognise the distinctions between 'teaching for creativity' and 'creative teaching' (Loveless 2002). In 1999 the publication of the influential *All our Futures* report by the National Advisory Committee on Creative and Cultural Education (NACCCE) characterised creativity as working imaginatively and with a purpose, judging and reflecting on the value of one's contributions to solving problems and fashioning critical responses. The report strongly concluded that creativity should no longer be associated solely with particular 'arts'-based disciplines, but rather as a process that can be mobilised across much wider domains. Others (Overton 2004; Harlen 2004; Howe 2004) have emphasised that creativity is not composed of uniquely creative events but is rather a process for learners of bringing together existing ideas, information and evidence to produce new combinations of ideas. This process, it is argued, is an integral function of learning, and while its activities are slower than more traditional classroom exercises, they facilitate learning that is more meaningful, more likely to 'stick' and more likely to satisfy children and motivate them to continue to learn. In a review of the literature in creativity, Loveless usefully provides a summary:

> Creativity in education can encompass learning to be creative in order to produce work that has originality and value to individuals, peers and society, as well as learning to be creative in order to support 'possibility thinking' in making choices in everyday life.
>
> (Loveless 2002: 3)

A recent special issue of *Primary Science Review* featured a number of practical examples of creative, cross-curricular teaching and learning intended to promote such 'possibility thinking', including field trips to old coal mines and dramatic role-play activities that illustrate such processes of discovery and exploration.

Indicative of the recognition of the importance of creativity to learning across subject domains, the Qualifications and Curriculum Agency (QCA) has established a 'creativity working group' which promotes creativity based on the model of constructivism as extended knowledge building tasks (QCA 2002) and recently launched the 'Creativity: Find It,

Promote It' website (http://www.ncaction.org.uk/creativity) to support good practice. The current government's strategy for primary schools (DfES 2003) also underlines the value of creativity as a broad and cross-curricular concern rather than a discrete specialisation. Such work makes it clear that science lessons in primary schools should have a creative, collaborative and cross-curricular emphasis that does not characterise science as an isolated, meaningless discipline, which students find de-motivating.

In science education we might, then, characterise 'being creative' as: working imaginatively with existing ideas, information and evidence; sharpening one's interpretation of them, often by sharing and working on ideas with others; and constructing expressions of the meanings of these ideas, information and evidence that accurately articulate one's personal understanding of what has been achieved.

Creativity and digital technology

As McFarlane (2003) has identified, working with ideas is characteristic not just of creativity but of the ways in which ICT can be used most effect-ively in schools. Loveless (2002: 12), too, identifies that key features of ICT applications such as interactivity and provisionality 'enable users to make changes, try out alternatives, and keep a "trace" of the development of ideas'. A compelling example of this is provided by McFarlane, who sug-gests that dynamic simulations offer opportunities for children to interact with and manipulate complex systems:

> The value of dynamic representation is likely to reside in the rendering of the abstract as concrete. For example, it is possible to see, and interact with, a representation of the molecules in a gas [. . .]. By experimenting with the behaviour of these virtual systems it is possible to infer, and understand, the principles underlying often complex and otherwise abstract systems.
>
> (McFarlane 2003: 223).

Such simulations, of course, must be built on adequate models or algo-rithms of the reality being simulated, which is not always the case: some oversimplify or even misrepresent the phenomena under simulation (McFarlane and Sakellariou 2002). Similarly, caution should be taken with computer simulations since they represent 'cleaned-up' versions of the complex and messy real world (Osborne and Hennessy 2003), and they do need to present viable and convincing alternatives to children's everyday beliefs if their thinking is to develop (Hennessy *et al.* 1995).

One other notable line of enquiry in simulations is recent work in metaphor, particularly visual metaphor. Cameron's (2002) work on metaphors in the learning of science describes a metaphor as bringing together two distinct domains whose juxtaposition activates the possibility of interpretation. These domains, she suggests, are the Topic and the Vehicle, where Topic refers to the actual concept under scrutiny and Vehicle to properties from a related area; operating together, the two domains help to activate the meaning of each distinctly and complement each other. In a development of a science simulation reported by Sweedyk (2005), visual and textual metaphors were recruited to explain the concept of protein synthesis, with the Topic of proteins represented by the Vehicle of elixirs and protein synthesis described in terms of elixir production techniques. The juxtaposition of the Topic with its metaphorical Vehicle, then, may be both visual and verbal, with images and words complementing one another to support the construction of meaning by learners.

However, Osborne and Hennessy (2003) note that the value of interactive computer models such as simulations is not just in representing scientific ideas or phenomena. They can also encourage pupils to pose exploratory 'what-if' questions, to try out and observe what happens when variables are manipulated and to revise both their hypotheses and their investigative practices if they have made mistakes. The capacity to interact with systems that support provisionality, to be iterative in this fashion and to receive immediate feedback can, then, support the development of young people's repertoire of creative and scientific methods. Further, Loveless (2002) suggests that a characteristic of creativity with new technology is the recognition of how the features of particular applications can be manipulated and exploited. The implications for science are that different predictions can be recorded, experiments designed, data and variables can be manipulated, results observed and a range of inferences made. In a classroom equipped with these tools and techniques, then, children can keep a trace of the development of their ideas as they interact with and explore dynamic systems, access information sources, record data, create meaningful representations of their ideas and communicate conclusions or inferences through appropriate media and modes. Loveless (1995) and Claxton (2000) have both suggested that being capable with new technology, however, is more than just competence with a set of skills and techniques; it is subject to an individual's ability to recognise and evaluate the distinctive contributions that new technologies can make to specific tasks and working processes. The use of new technology on its own cannot be described as creativity, then, but the right new technology can certainly be used to support the creative, imaginative and purposeful exploration of science concepts and phenomena.

Three recent examples that use ICT to promote creativity in primary science are the Blaise Castle Project, Savannah and the Bedminster Down Space Centre, all Bristol-based projects. The Blaise Castle Project is an annual fieldwork exercise which saw 700 Year 6 pupils use data-logging equipment, laptops loaded with databases and offline website resources and digital cameras to conduct a thorough survey of insect habitats in a historic park on the edge of Bristol. Pupils took on the tasks of data collection and analysis, documented their activities and discoveries, and afterwards collated their data into multimedia presentations and wall displays. The more experimental Savannah project conducted in March and April 2004 provided children from Year 6 with handheld computers (PDAs) with global positioning system technology to allow them to explore a physical playing field with a virtual map of the African plains superimposed on it. By taking on roles in a pride of lions, the children had to 'scent' their territory, protect their cubs, hunt for food and evade starvation in the dry season. The process of playing the game required them to make predictions about lions' lives on the savannah, to collect data from the field, conduct 'desk research' using websites, books and video and continuously modify their strategies for game-play as the demands of the virtual environment changed. The Bedminster Down Space Centre is a website developed by Bedminster Down Secondary School that hosts local primaries. Children at the primary schools log in to space missions that they are then able to track over a two-week period. The site beams them information about planets and space, and about their chosen space rocket, so that they can then use this information to carry out experiments on electrical circuits, to make presentations about aspects of planetary science and the solar system and to work with others to make sense of complex data sets.

It is not the technical or pedagogical innovativeness of these applications that uniquely positions them as 'creative'. Rather, it is the modes of interaction that they promote which stimulate pupils' creativity. In all three examples, children are encouraged to imagine, suppose and generate ideas; to shape, refine and manage those ideas; to purposely produce tangible outcomes and to act alongside their peers as reflective, critical reviewers. The capacity to manage these disciplines is what makes a learner a creative practitioner and pursuer of meaning.

Many other applications relevant to primary science are discussed elsewhere in this volume. Furthermore, the multimedia capacities of ICT mean that children's exploration and articulation of ideas about science need not be confined to words, but can be expressed in images, sound and action. This chapter will confine itself to examining the role of image-making software and the implications of such applications for strengthening the relationship between creativity and science.

NESTA Futurelab has been working with Soda Creative Ltd to create a tool to support children to work creatively with simple science concepts at Key Stage 1. 'Moovl' is designed as a dynamic doodling environment where it is possible to create interactive drawings that can be animated according to simple rules of physics. Users draw directly on to a tablet

PC using a digital stylus, on to an interactive whiteboard using a stylus or finger (depending on the system) or with a mouse on a PC. Images can be assigned properties which affect how they behave and interact with each other on screen. Each property can be manipulated along a slider scale:

- mass/density – weightless, light, heavy
- elasticity/springiness – very elastic, a little elastic, stiff
- air resistance – no air resistance, some air resistance, fixed
- hardness/collisions – solid, semi-solid, not solid

During trials of the prototype, it was clear that the software could only produce approximations of these physics, not accurate simulations. However, the purpose of the project was principally oriented towards encouraging young children to externalise and manipulate their mental concepts of dynamic phenomena and then to be able to present these to their teachers and to each other. In addition to the doodling functionality, Moovl also utilises the networking capacity of the tablet PCs to allow users to share their simulations through a 'scrapbook' function. The scrapbook allows users to simply 'drag and drop' their images into a series of 'bins' that are then visible to others working on the same local network.

In the study of Moovl being used by children in two classes at a primary school in Bristol (Figures 5.1 and 5.2), we were interested in how children articulate their understandings through their construction of dynamic drawings, how we can interpret their representations and models and, thus, what further work may be required to advance these understandings. In other words, what creative practices were being mobilised by the children in the use of the software? A group of Year 1 children (aged 5–6 years) began to demonstrate how the provisionality of the Moovl program allowed them to take a creative, iterative approach. In one example, pupils Maisie and Connor were illustrating how a group of elephants from *The Jungle Book* (the class reading text for the week) could get across a ravine (Figures 5.3 and 5.4):

Connor: [*quietly to Maisie*] Which one shall we do?
Maisie: Shall we draw a elephant, a aeroplane for the elephant to go in the aeroplane then we need to do a seat on the top
[*Connor drawing*]
Connor: I think they should, I think they should do another bridge
Researcher: Yeah?
Maisie: With lots of wood
[*Connor draws bridge spanning ravine. He tries to move the elephant but finds that it comes apart when moved*]
Connor: Oh. I'll rub him out
[*Maisie takes pen, re-draws elephant*]
Connor: [*takes pen*] Let's see if it works

Figures 5.1 and 5.2 Moovl on whiteboard in Year 1 classroom.

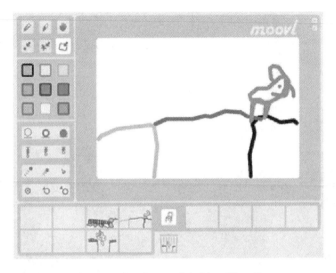

Figure 5.3   Connor's bridge (Year 1).
© Copyright NESTA Futurelab/Soda Creative 2004

Figure 5.4   Maisie's aeroplane (Year 1).
© Copyright NESTA Futurelab/Soda Creative 2004

[*Connor moves elephant across bridge*]
Connor: We did it, we did it already

The process of drawing and trying out shapes with different properties, and then of reviewing the effectiveness of those representations and iteratively redesigning them, is a creative enterprise that could not so easily be accomplished with a pen and paper. This provisionality and the iterative working it promotes is a core creative competence and the software allowed the children to complete the exercise by creating 'workscratchings' and then discarding or elaborating these.

One of the key aspects of creativity in science that has been identified is the ability to be able to ask exploratory 'what-if' questions and then to explore the consequences of taking certain actions or manipulating certain variables in an experiment. The flexibility of Moovl was intended to encourage children to ask such 'what-if' questions, particularly when they are manipulating the properties they have assigned to their images. In the trials of the software a number of the children's questions emerged. These tended to fall into two distinct types of question: those that asked why the software had behaved in certain ways and those that asked whether the software could simulate certain behaviours. The first set included these examples:

'Hey why did it fall down?' (Hanna, Year 1)
'How do you get this to bounce?' (Eloise, Year 3)
'Do you think it's extra springy?' (Marley, Year 3)
'Why's it still bouncing?' (Martha, Year 3)
'How did that happen? What's the mix like?' (Jack, Year 3)
'How come it isn't working?' (Jack, Year 3)
'Why did it go all up there?' (Jacob, Year 3)

The second set of questions included other examples which demonstrate the children beginning to ask more exploratory questions:

'It needs to be thinner, dunnit?' (unknown, Year 3)
'I thought, how do you get the river to move?' (Connor, Year 1)
'How do you make it fly?' (Maisie, Year 1)
'So now you see nothing happens . . . So now what they gonna do? . . . What's this one do I wonder?' (Marley, Year 3)

Another episode from the trial of the software indicated the value of children working together to share ideas, to show each other their drawings and then to make modifications of these based on each others' input. These children were regularly making predictions to one another about the actions the software would simulate if they manipulated their images and the variables in these. In this example, they were illustrating the forces of pushing and pulling and had chosen to picture this as a jumping cat leaping to knock a piece of fruit out of a tree:

Zoe: [*to Sam*] It's going to be a cat, as big as the tree
Kelsey: [*to Sam and Zoe*] We haven't done it yet
Sam: Ah, sucker, you can't do it
Kelsey: We can but we just keep doing it wrong
[. . .]
Sam: [*to Zoe*] Why are you rubbing out the cat?
Zoe: Because it's too big, it's as big as the tree. It may as well not jump if it's going to be as big as it
[. . .]
Sam: [*to Zoe*] Do it, make it bounce more
Kelsey: That was funny
Sam: [*pointing to screen in front of Zoe*] Do it on that one, that one's bigger
Zoe: I don't know what to
[*Sam takes pen*]
Zoe: [*pointing to screen in front of Sam*] I wonder if you make this thing really high up here. Rub that out and draw something really high
[*Zoe tries to take pen*]
Sam: No wait, get off a minute
Zoe: That makes it go really small
Sam: Then . . .
Zoe: Put something really high up there
Sam: You're up in the air . . . Eats something, gets the food [*hands pen back to Zoe*]
Zoe: Can I rub that out?

Sam and Zoe's dialogue accompanies an ongoing process of drawing, erasing and revising as they work out how to get their cat to jump into the treetop where it can push the fruit out of the tree. Throughout their dialogue, the pair conjecture about what features of the program will change the dynamics of the image they have created and they are able to try these ideas out iteratively.

In another example, three pairs of children sitting around the same set of tables launched into a longer dialogue during which a variety of existing understandings were articulated. Again, the children were experimenting with springiness and conjecturing about which sorts of animals they could draw that they could then simulate with the spring functionality:

Marley: What other animals could we possibly do?
Jack: Mmmm, a big blue whale
Marley: No, listen [*inaudible*]
Emily: [*whispers to Marley – inaudible*]

Marley:	An elephant? Elephants can't jump
Martha:	I might do an otter
Researcher:	An otter?
[. . .]	
Jacob:	The sea doesn't bounce
Martha:	It can jump
Jacob:	So? The sea doesn't bounce
[. . .]	
Martha:	Huh a dolphin can jump . . . [*louder*] a dolphin can jump

In this discussion, the children exchanged a variety of understandings. Marley recognises that elephants cannot jump and Martha realises that a dolphin can; Jacob states that the sea cannot bounce. As they discussed these ideas, the children were already in the act of drawing many of these items and manipulating the variables that determined their dynamics. The process of sharing ideas with one another, then, was complemented by the capacity of the software to allow the children to visualise these ideas.

Visual literacy

Osborne (2002: 206) identifies that in the professional domain 'science is a complex interplay of phenomena, data, theories, beliefs, values, motivation and social context both constituted by, and reflected in, its discourse. Science as a professional discipline, in short, is a process that relates the imaginative conjecture of scientists to an evidential base and to the work of others. This, as Osborne points out, is not purely to do with practical activities either. Rather, science is learned and expanded through its discourse – its practices, its representations, and its language, that is, the communicative modes in which ideas are articulated, considered, rejected or received. According to Gee (1996), being knowledgeable and familiar with these discourses leads to the development of 'scientific literacy', where being literate in this sense means developing fluency with the words, actions, values and beliefs of scientists. Even more particularly, it means being critically reflective about the practices of scientists, about the major scientific explanations, the beliefs which underpin them and the ways in which science is used and abused (Osborne and Hennessy 2003). If the emerging emphasis in science education is on how young people make meaning, then scientific literacy is the framework of content understandings and process competencies that will allow them to accomplish this. However, to take the social semiotic view of science literacy, science is bound in discourses and modes of representation which are far from exclusively lexical. Lemke (1998), for example, argues that science sometimes cannot be articulated in the language of words alone; it needs diagrams, pictures, graphs, maps and other visual forms of expression.

Given, then, that science is a multi-modal (Jewitt *et al.* 2001; Kress *et al.* 2001) or multi-semiotic (Lemke 1998) discipline – that is, it involves the negotiation and production of meanings in different modes of representation, from verbal text to image – many have begun to identify the important role that 'visual literacy' can play in science education. For Kress and colleagues, such a view of science education involves the understanding that when a sign-maker creates a representation of scientific phenomena it is to find 'the most plausible form for the meaning that (s)he wishes to express' (Kress *et al.* 2001: 5). In the primary curriculum for Key Stage 1 science there is already a requirement for children to communicate the findings of their scientific investigations in a variety of ways. This includes using ICT and producing drawings, tables, graphs and pictograms. It requires, then, a 'bringing together' of ideas in multiple formats, media and modes, not just for summation but in order to further develop understandings. The science classroom is already multi-modal and multi-semiotic, with emphasis placed on the visual as well as the verbal.

New technology is already beginning to allow children and educators to engage in complex science and dynamic systems (McFarlane and Sakellariou 2002; McFarlane 2003) in ways which are authentic to the actual experience of the observed or perceived world. The multiple modalities of representation that new technology increasingly offers do not just allow children to present creative interpretations of scientific concepts and phenomena; new technology should offer tools which afford children and their teachers the opportunities to think about science and to 'do science' (Osborne 2002) in meaningful ways. In short, it should allow us to be creative, inventive, imaginative and purposeful in science and to perceive science as a process of constantly making meaning.

During the study of Moovl, it was clear that many of the children were able to articulate their ideas in images, but that they were less confident in explaining what images and actions their images represented. Often the children involved in the trial drew images in silence, or spoke very quietly to themselves. What was apparent was that once they had seen others' pictures, many of them would duplicate this and produce very similar images themselves, as in Figures 5.5 and 5.6 which show how two children sitting near to each other had both drawn similar boat designs.

The children, then, appear to have been involved in the wordless exchange of representation, where the actual visual signs represented in their drawings and the dynamic movements afforded by the software allow them to communicate meanings that can then be shared with others. In the above examples, Hamera had been unable to identify how she planned to get her *Jungle Book* elephant across the ravine until she had seen Liam producing his image of the boat. The two pictures indicate

Figure 5.5   Liam's boat (Year 1).
© Copyright NESTA Futurelab/Soda Creative 2004

Figure 5.6   Hamera's boat (Year 1).
© Copyright NESTA Futurelab/Soda Creative 2004

strongly how her reception of these image-based ideas has influenced her thinking and thus influenced her image.

However, in some cases the children found that communicating in images was more difficult. In this excerpt two Year 3 children, Marley and Amy, had constructed an image of Mr Springylegs, an imaginary crab-like superhero with springs for legs, who they were using to illustrate the behaviours of springy objects:

Marley: Look yeah look I we did it, we did it
Amy: Oh yeah oh yeah
Researcher: Did it work?
Marley: Not exactly how I wanted it
Researcher: Not bad though is it
Everton: [*standing and looking over*] How come it walks?
Marley: It's isn't it's jumping

For Marley, the capacity of the software has limited his ability to represent his idea as well as he hoped. However, being able to illustrate the dynamics of springs seemed to free his imagination so that his representation of this phenomena is framed as an imaginary character who jumps across the screen. Moovl provides the potential for children to create visual, representational models of observable phenomena therefore, to an extent, offering the modalities of animation as a means of describing their perceptions of those phenomena. For this reason the actual images the children create in Moovl can be seen as important visual statements and models of their understandings. These understandings might also be beyond their linguistic grasp to explain, or may provide a better foundation for interpersonal understandings where language alone would be insufficient for articulating their meanings. Clearly, then, the children's representations created in Moovl should be seen as statements of their understanding of phenomena, although we may want to caution against assuming that their production of images accurately depicts their perceptions of the represented objects. As Dove *et al.* (1999) have warned in their study of young children's science drawings, many young children struggle with concepts such as scale, may tend to portray objects such as mountains and rivers according to stereotypical or idealised representations and sometimes their drawings display plain misconceptions. It is likely, then, that science educators in the near future will have to negotiate and interpret the representations created and articulated by children and the meanings articulated in them. These will come in a variety of modes, created in different media, and will be represented through the multi-semiotic discourses that constitute science and through which science constitutes itself. The images that children create in science emerge as purposely motivated signs of what children perceive to be the meanings in the world surrounding them. It is in this fashion that children are able to begin making sense of the science concepts which comprise the science curriculum.

Concept development

Children's conceptual development and their creativity in science are closely aligned. Much of the current emphasis on promoting creativity in primary science stems from two influential projects carried out in the 1990s that emphasised constructivist views of teaching and learning. The STAR (Science Teaching Action Research) project studied classroom practice in relation to process skills (see Russell and Harlen 1990), while the SPACE (Science Processes and Concept Exploration) project investigated children's own ideas about science (for example, Schilling et al. 1993). The SPACE project has subsequently informed the foundation for Nuffield's primary science scheme. It approaches the subject through the 'elicitation' of children's ideas about science and then through further activities and 'intervention' helps them towards better understanding of the topics under analysis. In a review of the research literature on children's conceptions in science Wandersee et al. (1994) notes that children have a variety of alternative frameworks arising from their personal experiences, observations and social interaction and that these can interact with formal school science learning in unintended ways. Similarly, Duit (1991) has found that children's pre-existing conceptions influence and guide their science learning throughout school. These alternative frameworks, then, need to be elicited by teachers not just so that they can be 'corrected' but so that teachers can design effective curricular and instructional strategies and materials.

The danger that such explicit elicitation of existing ideas – and the subsequent challenging of these ideas – may lead to demoralisation in the classroom (Asoko 2002) needs, however, to be recognised. What is required are classroom strategies which promote surprise and puzzlement and which can then be worked upon by teachers to raise the status of some ideas at the expense of others (Hewson et al. 1998). These approaches are broadly constructivist, that is, based on the assumption that children construct or build their own understandings about how the world works and that any misconceptions they have developed are best addressed by engaging them in activities that allow them to re-construct those conceptions. These approaches, then, are equally concerned with children's abilities to communicate to explain science as they are with practical science activities.

It is acknowledged that the creation of graphical images is important in science in allowing children to articulate their understandings of concepts (Cox 1999), as well as for young children's wider development of comprehension about the everyday world that they perceive (Browne 1996;

Kress 1997; Coates 2002). Previous studies of drawing activities in primary science have suggested that it taps children's holistic understanding of phenomena and concepts and that it therefore prevents them from feeling that their understandings are inferior to those of teachers or researchers. Further, many scientific phenomena, such as cloud types or leaf shapes, are better suited to visualisation than verbalisation (White and Gunstone 1992; Dove *et al.* 1999). Research into how children represent scientific concepts through drawing has focused both on specific concepts such as 'insects' (Shepardson 2002), 'the water cycle' (Dove *et al.* 1999), and 'evaporation' (Schilling *et al* 1993), and on abstract concepts such as 'technology' (Rennie and Jarvis 1995) and 'Earth viewed from space' (Arnold *et al.* 1995). Many of these have been intended to probe and detect levels of understanding.

In the Moovl study, the software was being used to elicit from children – through the externalisation of their mental images of phenomena – a range of understandings about dynamics, particularly how the weight and elasticity of objects affects their motion and their behaviour when they collide with or land on top of other objects. In this example, three children from Year 3 were demonstrating one of their images to the researcher:

Jack:	This is a way to cheat – you can't actually go to the bottom, or, a way so that it, you get not that many bounces
Jack:	Oh they're pushing it up
Marley:	Oh cool
Jack:	No they're whacking it and pushing it up
Marley:	Cool
Jack:	And one went through it
Sarah:	There it goes. Make it so it bounces on top of it, like that one does
Jack:	Awesome. Ah it's pushing it down now
Sarah:	Yeah but they will push it up
Researcher:	That's one's bouncing a lot isn't it
Jack:	Ah cool, wicked

During this session, the children were able to articulate their existing understandings about how objects would behave given certain degrees of elasticity and weight and also found that some objects behaved somewhat unpredictably, leading them to manipulate the image further and to conjecture about the likely consequences of doing so. Although a technical problem prevented it from occurring effectively, it was anticipated that the children would also be able to 'upload' their images to their teacher's machine so that they would then be able to present their creations from the whiteboard at the front of the classroom. The availability of such functionality, it is proposed, would have allowed the pupils to present their ideas to their classmates and to their teacher and to stimulate a longer discussion in which the teacher could have guided the development of their understandings by asking them probing questions. However, without this opportunity, the children instead adapted to conjecture and speculation about the affordances of the software and the effects of manipulating it:

Marley: I know what these do. [*points*] That means it's soft
Sarah: What does that do then? [*points to feature on screen*]
Jack: I don't know
Marley: I know
Sarah: What does it do?
Jack: What?
Marley: Squishes the [*inaudible*] underneath [*giggles*] . . . No it means . . .
Jack: That or that's got to be the speed of it
Sarah: What's it really do?

If one problem of using such drawing activities to elicit from children their existing understandings in science is that many of these are based on idealised or stereotypical forms and are hard to displace, or are based on plain misconceptions, then how can a program such as Moovl be used effectively to support the transformation of these understandings? The direct feedback it provides may begin to demonstrate if a particular conception is wrong, but this could just as easily be rejected if children do not understand it or if it is not consonant with their existing frameworks. The mechanism for tackling the issue of alternative conceptions in the Moovl project was to attempt to use the networked, public scrapbook functionality to promote collaboration. By this is meant collaboration between children, but also between the children and their teachers.

Recent work on changing the practices of school science has particularly highlighted the importance of the role of the teacher and the idea of cognition as a product of social interaction (see, for example, Asoko 2002; Watt 2002). Drawing on Ogborn *et al.* (1996), who call for classroom methods that facilitate 'talking ideas into existence', Warwick and Stephenson (2002: 145) state that 'if we are to encourage children to develop an understanding of the meaning of their work in science, there is at least one prerequisite – structured talk that acknowledges that pupils have pre-existing ideas'. In this 'social constructivist' model of learning, in which children and teachers are all collaborators, the curriculum may be subject structured but subject boundaries are often crossed by the teacher's approach as she/he looks at 'ways of making learning meaningful to the pupil by connecting knowledge that is presented in meaningful contexts' (Warwick and Stephenson 2002: 149). This statement, it seems, calls for a modified emphasis in science teaching that treats science as a collaborative subject in which teachers and pupils jointly construct meanings through social interaction, and as a nexus for cross-curricular links with other subjects and, indeed, non-curricular areas.

It was not possible adequately to trial the collaborative functions of Moovl, but the trials did begin to indicate how the software could prompt the kind of surprise and wonder that leads to talk in a creative classroom. The kinds of talk that many of the children were spontaneously engaging in whilst exploring the functionality of Moovl to complete the challenges set by their teachers were often characterised by exploratory questioning, conjecture and speculation. Arguably, the key function of the program is that it allows children to pose such questions and speculations and to simultaneously try out the ideas that emerge. Moovl is not alone, of course, in leading to such inquisitiveness. What we can learn from studying children's use of the program, however, is that multimedia and multi-modal tools provide engagement with ideas at many levels that appeal to many of the senses simultaneously. A box of plastic objects, or a collection of objects that create unique sounds, can have the same effect and be used effectively in the science classroom. The stimulation these tools can encourage in children should be seen as the starting place for the entire creative process of structured exploration and talk.

Conclusions

The research that has been carried out on Moovl and its uses in the primary science classroom is far from conclusive, nor is it intended to be. The purpose of the project was to investigate ways in which more creative and collaborative approaches might be made to scientific investigation to help to promote children's curiosity and enjoyment of science. It is clear that there are problems with Moovl that still need to be properly addressed. Likewise, there is much work still to be done to ensure that schools and the children in their care are using appropriate new technology resources and tools that can expand children's abilities to think and act creatively in science, rather than using resources which simply replicate the textbook question-and-answer standard or which misrepresent science as a field of static knowledge.

As a broad approach, it is critical that science educators understand the value of acknowledging children's pre-existing ideas and of working with children and their multi-modal representations of the world. By working with their existing conceptual frameworks it will be possible to transform these from naïve assumptions to understandings that have meaningful connections with the wider world of experience and of learning.

What the research using Moovl has confirmed is the value of providing young children with tools that can broaden the repertoire of communicational and representational facilities they have available. The process of being able to draw and revise images of dynamic phenomena allowed them to construct simple simulations or representations of real-world behaviours, and to use these illustrations to communicate their understandings. These facilities can act as a prompt to further discussion and have been shown to encourage children to begin asking exploratory questions about dynamics, materials, objects and the relationships between those things. The capacity for children to swap and share images using the scrapbook functionality, too, can promote their ability to review each others' contributions to solving a problem, to assess the suitability of images to fit their purpose and, finally, to collaborate on jointly agreeing on representations that adequately answer the challenges they have been set.

Moovl is fairly unique in allowing children to work with the modalities of the visual in order to begin investigating simple science concepts such as physical properties and dynamics. However, that is not to suggest that it is the only tool capable of being mobilised in the primary science classroom to promote such creative exploration of ideas. Many of the conclusions from the trial study of the software reported here are more widely applicable across the primary science domain. The study has confirmed the value of enabling young children to be creative by becoming actively involved in the construction of meaning. It suggests that children need to be able to articulate their existing understandings of scientific phenomena and then, through multiple modalities including image-making, performing actions in motion, and talking, review those understandings. Children's creativity in science is now recognised as the process of 'bringing together' ideas in multiple modalities, of being exploratory and purposeful while 'playing' with those ideas and of being critical and reflective about the value of those ideas and the ideas of others. In terms related to ICT, creativity can be promoted through tools which allow pupils to manipulate and edit, to juxtapose, to erase and to begin again; in short, actively and critically to construct content rather than passively consume it. Although 'creation' does not necessarily have anything to do with creativity, the ability to make meaning from the world of objects and phenomena from an early age has everything to do with it.

References

Arnold, P., Sarge, A. and Worrall, L. (1995) 'Children's knowledge of the earth's shape and its gravitational field', *International Journal of Science Education*, 17: 635–642.

Asoko, H. (2002) 'Developing conceptual understanding in primary science', *Cambridge Journal of Education*, 32 (2): 153–164.

Ball, S. (2003) 'ICT that works', *Primary Science Review*, 76: 11–13.

Browne, A. (1996) *Developing Language and Literacy 3–8*. London: Paul Chapman.

Cameron, L. (2002) 'Metaphors in the learning of science: a discourse focus', *British Educational Research Journal*, 28 (5): 673–688.

Claxton, G. (1999) *Wise Up: The Challenge of Lifelong Learning*. London: Bloomsbury Publishing Ltd.

Coates, E. (2002) ' "I forgot the sky!": children's stories contained in their draw-ings', *International Journal of Early Years Education*, 10 (1): 21–35.

Cox, R. (1999) 'Representation construction, externalised cognition and individual differences', *Learning and Instruction*, 9: 343–363.

Department for Education and Skills (2003) *Excellence and Enjoyment: Strategy for Primary Schools*. Nottingham: DfES Publications.

Dove, J.E., Everett, L.A. and Preece, P.F.W. (1999) 'Exploring a hydrological concept through children's drawings', *International Journal of Science Education*, 21 (5): 485–497.

Driver, R., Leach, J., Millar, R. and Scott, P. (1996) *Young People's Images of Science*. Buckingham: Open University Press.

Duit, R. (1991) 'Students' conceptual frameworks: consequences for learning science', in Glynn, S.M., Yeany, R.H. and Britton, B.K. (eds) *The Psychology of Learning Science*. Hillsdale, NJ: Lawrence Erlbaum Associates.

Facer, K. and Williamson, B. (2004) *Designing Technologies to Support Creativity and Collaboration*. Bristol: NESTA Futurelab.

Gee, J.P. (1996) *Social Linguistics and Literacies* (2nd edn). London: Taylor and Francis.

Harlen, W. (2004) 'Editorial: creativity and science education', *Primary Science Review*, 81: 2–3.

Harrison, A.G. and Treagust, D.F. (2000) 'A typology of school science models', *International Journal of Science Education*, 22: 1011–1026.

Hennessy, S., Twigger, D., Driver, R., O'Shea, T., O'Malley, C.E., Byard, M., Draper, S., Hartley, R., Mohamed, R. and Scanlon, E. (1995) 'A classroom intervention using a computer-augmented curriculum for mechanics', *International Journal of Science Education*, 17 (2): 189–206.

Hewson, P.W., Beeth, M.E. and Thorley, N.R. (1998) 'Teaching for conceptual change', in Fraser, B.J. and Tobin, K.G. (eds) *International Handbook of Science Education*. London: Kluwer Academic.

Howe, A. (2004) 'Science is creative', *Primary Science Review*, 81: 14–16.

Jarvis, T. and Rennie, L. (1998) 'Factors that influence children's developing per-ceptions of technology', *International Journal of Technology & Design Education*, 8 (3): 261–279.

Jewitt, C., Kress, G., Ogborn, J. and Tsatsarelis, C. (2001) 'Exploring learning through visual, actional and linguistic communication: the multimodal environment of a science classroom', *Educational Review*, 53 (1): 5–18.

John-Steiner, V. (2000) *Creative Collaboration*. New York: Oxford University Press.

Kelly, P. and Waters-Adams, S. (2004) 'Primary reaction', *TES Teacher*, 20 February: 14–15.

Kress, G. (1997) *Before Writing: Rethinking the Paths to Literacy*. London: Routledge.

Kress, G., Jewitt, C., Ogborn, J. and Tsatsarelis, C. (2001) *Multimodal Teaching and Learning: The Rhetorics of the Science Classroom*. London and New York: Continuum.

Lemke, J. (1998) *Teaching All the Languages of Science: Words, Symbols, Images and Action*, http://academic.brooklyn.cuny.edu/education/jlemke/papers/barcelon.htm.

Loveless, A. (1995) *The Role of IT: Practical Issues for Primary Teachers*. London: Continuum.

Loveless, A. (2002) *Literature Review in Creativity and Learning*. Bristol: NESTA Futurelab Series.

McFarlane, A. (2000) 'The impact of education technology', in Warwick, P. and Sparks Linfield, R. (eds) *Science 3–13: The Past, The Present and Possible Futures*. London: Routledge Falmer.

McFarlane, A. (2003) 'Learners, learning and new technologies', *Educational Media International*, 40 (3/4): 219–227.

McFarlane, A. and Sakellariou, S. (2002) 'The role of ICT in science education', *Cambridge Journal of Education*, 32 (2): 219–232.

Millar, R. and Driver, R. (1987) 'Beyond processes', *Studies in Science Education*, 14: 33–62.

Millar, R. and Osborne, J. (Eds) (1998) *Beyond 2000: Science Education for the Future*. London: King's College.

Murphy, C. (2003) *Literature Review in Primary Science and ICT*. Bristol: NESTA Futurelab Series.

NACCCE (National Advisory Committee on Creative and Cultural Education) (1999) *All our Futures: Creativity, Culture and Education*. Sudbury: DfEE Publications.

Newton, D. and Newton, L. (1992) 'Young children's perceptions of science and the scientist', *International Journal of Science Education*, 14 (3): 331–348.

Ogborn, J., Kress, G., Martins, I. and McGillicuddy, K. (1996) *Explaining Science in the Classroom*. Buckingham: Open University Press.

Osborne, J. (2002) 'Science without literacy: a ship without a sail?', *Cambridge Journal of Education*, 32 (2): 203–218.

Osborne, J. and Hennessy, S. (2003) *Literature Review in Science Education and the Role of ICT: Promise, Problems and Future Directions*. Bristol: NESTA Futurelab Series.

Osborne, J.F., Duschl, R. and Fairbrother, R. (2002) *Breaking the Mould? Teaching Science for Public Understanding*. London: Nuffield Foundation.

Overton, D. (2004) 'A creative science experience', *Primary Science Review*, 81, Jan/Feb: 21–24.

Qualifications and Curriculum Authority (2002) *Designing and Timetabling the Primary Curriculum – a Practical Guide for Key Stages 1 & 2*. London: QCA.

Prentice, R. (2000) 'Creativity: a reaffirmation of its place in early childhood education', *The Curriculum Journal*, 11 (2): 145–156.

Rennie, L.J. and Jarvis, T. (1995) 'Children's choice of drawings to communicate their ideas about technology', *Research in Science Education*, 25: 239–252.

Robinson, K. (2001) *Out of our Minds: Learning to be Creative*. Oxford: Capstone.

Russell, T. and Harlen, W. (1990) *Assessing Science in the Primary Classroom: Practical Tasks*. London: Paul Chapman.

Schilling, M., McGuigan, L. and Qualter, A. (1993) *The Primary Science and Concept Exploration (SPACE) Project: Investigating*, 9: 27–29.

Shepardson, D.P. (2002) 'Bugs, butterflies, and spiders: children's understanding about insects', *International Journal of Science Education*, 24 (6): 627–643.

Sweedyk, E. (2005) 'Games, metaphor and learning'. Paper presented at DIGRA 2005 Conference, Vancouver, Canada, 17–20 June.

Vosniadou, S. (1997) 'On the development of understanding of abstract ideas', in Harrnqvist, K. and Burgen, A. (eds) *Improving Science Education: The Contribution of Research*. Buckingham: Open University Press.

Wandersee, J.H., Mintzes, J.J. and Novak, J.D. (1994) 'Research on alternative conceptions of science', in Gabel, D.L. (ed.) *Handbook of Research of Science Teaching and Learning*. New York: Macmillan Publishing.

Warwick, P. and Stephenson, P. (2002) 'Reconstructing science in education: insights and strategies for making it more meaningful', *Cambridge Journal of Education*, 32 (2): 143–151.

Watt, D. (2002) 'Assisting performance: a case study from a primary science classroom', *Cambridge Journal of Education*, 32 (2): 165–182.

White, R. and Gunstone, R. (1992) *Probing Understanding*. London: Falmer Press.

DO COMPUTER CATS EVER REALLY DIE? COMPUTERS, MODELLING AND AUTHENTIC SCIENCE[1]

Patrick Carmichael

Introduction

In this chapter, I will explore how information and communications technology can contribute to the participation of primary age children in 'authentic' learning activities in science and will discuss how in certain circumstances ICT can be a medium with sufficient 'analogical capability' (that is, the ability to express ideas) to allow even young children to engage in tasks in which they 'think like scientists'. In other words, I will discuss whether the integration of ICT into young children's learning environments makes the activities in which they take part resemble more closely the activities of core members of the scientific community. In doing so, I will introduce some of the ideas of Mary Hesse, a philosopher of science whose work on the nature and role of analogical modelling in science may illuminate the thinking and learning of young children.

The chapter draws on the developing field of research into modelling in science and describes some of the features of particular kinds of ICT-based analogical models that might be used to support and stimulate children's learning. The account is illustrated with excerpts from transcripts of interviews and conversations collected in the course of a small-scale research project in which young children (aged 4–10) used a variety of computer programs designed to represent individual animals, communities and whole – albeit simplified – ecosystems. This was initially stimulated by the work of Amy Bruckman, who developed a novel collaborative online environment for children called 'Moose Crossing' (Bruckman and de Bonte 1997; Bruckman 1998), described as a place where children 'can create objects ranging from magic carpets to virtual pets' using a simple programming language. While Bruckman's work was largely concerned with patterns of social interaction, and with knowledge construction and exchange in this online environment, I was more interested in the relationship between children's 'real-world' experience and the representation of objects, particularly living organisms, in what Papert calls 'microworlds', such as Moose Crossing and other virtual environments (Papert 1980: 38).

Science, authenticity and modelling

Following from the radical reassessment of the nature of science by, among others, Hanson (1958), Kuhn (1996) Lakatos (1970, 1974) and Feyerabend (1978, 1987), *authentic science* has been characterised as involving, or at least allowing, the following elements: working and learning in contexts constituted by ill-defined problems; the tolerance of ambiguity and uncertainty; and the expectation that theories may be challenged and ultimately discarded. Individual learning of science is characterised as a 'sense-making' activity predicated on current knowledge, with learners participating in communities of enquiry in which they have opportunities to draw on the expertise of more knowledgeable others (Roth 1999). Roth associates 'authenticity' of learning activities with a view of learning as a 'situated' activity and contrasts this to the artificial nature of most school 'problems'. 'Out of school problems', he argues, 'are not "set" . . . [and] have to be framed as problems before they can be solved. In many cases, there are no prospects to get a "right" solution' (Roth 1999: 14).

For learners' experience of learning and doing science to be authentic, then, they must be involved in the development and application of theory (taken here not necessarily to mean the formalised predictive theories of science, but concepts, models and counterfactuals) and the 'ways of thinking and practising', the 'particular understandings, forms of discourse, values or ways of acting' (Hounsell and McCune 2004) of professional scientists. One of these particular and characteristic forms is modelling. Scientists and science teachers use a range of types of model (verbal, visual, gestural and concrete, amongst others) as they conceptualise, problematise and discuss complex concepts, processes and relationships. In this respect, modelling represents a characteristic 'form of discourse' but they also represent a pattern of engagement with 'real world' domains and problems rather than with a curriculum of predefined problems with 'right answers'. There is no 'right' model for helping to understand a given situation or problem – just a 'currently-best-in-my-opinion' one.

Models have a role to play at every stage in the scientific process from prototyping and 'what-if' statements through to 'textbook' reifications of concepts or processes. Boulter and Gilbert (1998) differentiate between notions of 'mental' and 'expressed' models – for them a model is a representation of an object, event, process or system; mental models are personal, private representations of the target; expressed models are placed in the public domain. Aspects of science, and of the science curriculum, are characterised by different kinds of model and different modes of expression (see Boulter and Buckley 2000 for a useful typology) and any attempt to foster authentic learning in school science needs, therefore, to involve the incorporation of appropriate models. Modelling's claim to a place in the curriculum, however, is not based solely on its being an authentic activity; there is a body of evidence (DiSessa 1986; Mellar 1994; White and Fredricksen 1998) which suggests that a modelling-based curriculum also has the potential to leverage important changes in classroom culture and levels of learner engagement and autonomy. Interestingly, it has been argued that, in most current school contexts at least, Design Technology, with the patterns of modelling it involves and the opportunities for the learner to be designer, maker and evaluator, presents greater opportunities for an authentic role for modelling than does school science (Gilbert *et al.* 2000).

Models also have a role to play in learning beyond merely acting as illustrations or as simplifications of complex situations. Johnson-Laird (1983) describes how inferential reasoning (another key 'way of thinking' for scientists) involves an iterative process in which mental models are progressively elaborated and new ideas generated. This view is advanced by Gentner and Gentner (1983) who, in their work with high school and college students, demonstrated how analogical models are conceptual tools capable of generating new understanding through a process of mapping of features from one domain to another. Nersessian (1992) goes further still by arguing that, in the work of professional scientists, it is analogical reasoning that 'do[es] the work' of problem solving, rather than simply acting as a guide or a heuristic device.

Mary Hesse and analogical modelling

Mary Hesse's view of the role of models in science, advanced in her book *Models and Analogies in Science* (Hesse 1970) is, like that of Nersessian, of models not only as heuristic methods but as a key element of scientific reasoning. She links the process of model building explicitly to the construction of strong but falsifiable scientific theory, a key element of authentic scientific activity according to Popper (1963). Hesse's view of models includes three kinds of analogies – positive, negative and neutral. Positive analogies are aspects of the model in which properties of the model are identical with those of the system it models. So in the context of the kinetic theory of gases, particles may be modelled as being like billiard balls and there is a positive analogy in that both particles and billiard balls obey Newtonian mechanics. There may be, however, some negative analogies as there are some aspects of billiard balls (colour, for example) we do not want to ascribe to particles. There are also neutral analogies – features of the model which cannot yet be reliably classified as positive or negative; these are frequently the basis of fruitful research for scientists, and in the case of the kinetic theory of gases, led scientists to investigate the effects of temperature and pressure on gases. Hesse's model of scientific progress, then, involves identification and subtraction of negative analogies, together with efforts to identify (as positive or negative) any neutral analogies, through a process of systematic enquiry. Neutrality is tolerated – encouraged, in fact – as an aspect of science which may be central to the generation of better understanding and new knowledge, and as such is an 'authentic' concept with which learners of science should be personally and collectively engaged.

What role for ICT?

This discussion of models and modelling raises a number of questions as to the specific role for ICT. Is it, for example, just one of a number of media for the 'expression' of models, or can it act as a bridge between the mental and the expressed models of learners? If we consider the first of these options, it can certainly be argued that ICT has considerable analogical capability by virtue of the range of media it can encompass and the ways in which they can be combined. The fact that learners can interact with highly realistic on-screen environments appears to present opportunities for learning in highly authentic environments – even to the point, as with 'virtual fieldwork', where it is seen as an alternative to working in the 'real world' where distance or danger preclude actual visits. Another argument for the use of ICT in teaching and learning has been put forward by Papert (1980), Resnick (1994) and others; namely, that the availability of 'microworlds' allows a range of patterns of interactions on the part of learners and, critically, the support for learners' risk-taking encourages the authentic behaviour of building and testing hypotheses.

If Papert's 'microworlds' provide a supportive and forgiving context for learners to try things out, then Hesse's view of analogies and models, and of neutral analogies in particular, provides a framework for learning and thinking about the elements of those microworlds. What an appropriate ICT application can provide, then, is a context in which learners are exposed to models in which they are encouraged to identify positive, negative and neutral analogies and provide scaffolding for even young learners in the exploratory, theory-building processes associated with authentic 'thinking like a scientist'.

There are some aspects of computer models that need to be kept in mind, however; they differ from many of the other kinds of models in Boulter and Buckley's typology (2000) in that they are not only the 'expressed models' of individuals other than the learner but also that they may not reflect the consensus views of the scientific community. Many computer models are highly 'edited', but the rationale for, and nature of, this editing may not always be made explicit. There are some notable exceptions, such as simulations written in the Logo programming language, the program code of which may be inspected and adapted (Collela *et al.* 2001), but many more are proprietary products the program codes of which are not exposed to users (see Carmichael 2000 for further discussion of this issue).

What this means is that many computer models have considerable potential to mislead or over-simplify the entities and processes they 'model'. In some cases this is due to decisions being made by designers or programmers as to the content of the program and may be related to perceptions of what is appropriate for the intended audience. In others, the simplification may reflect the difficulty of modelling complex situations and as such a stochastic model may come to be represented in what Boulter and Buckley (2000) call a 'determinative' way. It is difficult to model random motion of particles in a computer model of a gas, for example, and programmers might well use an algorithm to calculate their positions which in fact is deterministic, rendering the model a complex *animation* and, in Hesse's terms, increasing the negative analogy of the model.

Children thinking and practising science with ICT

I interviewed and observed a group of children aged between 4 and 10 (in two groups, 4–5 and 7–10) over a period of about six months, during which time they were able to use a number of software applications in which living things were represented in a number of forms. These applications[2] varied in their scope and complexity, but all involved representations of living things which were to some extent interactive – that is, the children were able to control or influence the behaviours of the simulations of animals within them, so these were more than simply animations over which they had no control. The applications were *Catz* and *Dogz*, 'virtual pet' applications from Mindscape Software (http://www.mindscapeuk.com); *SimAnt*, an interactive simulation in the form of a game from Maxis Software (http://www.maxis.com) and *Vivarium*, an independently produced freeware application developed by Ryan Koopman, which allowed modelling of predator–prey relations in 'microworlds' created and populated with a variety of living things by the children.[3]

The semi-structured interviews that took place involved me sitting alongside the child or children as they used the applications. Initially, the interview structure was limited to the children talking aloud as they 'demonstrated' the applications while I offered some stimulus questions which were, at least initially, based on the expected knowledge about living things from the Foundation Stage and Key Stages 1 and 2 of the Science Curriculum for England and Wales (http://www.qca.org.uk). What I was particularly interested in, following Hesse, was whether (and on what basis) the children identified positive, negative and neutral analogies in the computer applications. However, as we shall see, the interviews, while they remained focused on the applications and the simulated organisms within them, were to range over a rather broader range of issues than curriculum content alone and the 'point-for-point' comparison of simulations with real animals proved to be only one aspect of the children's modelling and learning.

Virtual pets

The youngest children worked primarily with *Catz* and *Dogz* running on Apple Macintosh Powerbooks. They were able to select a cat or dog to be their 'pet' and could choose a template which they could then adapt by adjusting colour and other aspects of its appearance. From the outset, the children referred to 'their' pets and they were regularly 'fed' and 'played with'. The application provides a variety of pet foods, grooming equipment and toys which can be manipulated with the computer mouse, allowing interaction with the virtual pets – the cats, for example, responding to grooming by purring. Even before interviews took place, the children were able to draw parallels between the simulations and real animals of which they had personal experience. They rapidly became familiar with the features of the application and discovered and shared knowledge of undocumented features. In this extract two of the children (A – 4 years old and B – 5 years old)[4] have discovered that it is possible to catch a mouse that periodically runs across the cat's living area and are attempting to feed it to the cat; this involves clicking the computer mouse while the cursor is over the mouse on the screen and holding the 'Shift' key (no easy task and one not documented in the user guide):

A: Got him. Come on mousie, time to die . . . [*drops mouse on to cat's head.
    Cat ignores it and mouse runs away. B takes control of computer mouse*]
A: Here's the mouse . . . grab him . . . use shift like for the cat
B: Got him . . . wiggle wiggle. Oh . . . he got away again

Those children who had pets of their own, or who had spent time with pet animals of friends or neighbours, were quick to make comparisons between their behaviour and that of the simulations. Here, A describes how Willow (a cat belonging to a neighbour) and the simulation differ in their behaviour – in Hesse's terms, negative analogies – also identifying how the application constrains her behaviour as a user:

A: I wouldn't pick Willow up like that. I'd cuddle him. Not by the leg or tail [*tries to use cursor to pick up simulation by tail*]. Oh . . . oh . . . I can't. You can't pick him up 'cept like this [*uses cursor to pick up simulation by neck*]

R: Maybe you can't pick him up so as you'd hurt him.

A: I can pick Jester [the simulation] up like this [*uses cursor to pick up simulation by neck again. Cat rotates slowly on screen and glares*]

R: Yes, but he doesn't like it, does he?

A: Look . . . look! He's really grumpy!

Other children who had less experience of playing with or caring for real animals were characteristically more cautious in making judgements about the extent to which the simulations were realistic and to identify positive and negative analogies. At the same time, faced with neutral analogies, they were more willing than others to experiment in order to establish the behaviour of the simulations, only stopping to reflect on the realism of the simulations when prompted by an adult. Here, C (5 years old) who has little experience of real dogs, begins by spraying a simulated dog with water – the only sanction, other than denial of food, available:

C: [*Sprays dog nine times. Dog looks depressed, edges away*] He doesn't like that! [*Dog goes to bowl and eats food*] Look at him! He likes that!

R: Is he like a real dog?

C: Mmm . . . yes.

R: If you squirted a real dog, what'd he do?

C: Roar at you . . . Rooaaarrrrr . . . 'cos he's so fierce

R: Do you think this dog ever gets fierce, or cross?

C: No . . .

R: Not ever?

C: [*Sprays dog a further four times. Dog yelps and moves away*] He just gets sad . . .

Even the youngest children were able to identify negative analogies in the simulations, most relating to the lack of realism in potentially danger-ous and injurious behaviours. In the *Catz* application, for example, the cats never kill the mouse and they are able to fall from the top of the application window to the bottom without injury. The analogy, initially a neutral one, which most interested the children, however, was the question of whether the simulations could survive without care and food and there were a number of discussions around the issue of whether they would eventually die if left unattended for a long period. A, who had by this time used the application and maintained her simulated cat 'Jester' for several months, describes her experiences and demonstrates an emerging awareness of the analogical limitations of the application.

A: If you don't feed them they d-i-e [emphasis]

R: Have any of your computer cats and dogs ever died?

A: No . . . oh . . . what happens when they die? Do they die like a real cat? Do computer cats ever really die?

R: What happens if you don't use the computer and leave them for a long time? Have you ever done that?

A: I didn't wake Jester [the simulation] up for ages and ages and when I did he was really hungry. His bowl was all empty.

R: Did he look sick, or thin?

A: No . . . no, he was grumpy and meowed a lot like 'feed me, feed me' so I gave him food and biccies and he ate and ate and ate like 'snarf snarf' [*laughs*] . . . like me!

While the animals were perceived as being 'really hungry' (a positive analogy in Hesse's terms) the issue of whether a simulation could 'die' remained unresolved and thus neutral for some time. Despite some of the children leaving their simulations for longer periods (up to six weeks in some cases), no simulations underwent virtual 'death' and the consensus was established among groups of children that while the cats and dogs became hungry, they seemed immortal – a negative analogy recognised by all the children. Only one of the older children (E, 7 years old) recognised the hidden hand of the application designers and developers at work in this, however, and suggested that the negative analogy was imposed to prevent 'upsetting little children if their cat dies', recalling a 'real fuss' a friend had made when another virtual pet had 'died'.

Ecological simulations

Software applications which represented more complex situations (such as Maxis's *SimAnt* which represents an ant colony and Ryan Koopman's *Vivarium* which allows modelling of population growth, competition for resources and predator–prey relationships) were less immediately appealing to the younger children, and even older children had a tendency to misinterpret the purposes of the applications which they regarded as 'games' to be mastered. Lack of familiarity with the subject matter led to children being initially more tentative and subsequently exploring neutral analogies through experimentation, leading to assertions such these:

'The mice were better than the bugs because we put more in and they got to the food quicker and had more babies.'

(F, 8 years old)

'You got to keep your queen safe 'cos she lays the eggs, and no more eggs, no more ants.'

(E, 7 years old)

'The yellow ant (controlled by the computer user) has to get help to carry all that food so she can call up her friends to carry for her.'

(G, 8 years old)

'If the slugs' food ran out they eat each other . . . but they never found each other, they just went on and on. The slugs couldn't have babies so they slowly went down and down.'

(H, 7 years old)

As with the virtual pets, the representation of mortality was a point of discussion amongst the children. In *SimAnt*, it was possible for the user's 'representative' (the 'yellow ant') to be trodden on, be eaten by predators or starve to death, but it is 'reincarnated' (the word used by in the application's documentation) back at the nest. Some children chose to interpret this as a negative analogy: 'If you was a real ant, right, and you got squashed, that's it, you've had it. But that wouldn't make much of a game, and you'd get fed up' (E, 7 years old).

Others disagreed and offered the interpretation that the 'reincarnated' ant was in fact a new individual, thus avoiding a negative analogy: 'Ants all live like a family, and the new ant takes over and becomes the boss ant' (F, 8 years old).

Interaction analogies and the 'real world'

The children were able to identify positive analogies (the computer cats were like real cats in terms of appearance, behaviour, appetite) and negative analogies (they were immortal, passive and did not excrete). The children also discussed and explored areas of neutral analogy – a characteristic and authentic activity of science. Hesse, however, identifies 'a further role for analogies' beyond the 'literal, point-by-point comparison of two systems' and the identification of positive and negative analogies – between model and 'target', computer cat and real cat. As we have already mentioned, Johnson-Laird (1983) and Nersessian (1992) argue that analogies can themselves 'do the work' of changing conceptualisations and solving problems.

Hesse, too, developing ideas first advanced by Black (1962), describes (Hesse 1970, 1980) how analogies can be 'interactive'; this involves the transfer of ideas and implications from the secondary system to the primary, involving selection, emphasis, suppression and illumination. As a result of this interaction, 'the two systems are seen as being more like each other, they . . . interact and adapt to each other' (Hesse 1980: 163), even to the point where they may lead to mental models of either system, or both being reassessed. What this means in the context of children's learning is that they have opportunities to 're-experience the world' (DiSessa 1986) and to take part in activities which are authentic science, but which are also personally authentic in that they are relevant to their own development as learners, not just as 'proto-scientists':

E: Look, look, they're making a trail. What have they found?
F: Must be food . . . where's the food?
G: In the hole?
F: Where's the yellow ant?
E: That's in the game. These are all black.
F: Where's the boss ant?
G: In front . . . that must be it . . . no . . . that one.
E: It must have found the food and told the others.

The area in which this kind of thinking and critical reassessment was most evident was in the children's discussions of the relationship between the target domain and the computer application itself, rather than between the target domain and its visual representation on the screen. On the whole, the children found it hard to articulate their understanding of how the simulations worked; only one, E (7 years old) recognised the critical role of the programmer in pre-defining behaviours: 'There's nothing in the computer to say "if you don't eat for ten or twenty or some days then you die", it just says "if your bowl's full then eat some food".'

Lack of technical insight did not prevent children from drawing parallels between the functioning of the computer and living things and, in doing so, going beyond comparative analogies. The question of whether 'computer cats ever really die' is interesting, then, not only as evidence of thinking about the death of a living thing (the *cat* represented by the model), but also because it signals the emergence of thinking about what 'death' might mean in the context of a computer-based model (the *computer cat*), and even of electronic devices more generally (the *computer* through which the model is expressed).

This was also explicitly addressed in discussions of other concepts including intellectual capacity, memory and sleep. In relation to *SimAnt* children started to refer to the computer as 'the yellow ants' brain' and then began to question how the computer could make all of the ants represented onscreen apparently function independently of each other. In the *Vivarium* program, children noticed that smaller 'worlds' appeared to run more quickly: 'The computer's got more work to do and it has to think for all the bugs . . . if you give it too many bugs and things it has to share its brain out and it can't think that much' (D, 7 years old).

In another example, E (7 years old) compared the 'brain power' of different computers and of the cats represented in *Catz*:

R: Do you want to put your cat on a disk and take it home?

E: Mmm . . . yes. Will it work on my computer?

R: Should do.

E: It might be really slow though like [*mimes walking in slow motion*] 'cos it's old [. . .] I don't think it's got enough brain to be a cat. It's not as smart as this one.

Once they had learned to start up and shut down the computers, locate the icons with which programmes were launched and load and save programme files, familiarity with the hardware and software led the children to draw other parallels. The 'sleep' function, which allowed the laptops to conserve battery power, led to comments such as: 'I've put the cats to sleep now. The computer's sleeping so the cats are sleeping too' (F, 7 years old). The question of what became of the cats was discussed by some of the children once they had become familiar with the process of 'minimising' windows. Here, A (4 years old) and C (5 years old) discuss switching between cats:

C: I want to see my cat now. Can I see my cat please?

A: [*speaks into microphone on computer*] You eat your food, and I'll go and talk to the other cats. I'll be back in a moment. [*minimises window on screen, no cats are now visible*]

C: Where's my cat?

R: A, where is your cat now?

A: I don't know, just hanging about. He's OK. He's got food to keep him going.

C: Is he OK? My cat's OK and he was switched off all week.

A: Yes, yes . . . the computer keeps them going. It remembers them.

The interactions illustrated here have the potential to act as starting points in discussions which address questions such as: in what way is a computer's sleep like that of a cat? Or like that of a human? Does thinking of the computer as 'like a human brain' help us understand what it means for the computer to 'sleep'? And conversely, does thinking of the human brain as 'like a computer' help us to understand what it is for *us* to 'sleep'? In the same way, how might thinking of our brains as computers shape our conceptualisation of memory, or our interpretation of the act of forgetting, or of the tendency to be forgetful? As Hesse suggests, a powerful analogy can alter our thinking about both of the concepts or domains that it involves.

Conclusions

The increasing role of ICT in the lives and education of young children makes it necessary for us to develop more sophisticated frameworks for analysing their thinking and learning. Piaget's notion of young children's 'animism' – the attribution of life-like processes such as intention to inanimate objects – remains relevant, up to a point. However, the strategies and complex reasoning demonstrated by the excerpts of children's talk in this chapter suggest that they are able to apply and adapt models as a process of conceptual change (rather than being based on a 'deficit model' in which a crude 'animism' results from incomplete understanding) and that ICT can play a part in enabling this process.

ICT applications can still be seen as addressing curriculum content, but more critical is the potential for learners to identify and explore neutral analogies in specific domains. Ideally, any neutral analogy identified by a learner within a computer application could be suggestive of some kind of virtual experimentation and a review of understanding of the real-world phenomena modelled. Of course, it is when this extends or 'blends' into observation and experimentation of the real-world domain modelled that children's learning becomes more apparent and can be said to have transferred across contexts. In crude terms, it is when knowledge is applied to a real-world phenomenon that the 'learning gains' of the computer application become obvious. But we can take a further step beyond seeing computer-based learning in terms of curriculum content or as a 'microworld'; what the activities reported here promoted through the interaction process described by Hesse was a 'meta-level' of learning about *the value of modelling itself*. What the children were doing was not only comparing a computer model with a real-world situation, but also – when they were talking about the relationship between computers and living things – beginning to address questions about the nature of the medium in which the models were presented.

At the heart of this argument is that view that models, rather than being imperfect mirrors, are opportunities for higher-order thinking and learning even in young children. The challenge for teachers is to stimulate and support this level of discussion; questions of the form 'how is this toy animal similar to and different from a real one?' suggest that, at most, a point-for-point comparison is required. Far more challenging and potentially rewarding are questions which address the iterative processes of model-formation, model-use, model-elaboration and model-abandonment, and which involve 'immersion' not just in an interactive computer environment but in a 'blended' learning experience.[5] Perhaps the greatest contribution that teachers and software designers alike can make is to collaborate in developing a culture of model-building and model-use that supports young learners as they make sense of a world that is, after all, far more immersive and interactive than any 'life on the screen'.

Notes

1 This title was posed as a rhetorical question by one of the children whom I interviewed during my research; the context is explored more fully in the text. It was only after I used it as the title of a conference presentation that people pointed out its resonance with the title (and, for that matter, the content) of Philip K. Dick's novella *Do Androids Dream of Electric Sheep?* Some of the children did indeed have discussions as to the content of the dreams both of real cats and their computer representations, so this chapter could well have been entitled 'Do Computer Cats Dream of Electric Mice?' (I also considered 'The Cats in the Machine'), but I decided to retain the title taken from the 'in vivo' quotation.
2 Throughout the remainder of the chapter the term 'application' will be used to describe the program and the on-screen environment with which the children interacted, whilst the term 'simulation' will be used to refer to specific living things represented within the applications.
3 Despite being listed on a number of web pages devoted to Artificial Life, Koopman's simulation no longer seems to be available online.
4 In the excerpts, A–G are the children; their ages are shown in years and months. R is the researcher – myself.
5 See Collela (2000) and Collela *et al.* (2001) for perhaps the closest approximation to date of this approach to curriculum design.

References

Black, M. (1962) *Models and Metaphors: Studies in Language and Philosophy*. Ithaca NY: Cornell University Press.
Bliss, J. (1994) 'From mental models to modelling', in Mellar, H., Boohan, R., Bliss, Ogborn, J. and Tompsett, C. (eds) *Learning with Artificial Worlds: Computer Based Modelling in the Curriculum*. London: Falmer Press.
Boulter, C. and Gilbert, J. (1998) 'Learning science through models and modelling', in Frazer, B. and Tobin, K. (eds) *The International Handbook of Science Education*. Dordrecht: Kluwer.
Boulter, C. and Buckley, B. (2000) 'Constructing a typology of models for science education', in Gilbert, J. and Boulter, C. (eds) *Developing Models in Science Education*. Dordrecht: Kluwer.
Bruckman, A. (1998) 'Community support for constructionist learning', *Computer Supported Cooperative Work*, 7 (1/2): 47–86.
Bruckman, A. and de Bonte, A. (1997) 'MOOSE goes to school: a comparison of three classrooms using a CSCL environment'. Proceedings of CSCL 97, Toronto, Canada, Dec 1997, http://www.cc.gatech.edu/~asb/papers/cscl97.html
Carmichael, P. (2000) 'Computers and the development of mental models', in Gilbert, J. and Boulter, C. (eds) *Developing Models in Science Education*. Dordrecht: Kluwer.
Carmichael, P. (2003) 'Teachers as researchers and teachers as software developers:

how use-case analysis helps build better educational software', *Curriculum Journal*, 14 (1): 105–122.

Collela, V. (2000) 'Participatory simulations: building collaborative understanding through immersive dynamic modelling', *Journal of the Learning Sciences*, 9 (4): 471–500.

Collela, V., Klopfer, E. and Resnick, M. (2001) *Adventures in Modelling: Exploring Complex, Dynamic Systems with StarLogo*. New York: Teachers' College Press.

DiSessa, A. (1986) Artificial worlds and real experience, *Instructional Science*, 14: 207–227.

Feyerabend, P. (1978) *Against Method*. London: Verso.

Feyerabend, P. (1987) *Farewell to Reason*. London: Verso.

Gentner, D. and Gentner, D.R. (1983) 'Flowing waters or teeming crows: mental models of electricity', in Gentner, D. and Stevens, A.L. (eds) *Mental Models*. Hillsdale, NJ: Lawrence Erlbaum.

Gilbert J. (1998) 'Explaining with models', in Ratcliffe, M. (ed.) *ASE Guide to Secondary Science Education*. Cheltenham: Stanley Thornes.

Gilbert, J., Boulter, C. and Elmer, R. (2000) 'Positioning models in science education and design and technology education', in Gilbert, J. and Boulter, C. (eds) *Developing Models in Science Education*. Dordrecht: Kluwer.

Grosslight, L., Unger, C., Jay, E. and Smith, C. (1991) 'Understanding models and their use in science: conceptions of middle and high school students and experts', *Journal of Research into Science Teaching*, 28: 799–822.

Hanson, N. (1958) *Patterns of Discovery*. Cambridge: Cambridge University Press.

Hesse, M. (1970) *Models and Analogies in Science* (2nd edn) Notre Dame, IN: University of Notre Dame Press.

Hesse, M. (1980) *Revolutions and Reconstructions in the History of Science*. Brighton: Harvester Press.

Hounsell, D. and McCune, V. (2004) 'The development of students' ways of thinking and practising in three final-year biology courses'. Paper presented at EARLI Second Biannual Assessment Conference, Bergen, 23–25 June 2004.

Johnson-Laird, P. (1983) *Mental Models*. Cambridge: Cambridge University Press.

Kuhn, T. (1970) *The Structure of Scientific Revolutions*. Chicago: University of Chicago Press.

Kuhn, T.S. (1996) *The Structure of Scientific Revolutions*. Chicago: University of Chicago Press.

Lakatos, I. (1970) *Criticism and the Growth of Knowledge* Cambridge: Cambridge University Press.

Lakatos, I. (1974) 'Falsification and the methodology of scientific research programs', in Lakatos, I. and Musgrave, A. (eds) *Criticism and the Growth of Knowledge*. Cambridge: Cambridge University Press.

Mellar, H., (1994) 'Towards a modelling curriculum', in Mellar, H., Boohan, R., Bliss, J., Ogborn, J. and Tompsett, C. (eds) (1994) *Learning with Artificial Worlds: Computer Based Modelling in the Curriculum*. London: Falmer Press.

Nersessian, N. (1992) 'How do scientists think? Capturing the dynamics of conceptual change in science', in Giere, R.N. (ed.) *Cognitive Models in Science*, XVI. Minnesota Studies in Philosophy of Science. Minneapolis: University of Minnesota Press.

Papert, S. (1980) *Mindstorms: Children, Computers and Powerful Ideas*. New York: Basic Books.

Popper, K. (1963) *Conjectures and Refutations: The Growth of Scientific Knowledge.* London: Routledge.

Resnick, M. (1994) *Turtles, Termites, and Traffic Jams: Explorations in Massively Parallel Microworlds.* Cambridge, MA: MIT Press.

Roth, W-M. (1999) 'Authentic school science: intellectual traditions', in McCormick, R. and Paechter, C. (eds) *Learning and Knowledge.* London: Paul Chapman.

White, B.Y. and Fredricksen, J.R. (1998) 'Inquiry, modelling, and metacognition: making science accessible to all students', *Learning and Cognition*, 16 (1): 3–118.

7

'IS THERE A PICTURE OF BEYOND?' MIND MAPPING, ICT AND COLLABORATIVE LEARNING IN PRIMARY SCIENCE

Paul Warwick and Ruth Kershner

It's quicker with everyone's ideas . . . one person can only think of one thing.

(Helen, Y1/2)

It helps to hear other ideas, even if you don't really understand . . . hearing another idea makes it easy to think of another one.

(Jenny, Y1/2)

With a Starboard everybody can see it, and if you make a mistake with spelling and it's a really easy word you're going to be a bit embarrassed if everybody sees that you've got it wrong.

(Nina, Y5/6)

It's always there on the big thing.

(Ewan, Y5/6)

Introduction

Diverse hardware and software are now employed in primary science classrooms and other chapters in this book reveal the various uses to which they have conventionally, and not so conventionally, been put. In many schools desktop computers can be found in every class in varying numbers, whilst in some they have been replaced by smaller, more versatile laptops. The advent of computer suites and laptop trolleys shared between classes has, some would argue, facilitated a more imaginative use of computer resources. The extensive introduction of interactive whiteboards (IWBs) – literally a 'big thing' in the primary classroom (Ewan, quoted above) – is now making a further contribution to the ways that we think about the impact of such resources on learning.

In this chapter we reflect on work carried out using laptop computers and IWBs in connection with a particular type of software used for 'mind mapping'. We draw upon evidence from our work with UK pupils in Year 2 (6–7 years) and in Year 6 (10–11 years), when we observed science lessons which involved the use of the IWB, laptops and other learning resources. The software used with the IWB was 'Kidspiration' (http://www.inspiration.com/productinfo/kidspiration), a tool designed for use by pupils of primary age. In carrying out our classroom observations and analysis, we were particularly interested in the ways in which the pupils' talk and activity related to their use of the hardware in combination with the mind mapping software and other classroom resources. We videoed teachers working with the whole class in producing mind maps on IWBs, laptop computers and on paper. We also videoed pairs of pupils working on laptops and small groups of pupils working at the IWB, focusing on the ways in which their developing understandings were expressed and negotiated during the activity. After the lessons we interviewed groups of children about their work in these lessons and about their general views on learning with the mind mapping software, the IWB and laptops.

Before going on to discuss the children's responses in these science lessons, we consider some general ideas about children's learning with ICT and the use of mind mapping for representing knowledge and thinking. The value of collaboration between pupils using computers is discussed in the next section, focusing particularly on the implications for learning in the classroom context.

ICT and learning in the primary classroom

As Crook (1994) points out, pupils collaborate and learn in several different ways 'with', 'around', 'through' and 'in relation to' computers. Whilst on some occasions pupils may interact directly with computers in a simulation of dialogue and guided learning, it is more common to see pupils and teachers interacting with each other in the presence of computers and with others beyond the classroom through the Internet. This provides a range of options for pupils' activity, participation and collaboration in the classroom and many teachers will make good use of the different possibilities in each lesson. Yet pupils' learning is not entirely predictable from the provision of certain learning resources and activities because of the individual ways in which each child may respond to the opportunities available in the classroom context. The concept of 'affordance' is useful here, referring back to Gibson's (1979) account of how the physical environment is perceived in terms of what actions it allows. Some objects in the environment are designed to be accessible and efficient to users (Norman 1998) – for example, a doorknob's use is intended to be easily evident to someone who wants to leave the room. The learning environment may seem to a teacher or classroom observer to provide similarly obvious affordances for activity and learning by pupils, but the key point is whether the pupils perceive them as such and respond accordingly.

The assumed connection between pupils' activity and their learning is based in the social constructivist model of learning outlined in Chapter 1. This model explains children's participation in classroom activities as the basis of the creation of knowledge and the development of the higher-level thinking involved in processes like investigation, problem-solving and creativity. The assumption is that learning depends on the collaboration of experienced learners and novices or peers engaged in what is seen to be a purposeful and worthwhile activity. Pupils' direct or peripheral involvement in classroom activities not only contributes to the completion of the task in hand but it also leaves 'residues' in the pupils' thinking which are taken forward to the next activity (Salomon 1996). As Sutherland *et al.* (2004) point out, this process implies three steps in learning where computer hardware and software may have influence:

- the involvement in the immediate learning process;
- the nature of the 'residues' left in children's thinking which affect future learning;
- the decoupling of computer use from a particular lesson so that it can be chosen in the future from the range of teaching and learning tools available in that setting.

These three steps reflect an increasing level of independence and conscious choice for pupils in deciding how best to use the learning resources available to them for different purposes.

The idea of a 'tool' for activity and learning is a central aspect of social constructivism. In the science lessons we observed, both the computer hardware and the mind mapping software can be understood as tools in this sense. A tool may be more than the pencil used for writing or the dictionary used for spelling. It is, broadly, any material or symbolic artefact which people use to carry out both ordinary and specialised activities: cutlery, maps, mathematical formulae, computers and human language are all tools which carry the cultural knowledge and skills of the inventors and previous users. Other people may be perceived as 'tools' when they are involved in assisting or directing activities. In this sense they act to mediate learning and support development by enabling learners to achieve with help what they could not do alone (Vygotsky 1978, 1935). Most tools are so familiar and embedded in daily life that it is hard to imagine what we would do without them. However, certain activities may call for the invention of new tools (ranging from swimming goggles to computer software) without which we could not achieve our goals (to swim in chlorinated water or to simulate the workings of DNA). It is worth noting that tools may both guide and constrain activity, depending in part on the immediate motivation and goals of the people involved in their use (Pea 1993). However, broader educational aims and intentions must also be taken into account. Sutherland *et al.* (2004) remark that ICT tools may facilitate what would otherwise be impossible for pupils, contributing in this way to democratisation, access and inclusion in education. Yet there is a dynamic aspect to the introduction of new educational tools which may lead to unexpected outcomes. One of the general questions that arises in investigating any computer hardware is whether it is just a new form of an old tool (such as IWBs interpreted as replacing blackboards) or whether it is a new tool which may afford fundamental changes in pupils' learning in school. The key question is whether the process is one of replacement or transformation in the classroom? As we see later, this depends at least in part on the teacher's aims and the pupils' responses. A particular issue arising from the research discussed in this chapter is how different tools may be combined in the classroom use and orchestrated by the teacher to best effect in the light of what we know about how children learn and the aims for their learning.

In considering the pupils' learning during this study we focused on both procedural and conceptual understanding in science. The mind mapping software was an important tool which allowed us to highlight both aspects of learning as the teachers attempted to scaffold the pupils' collective construction of knowledge. We were, primarily, interested in how such 'content-free' software might facilitate a genuine exchange of science ideas and how these exchanges and interactions might differ depending upon the hardware used. Before discussing the findings in detail, however, it is worth considering the terms 'mind mapping' and 'concept mapping' as both came up in planning the research and working with the teachers.

Representing knowledge and thinking: concept mapping and mind mapping

The terms 'concept map' and 'mind map' are used interchangeably in much of the literature and in recent years the tendency has been to talk of mind maps rather than concept maps. In trying to understand their nature and purpose, however, we need to consider the literature that refers to concept maps as well as that which relates to mind mapping. Indeed, perhaps the most interesting work exploring the intentions and possibilities of such tools is written referring to concept maps.

In educational settings in particular, 'concept maps' have been used as a strategy for developing metaknowledge and metalearning[1] and there has been much interest over several years in their use in primary science classrooms, both for developing learning and as a technique for formative assessment (Harlen *et al.* 1990; Comber and Johnson 1995; Stow 1997). Whilst the use of such maps always relates to specific content – for example, in connecting ideas in an area of science – an underlying intention in classrooms is usually to enable learners to reflect upon *how* they are coming to develop and understanding concepts and the connections between them.

Concept mapping derives from the early and influential work of Novak and Gowin (1984), who developed the notion of the concept map from Ausubelian learning theory (Ausubel 1968). Novak and Gowin (1984: 4) define a concept as 'a regularity in events or objects designated by some label'. For them, language and other symbol systems are the central tools for such labelling. In essence, a concept map provides a schematic for representing how concepts are perceived to be connected. Whilst there are many ways in which this might be done, the work of Novak and Gowin suggests that it is the ways in which meaningful relationships are drawn between concepts – in the form of propositions – that is the key to their worth in developing not only subject learning but also metaknowledge and metalearning. In Figure 7.1, some exemplars are presented that reflect different levels of propositional thinking.

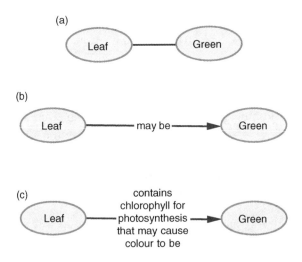

Figure 7.1 Concept maps showing different levels of propositional linking: (a) provides no indication of how the concepts might be connected; (b) suggests a simple propositional link; (c) suggests a more highly developed link in terms of science understanding.

Thus concept mapping 'is a technique for externalising concepts and propositions' (Novak and Gowin 1984: 17) primarily using language. In the simple maps presented in Figure 7.1 (b and c) there is a clear direction in the 'flow' of the map – represented by an arrow – and this is usually a feature of concept maps. As we can see from Figure 7.2, such a directional representation is not always possible to achieve, particularly for younger children. In addition, Novak and Gowin also point to the idea of developing notions of super-ordinate and sub-ordinate concepts within concept maps – this again seems to be only partially realised in the work of primary pupils.

Since the early 1960s 'mind maps' have been used in a variety of educational and business settings to summarise and consolidate information, as an aid to thinking through complex problems and as a means of presenting information (Buzan and Buzan 1993). Mind maps use a combination of different representational tools – pictures, diagrams, words etc. – to show concepts and the links between them. 'Mind mapping' therefore shares both the intention and the structures of concept mapping but there tends to be a greater emphasis on the use of combination of different representational tools to show concepts and the links between them. A further distinction that may be apparent is that concept maps tend to use as their starting point lists of words representing concepts, to be used as and where it seems appropriate to the learner. Though this is perfectly

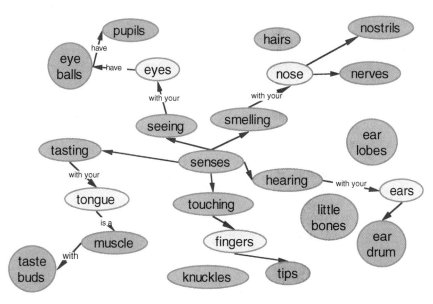

Figure 7.2   A 'typical' concept map produced by younger primary pupils
(Year 1/2).

possible with mind maps – and happened on occasions in both our research classrooms – such lists are rarely a *prerequisite* of working with mind maps.

There are now numerous mind mapping software products on the market ('Mindfull', 'Kidspiration' and Logotron's 'Thinking with Pictures' are amongst those appropriate for primary pupils). Most of these include banks of pictures that might represent ideas, the ability to manipulate colour and size, the possibility of creating 'word boxes' of different shapes and the inclusion of 'supergroupers' for clusters of concepts, as well as the organisational possibilities that might be seen in Novak and Gowin's concept maps (i.e. hierarchical structures and directional linkages). Advocates of the use of mind mapping software packages would suggest that because of their flexibility such tools have additional explanatory power beyond that of purely language-based models (Buzan and Buzan 1993).

We will now turn to the science activities that were undertaken in our research classrooms using the mind mapping package Kidspiration with groups working on laptops and at the IWB.

Learning in science: some classroom observations

In the following accounts of science lessons in Year 1/2 and Year 5/6, a number of themes emerge in looking at the pupils' and teachers' uses of the IWB, laptop computers and other tools for learning. One of the main areas of interest is the nature of the collaboration between the children and how they talked to each other during their work. We also became aware of several issues to do with the pupils' conceptual understanding – notably in Year 5/6 the distinction between what might be 'home knowledge' and 'school knowledge'. The representation of existing knowledge (both conceptual and procedural) was particularly highlighted in the use of software imagery and this related to the pupils' perceptions of the software affordances and the associated constraints and opportunities. The public nature of the IWB was important in two ways – not only in influencing the sharing of ideas but also in bringing elements of social evaluation into play (e.g. ensuring correct spellings). There were clearly some key factors relating to technical skill with the unfamiliar software, as well as the level of the pupils' typing and writing skills, which prompted the Y1/2 teacher to mediate and record the group discussion much more extensively than in Y5/6. Observing each whole lesson drew attention to the flow of activity in that time period and the combination of learning tools by the teacher and pupils. The 'orchestration' of learning tools is part of the process of mediation by the teacher and the pupils themselves – a process which not only enables the development of scientific understanding in each lesson, but also serves to connect learning in different lessons and different school and home contexts.

Using IWBs and laptops in Y5/6 and Y1/2

In the Year 5/6 classroom (with children aged 10–11 years) the first activity using Kidspiration was the creation of a mind map of concepts related to 'Planet Earth and Beyond'. The second was an attempt to create a mind map for a fair test of a balloon-powered 'jet'. One group – of between four and six children – worked on the IWB in each lesson. Groups of between two and three children worked on the same tasks at laptop computers.

In the work on 'Planet Earth and Beyond', collaboration between all the pupil groups was apparent during most of the lesson. With the laptop groups the influence of pre-mapping teacher-guided discussion was very clear in the initial stages of the work. Pupils took it in turns to input data, with initial discussions being focused on who should write what and whether terms were spelt correctly, rather than on what should be included. They placed a great deal of information on their maps very quickly, using 'school knowledge' to define the direction of some of their work – 'we need to write about the moon and the Earth and the sun'. As the lesson continued the nature of the activity changed. There was clearly a selection being made from group knowledge for inclusion on each map and sharing of information across groups occurred, with evidence of a subsequent 'filtering' process that determined what each group would adopt as part of their map (Figure 7.3). The pupils, who at this stage were quite unfamiliar with the software, were very concerned about the representation of ideas and the connections between them. Pictures were mainly used to illustrate text boxes, but we noticed discussions reflecting a concern that picture sizes should suggest, as far as possible, relative planet proportions. (As an aside, there was a charming moment when one child who had just found a picture of the Earth asked her partner 'is there a picture of Beyond?')

For the group working on the IWB, the most striking outcome was that the map created included a fraction of the information in those from the laptop groups (Figure 7.4). Why was this? Class procedures – such as checking spelling – were particularly important to the children on the 'public space' of the IWB. Group size and role decisions all used time and some technical issues with the wireless keyboard were apparent. However, it was noticeable that the discussions about what could and should be included on the map, and how the information should be represented and orientated, were at times extensive. For example, strong consideration was given to which type of concept 'holder' should be used to represent the importance of an idea. The group was focused on the board at all times, often gesturing to indicate approval, disagreement or a need to alter the ideas being expressed. Arriving at a consensus seemed very important to these pupils, with ideas often only used if 're-voiced' by more than one group member. Rules for map construction similarly had to be agreed – for example, it was decided that most links should be arrows, with a

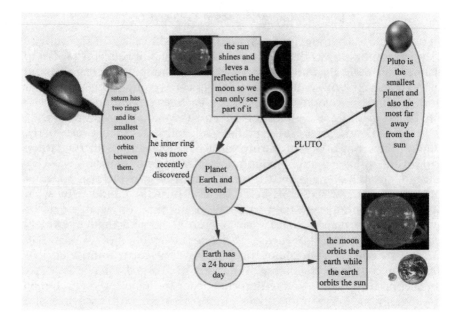

Figure 7.3   A 'Planet Earth and Beyond' mind map produced by a pair of pupils working at a laptop computer.

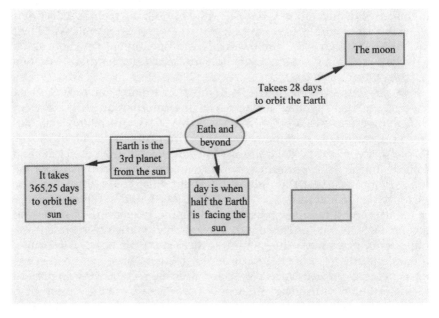

Figure 7.4   A 'Planet Earth and Beyond' mind map produced by a group of pupils working at an IWB.

directional meaning in linking concepts. Struggling with this construction seemed to help the process of deciding how best to show what was understood.

Many of the features noted above reappeared when the pupils were working on their design of a fair test investigation. Now more experienced in the use of the mind mapping software, the focus on procedural rather than conceptual categories led in some cases to a quite different approach by the children. All groups found the idea of a 'main idea' (which is part of the software presentation) impossible to interpret for this activity. The tendency was to group ideas connected to parts of the investigation, either through incorporating them within a 'super-grouper' or through the use of linking arrows (Figure 7.5). Here the affordances of the software were clearly being used by the pupils, yet it is noticeable that at least one of the groups working on the laptops used a simple list to define the experimental method, reverting to a familiar form of representation that might more easily have been achieved by other means. Here, one girl seemed to

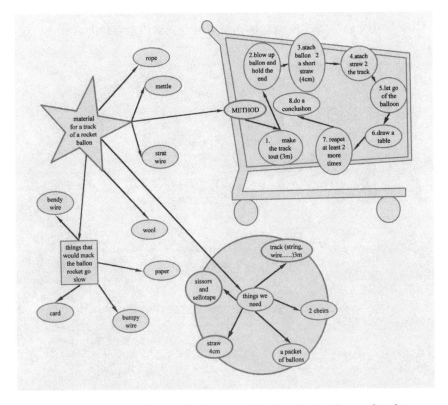

Figure 7.5   A concept map reflecting procedural understandings related to an activity with a balloon jet – laptop group.

be looking for ways to present her work as she would on paper – 'where's the bullet points?' (Figure 7.6)

Other affordances were, however, seized upon – modification of the content of concept boxes, or moving them to other parts of the map, happened regularly. Talk about *how* ideas might be represented was even more prevalent than in previous work – the idea that these representations had to mean something *to others* seemed to be at the centre of struggling to present the ideas clearly. For example, in re-organising and re-sizing concept boxes a child explained to her partner that it was 'so that everyone understands it'. In collaborating across groups, the pupils developed their own thinking – in one case a member of a group used another's map to pose serious questions about methodology, with the questioned pupil sufficiently convinced to say 'that's what I think' at the end of the exchange. For one group, teacher input using the flipchart was highly significant, providing a modelling of content that stimulated a complete re-working of the mind map. It could certainly be argued that the pupils would have been less willing to engage in this re-modelling if they had been working on pencil and paper.

For the IWB group, the negotiation of ideas again generally took longer

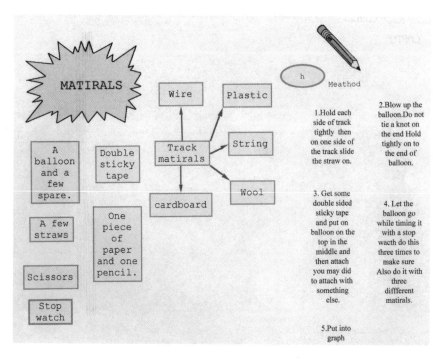

Figure 7.6   Using the software to produce a 'conventional' planning structure for the balloon jet activity – laptop group.

than with the laptop groups and consensus was usually, though not always, seen to be important. In many ways they were using the IWB to create a block of ideas and not in any substantive way using the affordances of the software. Overall, however, use of the software certainly led to a greater consideration of the relationship between representation and the meanings others may take from the completed group maps, hence the prevalence of discussions about the relationship of different forms of representation on the screen – pictures, words, symbols – and the links that should be made between them if an effective presentation of group thinking was to be created for others.

In the year 1/2 class (with children aged 5–7 years) the mind mapping software was used for three distinct purposes:

- to create a map of connected concepts about the human body, reviewed later from the perspective of work carried out in class;
- to allow the children to speculate about the concept of biological variation;
- as a basis for the construction and exploration of ideas related to a 'cars down ramps' friction investigation.

All of these activities were conducted with the whole class, with the teacher acting as an expert mediator of the pupils' ideas.

For the human body mind map, the teacher had pre-prepared the IWB screen to include key pictures and words to stimulate the children's thinking. This allowed the teacher to control the broad areas of discussion that might take place and so to focus the work on her curriculum objectives. She was able to mediate pupil responses, direct children to look at connections and probe their understanding where she felt they had more to offer. In reviewing this human body map, she focused the children's attention

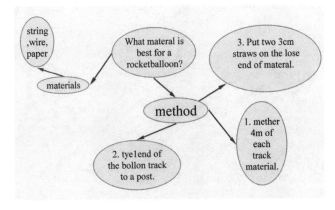

Figure 7.7 A mind map reflecting procedural understandings related to an activity with a balloon jet – IWB group.

on specific areas – such as healthy living – which had been the focus of classroom work. Here the children were concerned to introduce ideas and about exercise and about the kinds of drinks that might be considered to be healthy. Figure 7.8 presents the completed mind map.[2]

With such young children this guided, whole-class approach seemed highly effective in encouraging the children to think about school learning and to compare their thoughts with those of others. This recursive process of visiting and re-visiting information on the IWB is something that can be seen with other IWB software formats, for example notebooks. In this development work, the mind map allowed the children to revisit their initial thinking, to elaborate on their understanding of parts of the map and to draw connections between the major ideas presented.

In an ambitious later use of the software the teacher attempted to use the children's knowledge from their work on the human body to develop a wider conceptual framework related to the idea of biological variation. The teacher had placed words that she wanted the children to try to use – variation, same, different, humans and animals – in concept bubbles along the top of screen, together with pictures (in this case of people and

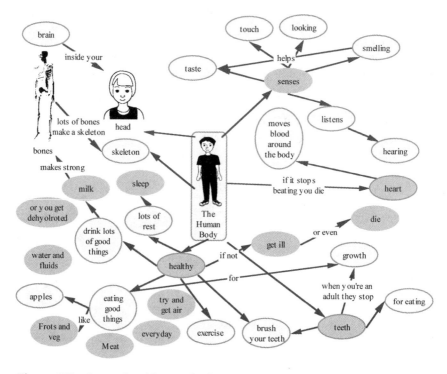

Figure 7.8   A completed human body mind map – Years 1/2 using the IWB with the teacher as mediator.

animals) that she felt the children might find helpful. Numerous ideas were elicited, from the simple – 'you could link animal with the cat' – to those that expressed more complex understandings – 'humans and animals are both living things . . . they eat food' (Figure 7.9). This public process of eliciting ideas (Howe *et al.* 2005) allowed the pupils to comment on the ideas of others and to develop what they knew. Using the software helped the teacher to physically point to the ideas noted on the screen whilst encouraging the children to express existing ideas and develop novel connections between them.

In the final lesson observed, this Year 1/2 class was engaged in developing a plan for a fair test investigation. Again, the teacher had pre-prepared a screen on which she had placed several areas of the consideration in devising a fair test investigation (Figure 7.10). During the lesson, she used the IWB to collect and orientate information from the children about how the investigation should be conducted. She used a range of additional tools to support the children's developing ideas, most notably a 'chest' containing all of the possible equipment the children might later use in their own investigations. By inviting the children to select objects from the chest and asking them how these might be used within the context of the proposed investigation the teacher was able to stimulate discussion, elicit ideas and build a framework of understanding on the mind map that could be used when the children engaged in their own investigation.

Physically, the teacher was in complete control of the IWB – she used

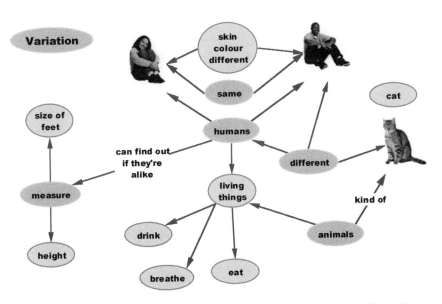

Figure 7.9 An ambitious attempt to map ideas related to biological variation with Year 1/2.

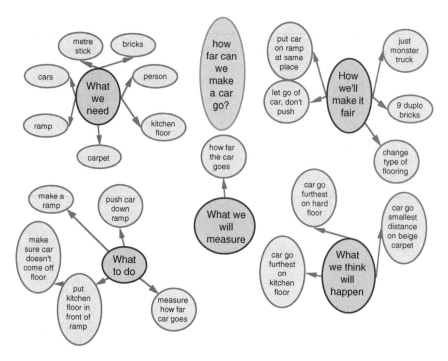

Figure 7.10 A teacher-created, pupil-adapted screen used on the IWB to
stimulate discussion about a fair test activity with Year1/2.

the keyboard to write the content of the concept boxes, controlled the
orientation of the concept boxes to one another and produced the links
between the concept boxes. But in doing this, she was merely the physical
operator of the technology – the ideas came from the children, as did the
reasons for making the links that were eventually made on the map. She
guided, questioned and challenged the pupils throughout the lesson,
skilfully mediating their ideas and creating the conditions by which the
children could see how they interrelated.

In orientating the children to the task, ideas from a previous lesson were
discussed. The teacher used familiar tools – just as the Year 5/6 teacher
used her flipchart – to focus initial interest and discussion. Each time a
child retrieved an item from the chest the children were asked about the
characteristics and possible uses of the objects: 'it's smooth'; 'we could use
it as a ramp'; 'to make the car stop'. As a result, they were asked to express
their thinking in relation to concrete objects and during the rest of the
lesson they were encouraged to try things out using the equipment that
they had retrieved from the chest. In discussing what would be the best
way to proceed they were encouraged to challenge one another's ideas

and to justify themselves, with the map being used as a public tool for orientating ideas that might work.

For teachers attempting to develop procedural understanding – the 'thinking behind the doing' of science (Gott and Duggan 1995: 26) – rather than to simply teach the skills associated with science investigations, this lesson illustrated how the visual presentation of ideas and the ways in which they are orientated to one another can help achieve this aim. In the following task, the children were led to a shared orientation. This seems fundamental to all that we have said so far and provides a clear example of how teachers might be involved in 'the creation of activities which necessitate learning dialogues' (Wegerif and Dawes 2004: 2).

In interview, these young children were clear that this was a very different exercise to the process of creating a map in which all of the boxes might in some way be seen as being conceptually related. At a simple level the concept groups created on the investigation map were understood as discrete elements – different parts of the process – that could be considered separately despite being part of the 'big picture' of the overall investigation. With respect to technical issues, there were obvious problems associated with the speed with which these young children could type on the computer keyboards, despite having received keyboard training. When the teacher herself used the keyboard the 'flow' of the lesson was much improved.

Some conclusions

The emerging findings from this research draw attention to certain key themes and issues relating to the pupils' classroom collaboration, their developing knowledge and understanding, their involvement in multiple aspects of learning in the primary classroom environment and the significance of the teacher's aims and strategies for learning and assessment. These are briefly discussed in the next sections.

Collaboration and talk, and learning to collaborate

One of the main principles informing the work discussed in this chapter is expressed well by Wegerif and Dawes (2004: 1), who argue that '. . . (l)earning with computers in school is a social activity in which the teacher plays a crucial role'. Yet we cannot assume that all primary pupils have the motivation and skills to collaborate in ways that promote their learning, even when provided with opportunities to be involved in tasks such as the ones that we have described. Perhaps to be truly effective, children need to experience something akin to the Nuffield Thinking Together project, which focuses on how children might be taught to interact and talk productively in the context of science. Dawes (2004: 685) remarks that the

Thinking Together project 'provided a core of talk skills lessons that enabled classes of children to generate and agree to use "ground rules" for exploratory talk'. We should note that, as in the case of our research, groups of pupils may be extremely adept at collaborating and respecting one another's views because of the ethos already established in the classroom and the whole school. However, this cannot be taken for granted and, as discussed by Kutnick and Manson (2000), certain pupils may need additional support to develop the social competence in relationships with others that allows them to take full advantage of the social interaction needed for collaborative learning.

Knowledge and understanding in science learning – making connections

The lessons discussed in this chapter draw attention to the importance of understanding how children develop and represent their knowledge in different contexts. In the Year 5/6 class the teacher explicitly told the pupils that for the purposes of that lesson she wanted to know what they had learned in school that term, not just the factual knowledge about the planets which they had gained largely through their homework. She used another classroom tool, the flipchart, to list some key concepts such as 'day and night' and this provided a visible representation of pupils' school learning to prompt them as they worked on their own maps. This tactic helped to mediate the pupils' 'home' and 'school' knowledge effectively and many were then able to begin to combine the two areas of their thinking. As Hart (2000) points out, the principle of making connections between the pupils' classroom responses and their wider learning experiences out of school is central to the thinking required of teachers and pupils and it is one of the fundamental ways to enhance learning and inclusion. The idea that teachers and pupils will combine the use of different classroom tools to make connections in learning draws attention to the need to place the use of any one resource, such as the interactive whiteboard, in the context of activity in the whole classroom environment. As we have already seen in Chapter 5, the work of Kress *et al.* (2001) extends this point in examining how pupils construct their understanding using a 'multi-modal' interplay of resources in speech, writing, gesture, action and visual images. Kress *et al.* (2001: 13) ask 'what constraints and possibilities for making meaning are offered by each mode present for representation in the science classroom, and what use is made of them?' The use of these different ways of representing knowledge is at the heart of the learning process, especially in attending to the connections that are made between them in science learning. Our lesson observations notably drew attention to the relevance of examining gesture, movement and other physical activity by teacher and pupils, in connection with the more familiar uses of speech, writing and visual images.

Multiple aspects of learning in the whole classroom environment – a question of control?

Individual tools such as the IWB do not stand alone in the classroom, but we do need to acknowledge that a particular resource may have specific and distinctive characteristics which can support, or hinder, different aspects of learning. For example, it was very clear from our study that the public nature of the IWB could have advantages and disadvantages. It could clearly help groups of pupils to share ideas with an easily visible point of reference. However, pupils were also aware of the possibilities for social evaluation as their work went up on the large screen and several were concerned about publicly demonstrating their technical skills including accurate typing and spelling. In discussing their review of research literature on ICT and pedagogy, Cox *et al.* (2003) identify one of the emerging themes as the control of learning. They note that work such as that by Hennessey *et al.* (2005) with teachers in secondary schools suggests that the use of ICT can be associated with a decrease of direction from the teacher and an increase in pupil self-regulation and collaboration. In our case both class teachers were concerned with involving the pupils in the learning, handing over as much as possible to them without withdrawing support all together. The idea that responsibility for learning can comfortably be shared by the pupils in the whole classroom environment, with all the prioritising, risk-taking and public errors implied, may be a goal to work towards as ICT tools become embedded in each primary classroom.

Teachers' aims and strategies for learning and assessment

This last point leads us to reflect on the centrality of the teacher's aims for pupils' learning in each lesson. The science lessons described in this chapter highlighted different views about whether the main focus would be on pupils' inclusion in the processes learning or on the assessment of what they had learned. This reflected alternative perceptions of what the mind mapping software could and should do in the given lessons. Yet these apparently alternative aims need not be contradictory. The classroom learning environment is a complex system which supports different aims and objectives for any one lesson. For example, Collins *et al.* (1996) propose the following framework of elements in the learning environment, expressed in terms of what teachers may want pupils to do:

- participating in discourse, for the purposes of active communication, knowledge-building and shared decision-making, as well as receiving information;
- participating in activities, in the form of purposeful projects and problem-solving, as well as practising exercises to improve specific skills and knowledge;

- presenting examples of work to be evaluated, which may involve both performing for an audience and demonstrating the ability to work out problems or answer questions.

These types of activity represent a mix of expectations and views about how children learn, including what may seem to be contradictory aspects of direct instruction and collaborative learning. However, Collins *et al.* (1996: 688) remark that most teaching and learning environments contain all these elements and that 'the most effective combine the advantages of each type'. Social constructivist models of learning emphasise the fundamental importance of the *processes* of participation, communication and active learning, but pupils are also asked to demonstrate their knowledge and achievements in relation to the science curriculum and more widely. Clarity about the priorities and multiple aims for pupils' learning is essential for developing the combined use of ICT tools in productive ways. Detailed classroom observations and further discussion with teachers and pupils can provide evidence of what they see as the opportunities for learning afforded by the computer software, hardware and other classroom tools in combination. However, there is more work to be done on what it really means for pupils to 'interact' with tools such as the interactive whiteboard and useful evidence may emerge as pupils continue to take on more responsibility and control in their use. This type of growth in pupils' involvement in learning is likely to be one of the main indicators of a fundamental transformation in teaching and learning as a result of new interactive technologies.

Notes

1 For the purposes of the discussion in this chapter, metaknowledge might be defined as knowledge about the nature of knowledge and knowing, whilst meta-learning refers, essentially, to learning about learning.
2 It's worth noting that, whilst the arrows on this map denote connections between ideas, they do not necessarily always represent the directional proposition or links proposed by Novak and Gowin (1984) for concept maps. This feature of the maps is much more prevalent in the work of the Year 5/6 class, where the teacher placed much greater emphasis on the *nature* of the links between concepts.

References

Ausubel, D.P. (1968) *Educational Psychology: A Cognitive View*. San Francisco: Holt, Rinehart and Winston.
Buzan, T. and Buzan, B. (1993) *The Mind Map Book: How to Use Radiant Thinking to Maximize your Brain's Untapped Potential*. New York: Penguin.

Collins, A., Greeno, J.G. and Resnick, L.B. (1996) 'Environments for learning', in De Corte, E. and Weinert, F.E. (eds) *International Encyclopedia of Developmental and Instructional Psychology*. Oxford: Pergamon/Elsevier Science.

Comber, M. and Johnson, P. (1995) 'Pushes and pulls: the potential of concept mapping for assessment', *Primary Science Review*, 36: 10–12.

Cox, M., Webb, M., Abbott, C., Blakely, B., Beauchamp, T. and Rhodes, V. (2003) *A Report to the DfES – ICT and Pedagogy: A Review of the Research Literature*. Norwich: HMSO.

Crook, C. (1994) *Computers and the Collaborative Experience of Learning*. London: Routledge.

Dawes, L. (2004) 'Talk and learning in classroom science', *International Journal of Science Education*, 26 (6): 677–695.

Gibson, J.J. (1979) *The Ecological Approach to Visual Perception*. Boston: Houghton-Mifflin.

Gott, R. and Duggan, S. (1995) *Investigative Work in the Science Curriculum*. Buckingham: Open University Press.

Harlen, W., Macro, C., Schilling, M., Malvern, D. and Reed, K. (1990) *Progress in primary science*. London: Taylor and Francis.

Hart, S. (2000) *Thinking through Teaching: A Framework for Enhancing Participation and Learning*. London: Fulton.

Hennessey, S., Deaney, R. and Ruthven, K. (2005) 'Emerging teacher strategies for mediating "technology-integrated instructional conversations": a socio-cultural perspective', *Curriculum Journal*, 16 (3).

Howe, A., Davies, D., McMahon, K., Towler, L. and Scott, T. (2005) *Science 5–11: A Guide for Teachers*. London: Fulton.

Kress, G., Jewitt, C., Ogborn, J. and Tsatsarelis, C. (2001) *Multimodal Teaching and Learning: The Rhetorics of the Science Classroom*. London: Continuum.

Kutnick, P. and Manson, I. (2000) 'Enabling children to learn in groups', in D. Whitebread (ed). *The Psychology of Learning and Teaching in the Primary School*. London: Routledge Falmer.

Norman, D.A. (1998) *The Design of Everyday Things*. London: MIT Press.

Novak, J.D. and Gowin, D.R. (1984) *Learning How to Learn*. Cambridge: Cambridge University Press.

Pea, R.D. (1993) 'Practices of distributed intelligence and designs for education', in Salomon, G. (ed.) *Distributed Cognitions: Psychological and Educational Considerations*. Cambridge: Cambridge University Press.

Salomon, G. (1993) 'No distribution without individuals' cognition: a dynamic interactional view', in Salomon, G. (ed.) (1996) *Distributed Cognitions: Psychological and Educational Considerations*. Cambridge: Cambridge University Press.

Salomon, G. (1996) *Distributed Cognitions: Psychological and Educational Considerations*. Cambridge: Cambridge University Press.

Stow, W. (1997) 'Concept mapping: a tool for self-assessment?', *Primary Science Review*, 49: 12–15.

Sutherland, R. with the InterActive Project Team (2004) 'Designs for learning: ICT and knowledge in the classroom', *Computers and Education*, 43: 5–16.

Vygotsky, L.S. (1978/1935) *Mind in Society: The Development of Higher Psychological Processes*. Cambridge, MA: Harvard University Press.

Wegerif, R. and Dawes, L. (2004) *Thinking and Learning with ICT*. London: Routledge Falmer.

8

EMERGENT SCIENCE AND ICT IN THE EARLY YEARS

John Siraj-Blatchford

As the Curriculum Guidance for the Foundation Stage (CGFS) makes clear, in the early years 'Children do not make a distinction between "play" and "work" and neither should practitioners' (QCA 2000: 11).

Within the CGFS the provisions for Knowledge and Understanding of the World provide the foundations for science education and the Early Learning Goals also suggest that, before children complete their reception year they should find out about and identify the uses of technology in their everyday lives and use computers and programmed toys to support their learning (QCA 2000).

The CGFS provides a series of 'Stepping Stone' statements that identify progression in science in terms of children's critical attitudes, their observation, recording and classification skills. My *Chambers Concise Dictionary* defines a stepping stone as 'a stone rising above water or mud to afford passage' and this seems highly appropriate in this case. The general principles or philosophy to be applied in providing an appropriate early education in science are not at all clear. The 'waters' are indeed murky. At Key Stage 2 the National Curriculum increasingly specifies the knowledge and understandings that are to be taught quite explicitly. As you move down through Key Stage 1 the orders tend to be less specific and refer to more general notions, like developing a respect for evidence and exploring similarities and differences. But there is no clear theorisation of the learning transition from early exploration to science education 'proper' (de Boo 2000).

Unfortunately, as we know, when educators are unsure of what they are doing they tend to keep very close to the script, or in this case to the stepping stones (so that they don't fall in!). In the circumstances the last thing we want is an early science education that is restricted to 'delivering' the stepping stones.

So what does it mean to support children's early learning in science? First, in understanding the nature of science education in the early years crucial distinctions have to be made between:

- natural phenomenon and behaviour;
- established scientific theories and explanations;
- children's individual scientific theories and explanations.

Learning science is not simply 'knowing about natural phenomena'; it provides a set of socio-historically established and agreed logico-mathematical constructions that explain these phenomena. But in the early years we cannot expect children to have experienced, or even to be aware of, all of the natural phenomena that they will later learn to explain in science lessons. A fundamental aspect of early science education is, therefore, to provide these awarenesses and experiences, to set the foundations for future science education. It is for this reason that provisions for sand and water play are very popular in the UK. However, the evidence suggests that without some form of scaffolding or instruction (e.g. demonstration, modelling etc.) the play involved may be repetitive, irrelevant and unproductive (Hutt *et al.* 1989, Siraj-Blatchford 2002a). Certainly, for this sort of play to be educational in terms of science, clear objectives need to be defined. Efforts should be made to draw children's attention to the workings of their own body and of the world around them. Imagine how difficult it would have been to understand atmospheric pressure if you had never gained confidence in conceiving of air as a substance beforehand! We can encourage 'air play' in the nursery, pouring it upside down in water, playing with bubbles and balloons, pumps and inner tubes, watching the effects of the wind and catching it in kites and sails.

To understand the problem of teaching 'established science' in the early years we need only consider the case of floatation. It is clear that any adequate understanding of the science of floatation must involve the concept of density and this will only be understood when children are able to consider the effects of proportional (and inverse proportional) changes in volume and mass – the intellectual equivalent of rubbing your stomach and tapping your head at the same time. At the Foundation Stage few (if any) children will be ready for this. Yet practical explorations of floating and sinking may be very valuable in the early years. Children can compare the buoyancy of small and large, heavy and light objects and we can encourage them to begin to develop hypotheses about floatation. But as Edwards and Knight (1994) have argued, in doing so we should only ever be trying to move children from their initial limited conceptions to 'less misconceived' ideas. The development of a practical recognition of the phenomenon of 'upthrust' might also provide a valuable support, if not a necessary prerequisite, for later understanding the scientific explanation.

A 'fact' is, as Margaret Donaldson (1992) has argued, something perceived and consciously noted. For scientific purposes it is also something described and recorded. But the business of perceiving and describing are quite different. For example, we don't consciously perceive everything that is available to our senses and there are many (perhaps an infinite number of) ways of describing what we perceive. Take the example of heat flow: science tells us that when we leave the warmth of our beds to stand bare-footed on a tiled floor, the excellent thermal conductivity of the tiles causes us to loose heat. But what we 'feel' is the sensation of the tiles being cold! A child may perceive and consciously note the fact that she feels warmer when she puts a coat on. But she will not have consciously perceived that heat was leaving her body before she did so and she therefore won't consciously perceive that the coat is providing an insulating layer that traps the heat around her. To the child the coat is simply warm. As Donaldson (1992: 161) says, 'theoretical preconceptions and reported observations are by no means independent of one another. Theories – or, indeed, beliefs not conscious enough to be called theories – guide the nature of the observations; and the guiding assumptions are often not recognized as being open to doubt.'

In the past many writers have referred to the child as a 'natural scientist' (Bentley and Watts 1994) because of their natural inclination to 'spontaneously wonder' (Donaldson 1992) about things. Driver addressed this directly in her book *The Pupil as Scientist*:

> The baby lets go of the rattle and it falls to the ground; it does it again and the pattern repeats itself . . . By the time the child receives formal teaching in science it has already constructed a set of beliefs about a wide range of natural phenomenon.
>
> (Driver 1985: 2)

As Driver (1985) went on to suggest, we now know that some of these beliefs differ markedly from accepted scientific knowledge and that they may be difficult to change. These are the 'misconceptions' that science educators in schools must later engage with. But the major difference between the scientific knowledge that every individual child builds up as an infant and the science constructed by professional scientists is not that one is 'right' and the other 'wrong'. It is related to the rigour with which every new 'scientific' idea is tested and to the benefits of professional collaboration and communication. 'Established' scientific knowledge is the product of a collective historical enterprise. When we refer to science as a 'discipline' we also draw attention to the fact that it constitutes an intellectual enterprise that has a distinct set of rules and that these rules are normally (or properly) adhered to by that particular academic community we know as 'scientists'. For a child (or for anyone else) to think 'scientifically' means to obey these rules and to keep an open mind, to respect yet always to critically evaluate evidence and to participate in a community that encourages the free exchange of information, critical peer review and testing. This latter point is crucial because, as Driver *et al.* (1996: 44) again put it, 'Scientific knowledge is the product of a community, not of an individual. Findings reported by an individual must survive an institutional checking and testing mechanism, before being accepted as knowledge.'

For all of these reasons it is important that we remain vigilant in our use of the term 'science' and discriminate clearly between 'scientific development' as itself a cultural phenomenon (and a knowledge base that children will be introduced to later in school), and cognitive development which, however analogous it may be to science, remains essentially an individual affair.

As Hodson (1998) has suggested, the contradiction that is often assumed to exist between the need to provide an enculturation into established science and the development of personal frameworks of understanding is in any event a false one. Even professional scientists who are working with the same theory while pursuing different purposes tend to apply different 'levels' of understanding. As Hodson (1998) goes on to argue, at whatever stage of education is being considered, the 'personalisation of learning' should involve the teacher in identifying and constructively engaging with the prior 'knowledge, experience, needs, interests and aspirations' of every learner.

Young children are naturally curious and we can encourage their explorations. We can also encourage an early interest in science, and the development of a respect for its achievements. As I have argued elsewhere (Siraj-Blatchford and McLeod-Brudenell 1999; Siraj-Blatchford 2002b), for all of these reasons it is important that we differentiate between a 'science education' that focuses on established conceptual knowledge (in the UK National Curriculum this currently starts in Key Stage 1) and an 'emergent science education' that focuses on hands-on experience, the development of emergent conceptions of the 'nature of science' and the development of positive dispositions to the subject.

In terms of learning theory and child development such 'emergent' approaches move us away from the simplistic notions of individual cognitive elaboration through 'discovery' to see effective practice in sociocultural terms involving the educator and the child engaged together in a 'construction zone' (Siraj-Blatchford and MacLeod-Brudenell 1999).

The large-scale and highly influential Effective Provision of Preschool Education (EPPE) project (Sylva *et al.* 2004) and the Researching Effective Pedagogy in Early Childhood (REPEY) project (Siraj-Blatchford *et al.* 2002) research suggests that adult–child interactions that involve some element of 'sustained shared thinking' are especially valuable in terms of children's early learning. These were identified as sustained verbal interactions that moved forward in keeping with the child's interest and attention. When children share 'joint attention' or 'engage jointly' in activities we know that this provides a significant cognitive challenge (Light and Butterworth 1992). Collaboration is also considered important in providing opportunities for cognitive conflict as efforts are made to reach consensus (Doise and Mugny 1984), and for the co-construction of potential solutions in the creative processes. Arguably, emergent science education provides the greatest curriculum potential for this sort of intellectual engagement.

An 'emergent' science curriculum is a curriculum responsive to children's needs as individuals; it accepts diversity of experience, interests and development. An emergent science curriculum is also a curriculum that respects the power and importance of play and that supports children in becoming more accomplished players, good at choosing, constructing and co-constructing their own learning. To sustain an interest in science is to sustain an interest in problem solving and exploration and for Bandura (1986) these processes begin with imitative learning which are subsequently internalised through identification and incorporated in the individual's self-concept and identity. So the real challenge is to provide children with strong models of science so that they develop positive attitudes and beliefs about the importance of the subject. A good deal of this can be achieved in small group work where children act as a 'collective scientist' under the direction of the adult (Siraj-Blatchford and Macleod-Brudenell 1999). The REPEY research (Siraj-Blatchford *et al.* 2002) found that the cognitive outcomes of the pre-school children whom they studied were directly related to the quantity and quality of the adult planned and focused group work. They also found that the most effective settings achieved a balance between the opportunities provided for children to benefit from teacher-initiated group work and the provision of freely chosen yet potentially instructive play activities.

It would be a nonsense to try to teach literacy by first teaching letter and words in isolation from stories and texts. It is also a nonsense to teach separate science 'skills' without providing a model of investigation. Early years teachers therefore need models, or 'recipes for doing' science if they are to model good scientific practice (Siraj-Blatchford and Macleod-Brudenell 1999). Teachers who teach emergent literacy provide positive role models by showing children the value that they place in their own use of print. In emergent science we can do the same by talking about science and engaging children in collaborative scientific investigations. We can tell the children many of the stories of scientific discovery. In doing so we will encourage children to develop an emergent awareness of the nature and value of the subject as well as positive dispositions towards the science education that they will experience in the future.

Socio-constructivist perspectives in early childhood education (Sayeed and Guerin 2000) recognise the importance of viewing play as an activity where children are developing their confidence and capability for interacting with their cultural environment. If we are to provide for an appropriate, broad and balanced education in the early years we must first think about children playing, but then we must also think about the particular subjects of that play. The clothing we provide for children to dress up in, and the props that we provide for their socio-dramatic play, should include resources to support emergent science. Many classrooms and play areas will already include resources to support playing doctors and nurses and there are usually plenty of resources to support measuring, but more resources need to be made available and I don't think we should be too worried about stereotypes at this age as long as they are not gendered. We might therefore consider providing young children with lab coats, extra large (plastic) test tubes and racks, flasks, burettes and coloured fluids and powders to play at being chemists. Toy manufactures could also do more to provide simple sensing equipment (I describe the Blatchford Buzz Box later in the section on data logging). For some early years practitioners this will all seem too prescriptive, but as Vygotsky (1978: 103) argued, 'In one sense a child at play is free to determine his own actions. But in another sense this is an illusory freedom, for his actions are in fact subordinated to the meanings of things and he acts accordingly.'

Screen-based activities have been shown to support the processes of verbal reflection and abstraction (Forman 1989). This is a theme specifically addressed by Bowman *et al.* (2001: 229) in the US National Research Council's report *Eager to Learn: Educating our Preschoolers*. The report strongly endorses the application of computers in early childhood:

> Computers help even young children think about thinking, as early proponents suggested (Papert 1980). In one study, preschoolers who used computers scored higher on measures of metacognition (Fletcher-Flinn and Suddendorf 1996). They were more able to keep in mind a number of different mental states simultaneously and had more sophisticated theories of mind than those who did not use computers.

The 'example materials' for the Foundation Stage produced for the Primary National Strategy (DfES 2004) provide concrete suggestions on how to use ICT to support the early learning goals within knowledge and understanding of the world:

> ICT resources can help children in 'developing crucial knowledge, skills and understanding that will enable them to make sense of their own,

immediate environment as well as environments of others'. Digital photographs, tape recorders, camcorders and webcams can allow children to investigate living things, objects and materials, some of which might not be accessible otherwise, for example with a webcam placed in a wildlife area.

As previously suggested, we need to begin by considering how this fits into a playful curriculum. In the following pages applications appropriate for supporting science in the Foundation Stage are illustrated under each of the following categories:

- Information sources
  e.g. ICT provides a wide range of resources, including the Internet and CD-ROM encyclopedias to support adults and children.
- Data handling
  e.g. especially providing support for 'counting' and in graphically displaying data from surveys.
- Data logging
  e.g. using sensors to observe changes more clearly.
- Sorting and branching
  e.g. identifying attributes and introducing classification.
- Simulation and modelling
  e.g. to investigate the effects of changing variables.

Play and problem solving

Most developmental psychologists treat play as either one, or some combination of three things:

1. Play as an exploration of the object environment
2. Play as an experience of an experimental and flexible nature, and
3. Play as a facilitator of the transition from concrete to abstract thought.
   (Adapted from Pepler 1982)

Exploration in this sense may be considered a necessary preamble to play, or as an initial stage within play. It may represent an integral part or a separate, although closely related, activity. In experimenting, the child moves beyond discovering the properties of objects, to determine what s/he can do with the object. This fits in well with Bruner's notion of 'mastery'. The importance here of the child being left free from the tensions of instrumental goals is often stressed. This allows for more novel, less inhibited, responses and applications of the objects of play. Play provides the opportunity for children to consider objects abstractly and this is an adaptive mechanism that facilitates problem solving. The folded paper that signifies for the child an aeroplane soaring through the sky becomes 'a pivot' for severing the meaning of 'aeroplane' from real aeroplanes. The focus of attention becomes what it is that the object signifies and can do, its properties and functions rather than its representation in the 'real' world. The objects of symbolic play thus provide important precursors for representational thought.

The importance of all this shouldn't be understated; pretend play has a major role in early cognitive development. The symbolisation that begins with objects goes on to be shared with the parent, then with peers and, as Piaget argued, the reciprocity in peer relations provides foundations for perspective taking and decentring. This in turn provides a model for symbolising 'the self' and the 'other' and supports the development of the child's 'theory of mind'. In the circumstances it isn't at all surprising that children's preference for socio-dramatic play has been shown to be correlated with intellectual performance (in terms of both IQ and ability scales).

Sylva *et al.* (1976) showed us that play facilitated problem solving. Divergent thinking is central to both play and creativity and longitudinal studies have also shown that creativity in pretend play is predictive of divergent thinking over time (Russ *et al.* 1999). As Edwards and Hiler (1993) argued in their teacher's guide to 'Reggio Emilia' (which is based in Italy and champions a particular approach to early years education), young children are developmentally capable of all the high-level thinking skills. We should therefore encourage them in their day-to-day practices of analysis (e.g. seeing similarities and differences); synthesis (e.g. rearranging, reorganising); and evaluation (e.g. judging the value of things).

In an evaluation of the Northamptonshire LEA Foundation Stage ICT programme (Siraj-Blatchford and Siraj-Blatchford 2006), we found that, given appropriate training, Reception teachers were able to make enormous progress in expanding the opportunities in their classrooms for play using ICT. In one example, Chrissie Dale, a Reception teacher at King's Sutton School, used Granada's Learning at the Vets software to support science and to encourage emergent writing (Figure 8.1).

1.11.04: The children have all been desperate to have a go at this one and demonstrating it on the whiteboard was a very effective way of showing the children how to use the program. However, when trying to use the program on the PC the children needed a lot of support. For each child to have a turn took a long time and has tended to initially interrupt the role play that has been established.

(Chrissie Dale, King's Sutton School)

This application was also developed further to incorporate a Listening Station (Figure 8.2) as an Answer Phone at the vets:

Some of the children have started pretending to write the messages down but I have not yet observed them taking these messages into their

Figure 8.1   Children working with Learning at the Vets.

play. I need to rerecord the messages as the volume levels are uneven and they need more careful thinking out to vary the play that they might develop. I need to buy a tape with the shortest running time that I can find or might even buy a cheap answering machine or ask parents to donate an old one.

(op cit)

Information sources

The value of taking children out of the classroom to learn from the environment is widely recognised in early education. CD-ROM talking encyclopaedias and the Internet extend the possibilities even further. A wide range of other early learning software is also available. One notable example is Percy's Animal Explorer (Figure 8.3).

Percy is a talking caterpillar who supports the children in learning about different animals and the sounds they make. Other games include finding the odd one out, matching pictures to sounds and learning where the animals live. The locations include a farm, a garden, the jungle and under the sea.

Figure 8.2   A child at the Listening Station.

Figure 8.3   Percy's Animal Explorer.

Data handling and display

Many early educators have found that 2Simple's 2Count and 2Graph provides a quick and effective data presentation program that supports children in reflecting upon their data and in answering their questions (Figures 8.4 and 8.5).

Graphs can also be used to summarise information collected over time for analysis, e.g. temperature, type of weather, the growth of a plant, etc.

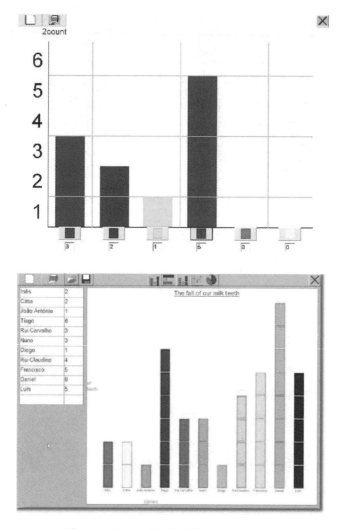

Figures 8.4 and 8.5   2Simple's 2Graph.

One application developed as a part of the Developmentally Appropriate Technology in Early Childhood (DATEC) project in Portugal (http://datec.org.uk) found that the 2Simple graphs can help the children to summarise and evaluate the data that they collected in different contexts as well as to communicate the information to others. One application involved the children studying their 'baby' teeth falling out. As Folque (2004) has suggested, this was a mathematical activity based on real and affective experiences. The children counted the teeth that they lost and associated this with a sign of their physical development. The experience was also the base for a range of other comparisons, measures and graphing activities. The 2Graph software allowed the children to save different graphs corresponding to different months and to explore each other's progress. The children also explored the different graph layouts in order to find the best one to communicate their central idea.

Data logging

The term 'data logging' may be applied in its broadest sense in the early years to denote all of those resources capable of supporting children in their observations of natural and humanmade phenomena. Early years data-logging resources, therefore, include a wide range of technologies, from digital cameras to technologies developed to support learning in much more discrete areas of the curriculum. One example of the latter is provided by the Blatchford Buzz Box (TTS) (CLEAPSS 2000; Siraj-Blatchford 2000). The Buzz Box provides an extremely sensitive electronic buzzer that will respond to minute current flows with an audible pitch proportionate to the current flowing in the circuit. It therefore provides a safe means of demonstrating the conductivity of the human body and of water. It is especially valuable in teaching young children about basic circuit principles and the dangers of electricity.

An example of a much more flexible data logging resource is provided by the Digital Microscope (Figures 8.6 and 8.7). At Gamesley Early Excellence Centre (Siraj-Blatchford and Siraj-Blatchford 2005) the staff have developed some excellent applications. In terms of the CGFS (QCA 2000), its particular value has been found to support the Knowledge and Understanding of the World and Communication Language and Literacy. A typical example of its use involved the children looking at mini-beasts found in the nursery environment. As the staff at Gamesley – and Feasey *et al.* (2003) – have found, with adult support even the youngest children can benefit from the use of this sort of equipment. Older Foundation Stage children have also been found to be capable of using the microscope independently.

Feasey *et al.* (2003) were commissioned by Becta to evaluate the use of the Intel Play QX3 Computer Microscope which was given to all schools

Figures 8.6 and 8.7   Early years children using a digital microscope, with adult support and alone.

in England as part of Science Year. They found that Foundation Stage children who used the microscope were highly motivated, and particularly keen to discuss their observations. They also found that it was often the children who became the instigators of using the microscope, showing the confidence to explore its potential. The study found that when most teachers and children first learn to use the microscope they cannot resist using it to view parts of their own body, but then they soon move on to discover the great potential that the technology has for supporting work in science, literacy and across the curriculum. Feasey *et al.* observed children using the microscope to view teeth, ears, skin, spiders and woodlice!

In our Northamptonshire ICT evaluation (Siraj-Blatchford and Siraj-Blatchford 2005) we found that digital cameras were being used to support children's reflection, for the purpose of display and documentation and to support communication with parents. At Aldwinkle School, for example, the Reception teacher (Shona Hall) found that the immediacy of the images produced by the digital camera were of immense value in her integrated activity associated with life cycles. The children used the digital camera to record the growth of their sunflowers:

> I believe that the digital camera provided the activity with more focus; because the children were taking their own pictures, they seemed to be looking more carefully for things to photograph. Most could provide an explanation of why they were choosing to take a particular shot and those who could not were given the opportunity to do so when we were viewing the images upon our return to the classroom. Most did this. In this way, I feel that the camera helped to clarify and consolidate the children's learning.
>
> (Shona Hall, Aldwinkle School)

At another Northamptonshire school (All Saints), a digital camera was used to support the children's investigation of bean growth. The children recorded the growth and made up their own 'Bean Diary' to record their findings (Figure 8.8).

> The children loved using the cameras . . . They enjoyed looking at their images after they had taken them and decide whether to re-take or if they were content. The children recorded their own serial numbers of photographs for printing purposes.
>
> (Mia Hobbs, All Saints)

Sorting and branching

Science began when people first recognised patterns. They recognised that there were patterns and regularities in nature that allowed them to

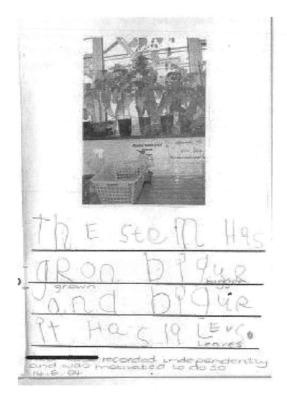

Figure 8.8   Using a digital camera to record a 'bean diary'.

predict the occurrence of natural events in advance. The most spectacular achievement of Thales of Miletus, who is often identified as one of the earliest scientists in Europe, was to predict the eclipse of 585BC.

In *Sammy's Science House: Sorting Station* (Edmark/Riverdeep) children sort pictures into categories, identifying similarities and differences as they begin to learn how plants, animals and minerals are classified (Figure 8.9).

Simulation and modelling

Whilst it presents an American environment, Acorn Pond from *Sammy's Science House* (Edmark/Riverdeep) provides an excellent example of how software can support a visit to a pond for some 'dipping', or other work associated with animal habitats (Figure 8.10). The CD-ROM supports children exploring animal habitats, seasonal changes and the effects of changing variables.

Figure 8.9 *Sammy's Science House: Sorting Station.*

Figure 8.10 Acorn Pond, from *Sammy's Science House.*

Off-computer activities might also include looking at butterfly books, colouring and printing butterfly wing designs. It might also include art work associated with the seasons, investigations of other animal habitats (under rocks, logs etc.), animal tracks and/or physical development sessions where the children 'fly' like butterflies, 'jump' like frogs, 'hop' like rabbits and 'slither' like snakes. *Sammy's Science House* also includes a 'Weather Machine' that allows the children to control the key variables of temperature, moisture and wind. Like a television presenter, Frederick the Bear then reports on the weather (Figure 8.11). The children learn how the changes in the key variables cause changes in weather conditions and also influence the dress and activity of the animated cartoon characters.

With adult support, there are a lot of other simulation and adventure CD-ROMs available that children will benefit from. One excellent example is provided by *Oscar the Balloonist Discovers the Farm* (Tivoli). The story line is described in an Amazon.com review as '. . . something along the lines of Doctor Doolittle meets Monty Python, meets Beatrix Potter'. Oscar tours the world in a hot-air balloon that enables him to travel through the seasons. In this adventure he crash-lands his balloon in a farm and meets an eccentric animal researcher, Balthasar Pumpernickel. In their explorations of the farm environment, children interview the animals and learn about them through a variety of virtual interactions and games. The

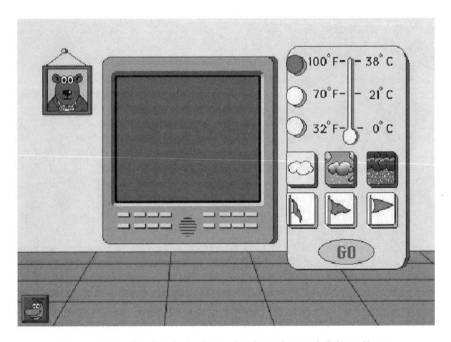

Figure 8.11 Frederick the Bear, also from *Sammy's Science House*.

software also provides a wealth of possibilities for investigations and activities away from the computer. One of the games invites the children to agree or disagree with suggestions such as 'Cows love to eat frogs'! Another title in the Tivoli series is *Oscar the Balloonist and the Secrets of the Forest.* Here, in addition to the changing seasons, the forest environment changes from day to night so that the children meet the nocturnal animals as well. The children will also discover whether squirrels are forgetful, whether ants freeze in the winter and why badgers get fat in autumn.

Conclusions: 'being a scientist'

I have argued that the best way to begin science education is for children to play together and, with adult support, at 'being scientists'. The fact that children in the Foundation Stage are too young to use many ICT applications on their own without adult support shouldn't trouble us at all. 'Science' in any event has never been the product of any individual. As Newton suggested, even the greatest contributions are only made by those 'standing on the shoulders of giants' – and that is just what adults are to young children.

References

Bandura (1986) *Social Foundations of Thought and Action: A Social Cognitive Theory.* Englewood Cliffs, NJ: Prentice Hall.

Bentley, D. and Watts, M. (1994) *Primary Science and Technology.* Buckingham: Open University Press.

Bowman, B., Donovan, S. and Burns, S. (eds) (2001) *Eager to Learn: Educating our Preschoolers.* Washington, DC: Committee on Early Childhood Pedagogy, National Academic Press.

Consortium of Local Education Authorities for the Provision of Science Services (CLEAPSS) (2000) The Blatchford Buzz Box, *Primary Science and Technology Newsletter*, Issue 20, http://www.cleapss.org.uk/priinfr.htm (The Buzz Box is available from: Technology Teaching Systems (TTS) Ltd, http://www.tts-group.co.uk).

de Boo, M. (ed.) (2000) *Science in the Early Years: A Teachers Handbook.* Hatfield: Association for Science Education.

Department for Education and Skills (DfES) (2004) *Learning and Teaching Using ICT: Example Materials from Foundation Stage* (Primary National Strategy/Sure Start CD-ROM) London: HMSO.

Doise, W. and Mugny, G. (1984) *The Social Development of the Intellect.* Oxford: Pergamon Press.

Donaldson, M. (1992) *Human Minds: An Exploration.* London: Penguin Press.

Driver, R. (1985) *The Pupil as Scientist.* Buckingham: Open University Press.

Driver, R., Leach, J. Millar, R. and Scott, P. (1996) *Young People's Images of Science.* Buckingham: Open University Press.

Edwards, A. and Knight, P. (1994) *Effective Early Years Education*. London: Routledge.

Edwards, C.P., and C. Hiler (1993) *A Teacher's Guide to The Exhibit: 'The Hundred Languages Of Children'*, Lexington, KY: College of Human Environmental Sciences, University of Kentucky.

Feasey, R., Gair, J. and Shaw, P. (2003) *Evaluation of the Intel Play QX3 Microscope*. Report to Becta, http://www.becta.org.uk

Fletcher-Flinn, C. and Suddendorf, T. (1996) 'Do computers affect the mind?', *Journal of Educational Computing Research*, 15 (2): 97–112.

Folque, A. (2004) *Graphs*. DATEC, http://www.datec.org.uk

Forman, E. (1989) 'The role of peer interaction in the social construction of mathematical knowledge', *International Journal of Educational Research*, 13: 55–69.

Hodson, D. (1998) *Teaching and Learning Science: Towards a Personalized Approach*. Buckingham: Open University Press.

Hutt, S., Tyler, S., Hutt, C. and Christopherson, H. (1989) *Play, Exploration and Learning*. London: Routledge.

Light, P. and Butterworth, G. (eds) (1992) *Context and Cognition: Ways of Learning and Knowing*. Hemel Hempstead: Harvester Wheatsheaf.

Papert, S. (1981) *Mindstorms: Children, Computers and Powerful Ideas*. New York: Basic Books.

Pepler, D.J. (1982) 'Play and divergent thinking', in Pepler, D.J. and Rubin, H. (eds) *Contributions to Human Development: Vol. 6 – The Play of Children: Current Theory and Research*. Basel: Karger.

Piaget, J. (1969) *Mechanisms of Perception*. London: Routledge and Kegan Paul.

Qualifications and Curriculum Authority/Department for Education and Employment (QCA/DfEE) (1999) *The National Curriculum: Handbook for Primary Teachers in England, Key Stages 1 and 2*. London: HMSO.

QCA/DfEE (2000) *Curriculum Guidance for the Foundation Stage: Stepping Stones*. London: HMSO.

Russ, S., Robins, A. and Christiano, B. (1999) 'Pretend play: longitudinal prediction of creativity and affect in fantasy in children', *Creativity Research Journal*, 12 (2): 129

Siraj-Blatchford, I. and Siraj-Blatchford, J. (2006) *A Guide to Developing the ICT Curriculum for Early Childhood Education*. Stoke-on-Trent: Trentham Books, http://www.trentham-books.co.uk/pages/guidetoict.htm

Siraj-Blatchford, J. (2000) *Nearly 101 Things to Do with a Buzz Box: A Comprehensive Guide to Basic Electricity Education*. Nottingham: SB Publications in association with Education Now Books.

Siraj-Blatchford, J. (2003) *Developing New Technologies for Young Children*. Stoke-on-Trent: Trentham Books.

Siraj-Blatchford, J. and MacLeod-Brudenell, I. (1999) *Supporting Science, Design and Technology in the Early Years*. Buckingham: Open University Press.

Siraj-Blatchford, J. and Siraj-Blatchford, I. (1998) 'Learning through making in the early years', in Smith, J. and Norman, E. (eds) *International Design and Technology Educational Research and Curriculum Development*. Loughborough: Loughborough University of Technology.

Siraj-Blatchford, J. and Siraj-Blatchford, I. (2002a) 'Discriminating between schemes and schemata in young children's emergent learning of science and technology', *International Journal of Early Years Education*, 10 (3).

Siraj-Blatchford, J. and Siraj-Blatchford, I. (2002b) 'Developmentally appropriate technology in early childhood: "video conferencing",' *Contemporary Issues in early Childhood*, 3 (2): 216–225.

Siraj-Blatchford, J. and Siraj-Blatchford, I. (2002c) *IBM KidSmart Early Learning Programme: UK Evaluation Report – Phase 1 (2000–2001)*. London: IBM.

Siraj-Blatchford, J. and Siraj-Blatchford, I. (2003) *A Curriculum Development Guide to ICT in Early Childhood Education*. Stoke-on-Trent: Trentham Books (published in collaboration with Early Education).

Siraj-Blatchford, J. and Whitebread, D. (2003) *Supporting Information and Communications Technology Education in Early Childhood*. Buckingham: Open University Press.

Siraj-Blatchford, I., Sylva, K., Muttock, S., Gilden, R. and Bell, D. (2002) *Effective Pedagogy in the Early Years: Research Report 356*. London: DfES.

Sayeed, Z. and Guerin, E. (2000) *Early Years Play*. London: David Fulton.

Sylva, K., Bruner, J.S. and Genova, P. (1976). 'The role of play in the problem-solving of children 3–5 years old', in Bruner, J., Jolly, A. and Sylva, K. (eds) *Play: Its Role in Evolution and Development*. New York: Basic Books.

Sylva, K., Melhuish, E., Sammons, P., Siraj-Blatchford, I. and Taggart, B. (2000) 'Effective Provision of Pre-school Education project – recent findings'. Presented at the British Educational Research Conference, Cardiff University, September 2000.

Sylva, K., Melhuish, E.C., Sammons, P., Siraj-Blatchford, I. and Taggart, B. (2004) *The Effective Provision of Pre-School Education (EPPE) Project: Technical Paper 12 – The Final Report: Effective Pre-School Education*. London: DfES/Institute of Education, University of London.

Vygotsky, L. (1978) *Mind in Society: The Development of Higher Psychological Processes*. Cambridge, MA: Harvard University Press.

9

USING ICT TO SUPPORT SCIENCE LEARNING OUT OF THE CLASSROOM

Nick Easingwood and John Williams

Introduction

Society in general, and children in particular, are becoming increasingly reliant upon 'virtual', rather than 'real' environments, both for entertainment and education. The latter unquestionably provide opportunities to develop the key scientific skills, based as they are on first-hand experiences of observation, hypothesis and recording; however, these could easily become lost in the virtual world. Although it could be argued that the use of new technologies as an integral part of scientific investigation negates the need for 'old and traditional' scientific investigative skills, the fact remains that practical, first-hand experience is crucially important in good primary science and primary practice in general (DfES 2003a; Chapter 3 in this volume). The key for teachers is to use new technology, and their pupils' sophistication in exploiting its power, to enhance the learning experience.

This chapter will examine how ICT can be used to support learning in 'out-of-school' contexts, focusing on some of the oldest scientific learning environments of all – museums. Because ICT provides interactivity, functionality, personalisation, speed, automation and instant feedback, it enables pupils to gain so much more from a museum visit than they might have done previously. Thus, we will review ideas about learning in museums, exemplify how a museum visit can enhance pupils' learning of primary science, suggest how ICT can enhance pupils' learning in a museum and examine and suggest how digital imaging can be used to enhance pupils' learning of primary science in a museum context.

Learning in museums

It seems clear to us that the focused use of a museum, in which a teacher helps pupils to learn by encouraging them to interact with exhibits in a structured and directed way, can provide a range of learning experiences that simply cannot be simulated accurately or meaningfully elsewhere. Indeed, Howard Gardner recommended that all children engage in museum learning, as it has considerable scope to stimulate their 'multiple intelligences' (Hawkey 2004). Museum learning also has the potential to break down some ideas about teaching and learning that are sometimes associated with schools by those who are not involved in education – namely, that learning must be constrained by a curriculum; that it is a simple acquisition of facts and skills; and that it involves transmission of knowledge from teacher to pupil. Museum educators – to some extent historically (*sic*) free from public scrutiny – have been at liberty to develop their ideas about learning and to engage in alternative approaches (Anderson 1999; MLA 2004). In so doing, many have been strongly influenced by the ideas set out below and drawn from Hawkey (2004).

Hawkey (2001) summarised how thinking about learning has been influential in enhancing museum learning opportunities. Bloom's (1984) *Taxonomy* suggests that learning may occur in any or all of three domains: cognitive, psycho-motor or affective. The cognitive domain is divided into several levels, the lowest of which is factual recall. Appreciating the low level of simple fact presentation has provided museum educators with an impetus to diversify their approach to learning. Gammon (2001) suggested a taxonomy into which museum learning experiences in particular could fall: cognitive, affective, social, skills development and personal. Hooper-Greenhill *et al.* (2003) established a similar taxonomy including the following categories: (a) knowledge and understanding, (b) skills, (c) values and attitudes, (d) enjoyment, inspiration and creativity, and (e) activity, behaviour and progression. These two analyses bear some relation to Gardner's multiple intelligences.

Wider definitions of learning and attempts to describe the learning process sequentially have also been useful. Sharples (2003) described learning as construction of understanding, relating new experiences to existing knowledge. Kolb (1984) attempted to develop such ideas by defining a model of experiential learning, which examines four components of a cycle of learning: immersion in concrete experience, observations and reflections, logical or inductive formation of abstract concepts and generalisations, and empirical testing of the implications of concepts in new situations. He suggested that learners often have strengths in particular components of this cycle, and defined learning styles (accommodator, assimilator, converger and diverger) accordingly. Serrell (1996) identified types of museum learning activities, and the outcomes to those activities, preferred and looked for respectively, by individuals with these learning styles. (Table 9.1)

Table 9.1 Learning styles and preferred activities and outcomes (Serrell 1996)

Learning style	Preferred activities	Outcome sought
Accommodator	Imaginative Trial and error	Hidden meaning
Assimilator	Interpretation that provides facts and sequential ideas	Intellectual comprehension
Converger	Try out theories	Solutions to problems
Diverger	Interpretation that encourages social interaction	Personal meaning

Given the increasing focus on classifying learning styles as visual, auditory and kinaesthetic (e.g. DfES 2003b), there appear to be intuitive and obvious ways in which museum experiences may target students with such styles. For example, the range of visual and 'hands-on' opportunities that could be offered within a museum would appear to appeal to pupils with kinaesthetic and visual styles (Stephenson and Sword 2004). It should be recognised that controversy exists about the applicability and reliability of the range of learning style models.

Another productive approach is to distinguish between theories of learning and theories of knowledge (Hein 1995, 1998), and to keep those classifications in mind when designing learning opportunities:

- Views of knowledge exist on a continuum, the extremes of which are:
  - knowledge is absolute truth
  - knowledge is the creation of the human mind.
- Views of learning exist on a continuum, the extremes of which are:
  - learning is passive with museums' purpose to pour learning into an 'empty vessel' of the mind
  - learning is actively assimilated into existing cognitive structures by the learner.

Table 9.2 shows the four ways in which these views of learning and knowledge can combine to yield four domains of learning.

Table 9.2 Domains of knowledge and learning

Domain	Knowledge	Learning
Didactic	Knowledge is absolute truth	Learning is passive
Heuristic	Knowledge is absolute truth	Learning is constructed from ideas and experience
Constructivist	Knowledge is constructed	Learning is constructed from ideas and experience
Behaviourist	Knowledge is constructed	Learning is passive

To design museum learning opportunities that respond to the ideas above cannot be done using a 'one size fits all' model. Learning from, rather than about, objects, providing the opportunity for a variety of active and enquiry-based learning activities – and structuring and coordinating a range of meaningful learning choices within a particular context – are essential components for success (Hawkey 2004; Johnson and Quinn 2004). Provision of motivating learning experiences that are stimulating, enjoyable and relevant is essential. Embedding such experiences in the interdisciplinary approach facilitated by museums is also more likely to enable pupils to make links between areas of learning (Hawkey 2004).

How can a museum visit enhance pupils' learning of primary science?

A recent case study shows very effectively how the ideas outlined above enabled primary pupils' learning of science during a museum-based investigation (OFSTED 2003; Stephenson and Sword 2004). This investigation concerned the challenges facing ancient civilisations and was cross-curricular, drawing upon science, history, and design and technology. The museum work involved unravelling the story of the granite sarcophagus of Rameses III in the Fitzwilliam Museum at the University of Cambridge. This was followed up with classroom-based investigations, which allowed pupils to test hypotheses developed in the museum.

Stephenson and Sword (2004) found that the museum experience can encourage pupils to think autonomously about science with versatility, imagination, individual creativity and tenacity. Their activity enhanced motivation, and encouraged the development of sophisticated information processing skills (OFSTED 2003). The museum's enhanced funding, by comparison to primary schools, enabled pupils to have access to appropriate authentic materials (inspiring 'awe and wonder'), which appeared to give learning more meaning, particularly as the tasks were situated within a contextualised cross-curricular approach with which pupils could empathise. By using open-ended problem solving, pupils were enabled to engage with learning at a level appropriate to them, facilitating differentiation. In the museum, pupils seemed less likely to pre-judge the 'correct' answer to investigations and to immerse themselves in the learning context.

Structuring the opportunity for focused discussion around museum artefacts, as part of a 'journey of enquiry' (OFSTED 2003) was key to facilitating learning. This 'journey' encouraged pupils to ask questions of those objects in relation to problem solving scenarios that required comparison and close observation, fostering learning and providing the opportunity for development of pupils' creativity. By exploiting the museum's own science educators and practitioners, pupils had access to specialists and specialist information, which could extend their discussions to broaden and deepen their learning.

Pupils' scientific principles and skills were also developed. In the museum, pupils had to search for evidence to support or refute their hypothesis. In the classroom, pupils subsequently carried out investigations, which were still 'situated' within the contexts developed in the museum. Across both locations, they were asked to think creatively to explain how things worked, to test and refine their ideas in the classroom and museum, to present their ideas to their peers and to interrogate each others' solutions; this developed their skills of reflection and evaluation and helped them to review their own learning. The group work involved in the whole process gave ample opportunity to assess pupils' learning.

How can ICT enhance pupils' learning in a museum context?

Museums have had a dual role in scholarship and education since their inception. These two roles have come together in recent years and this fusion is being facilitated by ICT, which is increasingly being used to enhance learning, both of schoolchildren and of lifelong learners (Hawkey 2004).

Of course, the lifeblood of the educative role of museums has traditionally been exhibited objects. Such artefacts can be awe-inspiring (such as the *Flying Scotsman*); alternatively they can challenge visitors to compare and contrast reality with popular image. For example, the popular image of George Stephenson's *Rocket* is of a large, yellow locomotive; yet modern-day visitors to the exhibit in the Science Museum in London will see a small, black locomotive, with a very rough iron outer casing.

ICT can enhance learning from such artefacts, both by facilitating and accelerating traditional learning approaches, but also by expanding the range of learning experiences available (Hawkey 2004). Museum education officers have increasingly tried to design exhibitions and experiences based upon defined learning objectives, interactivity, learner participation and collaboration, and the facilitation of learner initiative in interacting with exhibits. Although this has taken place both with and without the aid of ICT, ICT has certainly facilitated the process, and has helped to develop museums as places of exploration and discovery (Hawkey 2004). In fact, ICT may have become so pervasive so quickly in museum education because learning from museums and learning from digital technologies, share many of the same attributes, including learning from objects, rather than about them, and developing strategies for discovering information, rather than being presented with the information itself (Hawkey 2004).

ICT has considerable power to enable pupils' learning in museums because it can facilitate interactivity and participation, collaboration between learners (both onsite and online) to construct ideas and the personalisation of learning experiences to account for prior knowledge and learners' preferences (Hein 1990; Hawkey 2004); it may also exploit mobile technologies (Naismith *et al.* 2004). For example, personalisation of pupils' experiences may begin even before they reach the museum. Even using the museum's website (including maps, gallery information and virtual tours) to help plan their route around the museum, in response to activities suggested by the teacher, gives pupils immediate autonomy and enables them to make choices about their learning in the museum. However, other technology may also benefit learning in the museum context, including still and moving images (video and animations), simulations and presentations, games, and the increasing use of the Internet and the museums' own intranets (Littlejohn and Higginson 2003). Of these, the use of digital imaging appears to have very considerable potential.

How can digital imaging enhance pupils' learning of primary science within a museum context?

It is almost part of folklore that, in the past, pupils had to make a written description of every educational visit, with such writing often containing little reflection or analysis upon what had been seen. Although creative and imaginative teachers have always found alternative means of getting children to analyse, record and report educational visits, ICT can bring new opportunities that were largely unimaginable ten years ago. Central to this is the use of digital imaging.

Still images

Effective use of 'still' images must start with appreciation that the original capturing of the image or taking the photograph is no longer the end of the process, but the beginning of it. From here the images can be inserted into different types of applications; for example, they may be used in a multimedia presentation or web page, where children can point and click on hyperlinks to link to another slide. This means that information is not presented in a linear way, which in turn means that an element of creativity and imagination has to be employed by the designer and an element of choice by the user. Unsurprisingly, it is important for pupils to take account of their audience when designing such a website or presentation.

The STEM project at the Science Museum in London worked with primary pupils to help them design a website, based around their museum visit, which exploited digital images in this way. The children designing the website had to be able to analyse and synthesise what they had seen in order to present ideas in a manageable form and users commonly needed to be able to think laterally in order to exercise an element of choice. This means that both designer and user have to exercise higher-order thinking skills, an immediate validation of the use of ICT in this context. Previously the 'designer' would have written a simple report or description of what had been seen and the user would have simply read what was written. However, the use of web-page design or multimedia presentation ensures there is interactivity and engagement with both roles.

Much presentation software also enables further functionality to add to pupils' analysis and teachers' assessment of their learning. Examples include the embellishment and annotation of images by the use of callouts (speech bubbles) and draw tools and the embellishment of reports as a whole by the addition of a commentary, which itself could be recorded by the pupils in the museum.

Video images

Digital video is video that can be stored, manipulated and edited on computer. Digital video cameras can record museum experiences more effectively than still cameras and have advantages over analogue video cameras (Becta 2003) for the following reasons:

- digital cameras are smaller and lighter than VHS cameras, facilitating their use in a mobile museum context;
- picture quality is enhanced;
- digital video is easy to edit, enabling students to produce high-quality films in a short time;
- digital video can be integrated with other forms of technology, such as presentation software and the Internet.

Because digital video editing software is now so accessible and ubiquitous (for example, iMovie and Windows Movie Maker are both free with Mac and Windows operating systems respectively), primary school pupils now have access to functionality that until recently was restricted to professional television and film makers. This means that a child can video aspects of a museum visit, or indeed any 'out-of-school' work, and on returning to school can download the recorded 'footage' into a computer and then edit the movie into a manageable form for viewing. This could include adding a soundtrack such as narration, music and sound effects, as well as titles and transitions between clips. By dragging, cutting, copying and pasting into a storyboard at the bottom of the screen, children can edit and create a complete film in the same way that they can edit a piece of word-processed text. This flexibility means that the video can subsequently be used in similar ways to those images captured with a 'still' camera, as described above.

Making digital videos centred on museum visits appears particularly suited to enabling pupils' learning because many of the learning opportunities provided mirror – to some extent – those of the museum itself (Becta 2003). Digital video lends itself to cross-curricular activities (Becta 2004), facilitating the interdisciplinary approach highlighted as important earlier and exemplified in the case study (Stephenson and Sword 2004). Making digital videos can enhance motivation, enjoyment and self-esteem (Burn and Reed 1999; Ryan 2002), is more likely to draw on pupils' out-of-school interests (Parker 2002) and can enable self-expression and creativity (Becta 2002). Its motivational effects are exemplified by the length of extra time spent on digital video projects by students. Digital video also enables differentiation according to students' learning styles and attainment levels (Burn and Reed 1999) and removes literacy difficulties as an obstacle to learning. For example, rather than capturing and analysing data on paper, or recording their museum visit through words, pupils can now make simple records using moving images (Becta 2002, 2003). The process of working collaboratively in groups to produce and edit digital video encourages learning through discussion and problem-solving (Becta 2002) and encourages children to think about their learning (Swain *et al.* 2003). The flexibility afforded by digital video software and its timeline also allows students to draft and redraft sequences quickly and easily, encouraging creative experimentation (Buckingham *et al.* 1999; Burn and Reed 1999), and developing understanding of narrative and structuring of scientific argument (Becta 2002).

Despite the value that making a video may have for pupils' learning, the audience to the final product will usually take on the role of a passive viewer with little opportunity for interaction. Although still useful, enabling interactive engagement with the video can also help to maximise the learning of the viewer. For example, when inserted into a presentation, the user must point and click to select a video clip. The nature of this type of activity will mean that the clips will be shorter, and the investigative skills developed will engage the viewer with the material throughout, maintaining concentration more effectively.

To use digital video successfully within a museum visit will require planning and preparation by the teacher (Becta 2002). Hardware issues include the requirement to have modern computers with high-capacity hard drives and fast processing speeds (Yao and Ouyang 2001), enough digital video cameras for each group in the class and the facility to save products to external media, such as CD or DVD writers or USB memory sticks. Frequency of use by pupils is important, particularly if a specific subject focus is to be emphasised (Becta 2002), and the skills required to use the cameras effectively and edit the product clearly need to be taught. This might involve the addition of titles, soundtracks, fades in and out and special effects. Elements of production are also important. As the children become more familiar with the art of movie making, they will learn how to create movies that engage and keep the viewer interested, e.g. by using several different camera angles or 'cut-aways'. There is evidence that making a film for an audience, such as parents or peers, maximises the benefits to motivation and self-esteem (Buckingham *et al.* 1999). It is also important to ensure that the children record short clips of just a few seconds. This is because it is easier to edit short clips than longer ones and it will reduce download times. Key teaching questions include: 'Why are you recording that?' and 'How do you hope that it will fit with your final presentation?'

Exploiting the benefits of digital video in the context of museum learning of primary science requires a structured approach by the teacher. The stages of implementation include:

1. preparation, which is likely to occur before the visit;
2. identifying an audience;
3. producing storyboards and flowcharts;
4. making the film at the museum;
5. editing the film in school;
6. showing the film or using it as part of a presentation.

Evaluation of the outcomes by peers is an important part of the learning process (Becta 2003).

If we take primary science to include not only scientific observation and experimentation, but also aspects of role play and drama, then we can include not only the historical artefacts in the museum itself but also the scientists and engineers that first discovered or invented them. For example, in the past, personalities such as 'Doctor Who' have been used to take the audience back through time to meet such notables as Galileo, Newton, Faraday and Darwin (Williams 2000). Providing pupils with the opportunity to use museum websites to plan their approach to the science, to storyboard such a drama prior to the visit and then to use the museum for filming 'in role' and 'in situ' provides an excellent foundation for learning.

Another important example addresses the need to encourage and develop children's investigative skills. Many museums now have inter-active investigative galleries aimed directly at school pupils. A good example of this would be Launch Pad in the Science Museum in London. Using digital video and other relevant ICT to record quantitative and qualitative results of investigations enables pupils and teachers to exploit the learning benefits outlined above and to provide a springboard for fur-ther learning. For example, during the case study of Stephenson and Sword (2004), pupils could have used digital video to record information collected in the museum, and then to produce a film of the whole investi-gative process. Again, as with any use of ICT, it is essential in this example to provide pupils with tightly focused tasks, which stem from tightly struc-tured prior preparation in school and from regular progress reviews. Expecting pupils to arrive at a museum to record a digital video 'ad-hoc' is misguided and a waste of an effective learning opportunity.

Concluding and looking forward

Using digital imaging to record a museum visit enables pupils to learn during the visit, provides a record of what they saw after the visit and gives further opportunities for follow-up learning back in school. Such benefits would apply to other educational visits as well. Of course, the museum will often have its own website, or may produce a CD-ROM which can be used in the same way, but whatever type of ICT is used it is important to remember that all of the ICT mentioned in this chapter is relatively easy to use. Although time would be required to learn how to use the hardware and software, such skills can subsequently be reinforced and extended in much the same way as the other skills that a child learns and develops during the primary phase. This is a sound investment of time and resources. Combination of these new technologies alongside the oldest and most traditional of scientific environments, by an imaginative and creative teacher, can access levels of learning previously unimaginable.

So what opportunities will be available for exploitation by creative primary science teachers in the museums of the future? Mobile resources appear to have considerable potential to enhance personalisation and interaction in the way that visitors learn from museums. Indeed, the functionality of PDAs, mobile phones and digital cameras are already beginning to overlap and mobile resources of the future are likely to have ever-increasing computing power, enabling fully functional interaction with the Internet and access to communication networks, on the move. The integration of 'context-aware' functionality in mobile devices, which provides information to users according to their location, is becoming increasingly visible in museums and other centres of 'informal' learning, enabling some personalisation and direction to a visitor's learning experi-ence and enhancing the exploitation of novel learning experiences such as augmented reality gaming (Naismith *et al.* 2004). The ability to interact and collaborate both with other visitors (such as classmates in the museum) and extended groups of learners across more and more extensive learning networks (such as classmates back at school, or with pupils from other schools) is also ripe for exploitation (Naismith *et al.* 2004; Chapter 10 in this volume). Pupils are also likely to see an increase in what are becoming known as tangible technologies, part of the ubiquitous computing vision (Weiser 1991) in which technology becomes part of the environment and within which inputs (which conventionally were made via a mouse or a keyboard) become more physical, and more closely tied to outputs. Examples could include augmented museum displays, in which a soundtrack is initiated by moving a hand over some text, or simply by moving towards an exhibit; or exhibits in which visitors can manipulate physical objects to have a digital effect, for example on a simulation (O'Malley and Stanton Fraser 2005).

This chapter began with a statement about the importance of first-hand scientific experience for all pupils. We have described in detail the benefits of digital imaging, and examined some of the opportunities provided by ICT in museum education, but perhaps we should now remind ourselves of the primary reason for a museum visit. It is by visiting museums that most children will have direct contact with science and with the science that has led to the technological advances associated with the rise of numerous civilisations. Museums have changed considerably over the years. Not so long ago they were just collections of artefacts, models and specimens. Indeed, we still remember the first 'hands-on' exhibits, which caused much excitement because for the first time children could actually work machines, or by pressing a button actually observe some biological processes in action. Since then we have had specially designed ecological galleries which show the specimens in their natural environment. Today, we even have a dinosaur that moves (but only from side to side) and makes sounds (which are rather unlikely to be authentic!). Although more and more museums have become far more engaging, the balance between learning and entertainment may still need refining, and the work of Stephenson and Sword (2004) makes clear the very great continuing potential for developing science activity and engagement in 'traditional' museums. In all cases, however, the way in which teachers and museum educators exploit ICT will be a key feature in getting that balance right.

References

Anderson, D. (1999) *A Common Wealth: Museums in the Learning Age*. London: Department of Culture, Media and Sport.

Becta (2002) *Evaluation Report of the BECTA Digital Video Pilot Project*. Coventry: Becta.

Becta (2003) *What the Research Says about Digital Video in Teaching and Learning.* Coventry: Becta.

Becta (2004) *Using Digital Video Assets Across the Curriculum (CD).* Coventry: Becta.

Bloom, B.S. (1984) *Taxonomy of Educational Objectives.* Boston, MA: Allyn and Bacon.

Buckingham, D., Harvey, I., Sefton-Green, J. (1999) 'The difference is digital. Digital technology and student media production', *Convergence,* 5: 10–20.

Burn, A. and Reed, K. (1999) 'Digi-teens: media literacies and digital technologies in the secondary classroom', *English in Education,* 33: 5–20.

Department for Education and Skills (DfES) (2003a) *Excellence and Enjoyment: A Strategy for Primary Schools.* London: HMSO.

DFES (2003b) *Conditions for Learning 10: Learning Styles.* London: HMSO.

DfES (2004) *Excellence and Enjoyment: Learning and Teaching in the Primary Years: Understanding How Learning Develops.* London: HMSO.

Gammon, B. (2001) *Assessing Learning in Museum Environments: A Practical Guide for Museum Educators.* London: Science Museum.

Hawkey, R. (2001) 'Innovation, inspiration, interpretation: museums, science and learning', *Ways of Knowing Journal,* 1 (1): 23–31.

Hawkey, R. (2004) *Report 9: Learning with Digital Technologies in Museums, Science Centres and Galleries.* Bristol: NESTA Futurelab.

Hein, G. (1995) The constructivist museum, *Journal for Education in Museums,* 16: 21–23.

Hein, G. (1998) *Learning in the Museum.* London: Routledge.

Hein, H. (1990) *The Exploratorium: The Museum as Laboratory.* Washington: Smithsonian Institution Press.

Hooper-Greenhill, E., Dodd, J., Moussouri, R. *et al.* (2003) 'Measuring the outcomes and impact of learning in museums, archives and libraries'. End of project paper for the Learning Impact and Research Project. Leicester: Research Centre for Museums and Galleries.

Johnson, C. and Quin, M. (2004) *Learning in Science and Discovery Centres; Science Centre Impact Study.* Washington: ASTC.

Kolb, D.A. (1984) *Experiential Learning.* Englewood Cliffs, NJ: Prentice Hall.

Littlejohn, A. and Higginson, C. (2003) *A Guide for Teachers (e-learning series number 3).* London: Learning and Teaching Support Network.

MLA (Museums, Libraries and Archives Council) (2004) *Inspiring Learning for All.* London: MLA.

Naismith, L., Lonsdale, P., Vavoula, G. and Sharples, M. (2004) *Report 11: Literature Review in Mobile Technologies and Learning.* Bristol: NESTA Futurelab.

Office for Standards in Education (OFSTED) (2003) *Expecting the Unexpected: Learning and Teaching in the Primary Years.* London: HMSO.

O'Malley, C. and Stanton Fraser, D. (2005) *Report 12: Literature Review in Learning with Tangible Technologies.* Bristol: NESTA Futurelab.

Parker, D. (2002) 'Show us a story: an overview of recent research and resource development work at the British Film Institute', *English in Education,* 36: 38–44.

Ryan, S. (2002) 'Digital video: using technology to improve learner motivation', *Modern English Teacher,* 11: 72–75.

Serrell, B. (1996) *Exhibit Labels: An Interpretive Approach.* Lanham, MD: Altamira Press.

Sharples, M. (2003) 'Disruptive devices: mobile technology for conversational learning', *International Journal of Continuing Engineering Education and Lifelong Learning*, 12: 504–520.

Stephenson, P. and Sword, F. (2004) 'The ancients' appliance of science', *Primary Science Review*, 84: 22–25.

Swain, C., Sharpe, R. and Dawson, K. (2003) 'Using digital video to study history', *Social Education*, 69: 154–157.

Weiser, M. (1991) The computer for the 21st century, *Scientific American*, 265: 94–104.

Williams, J. (2000) 'Galileo after the trial: a short play', *Breakthrough*, 2 (3).

Williams, J. and Easingwood, N. (eds) (2003) *ICT and Primary Science*. London: Routledge Falmer.

Williams, J. and Easingwood, N. (2006) 'Possibilities and practicalities – planning, teaching, and learning science with ICT', in Warwick, P., Wilson, E. and Winterbottom, M. (eds) *Teaching and Learning Primary Science with ICT.* Maidenhead: Open University Press.

Yao, J.E. and Ouyang, J.R. (2001) 'Digital video: what should teachers know?' Paper presented to the Society for Information Technology and Teacher Education International Conference.

———— **10** ————

VIRTUAL LEARNING IN
PRIMARY SCIENCE

Helena Gillespie

Introduction

Since the mid-1980s the Internet has been connected with science teaching and learning. Originally the invention of professional scientists, the Internet was first used to communicate findings and ideas. Since then, its use in teaching and learning in schools has become common in all phases and subjects. Some would argue that the future of education will continue to be substantially affected by what the Internet can do.

However, the case for computer use in general, and Internet use in particular, has yet to convince some teachers and educationalists. Substantial funds have been invested in ICT in schools over the past few years (in UK schools, the total investment for 2005/06 is in the region of £700 million). Despite this investment, some studies (Harrison *et al.* 2002) have pointed to the difficulty of finding clear and conclusive evidence that ICT can enhance teaching and learning.

It is increasingly clear, however, that teachers, teacher trainers, academics, administrators, advisory teachers and members of government believe that computers can positively affect teaching and learning. In 2004, Charles Clarke, then Secretary of State for education in the UK, asserted that although the 'potential for transformation remains largely untapped',

ICT can 'undoubtedly' be beneficial for education (DfES 2003). This chapter will examine what this potential for transformation might be in the context of virtual learning environments (VLEs) and how the transformation might take place in primary science education.

In 2005 the Joint Information Systems Committee (JISC) set out their requirements for a VLE. Elements include a mapped curriculum, based on some sort of electronically delivered content, with the ability to track student activity with a communications system. However, whether this is a comprehensive and universal description is open to question. In essence, a VLE is a place where learning takes place via the Internet. It is different from a web page because it brings together resources, allows communication and has the tools to enable teachers to track how the resources are being used.

Very few primary schools are currently using VLEs, although they are widespread in higher and further education and are becoming increasingly common in secondary education. The intention of this chapter is to look to the future and to examine the potential of VLE use in primary schools, drawing upon what we have learnt about VLEs from their use in higher and secondary education. The 'real transformation' (DfES 2003) should be to the benefit of real, authentic, meaningful primary science teaching and learning.

Identifying the potential of virtual learning environments

Since the end of the 1990s, there have been a number of enthusiasts who have suggested ways to bring together content and communication electronically, not only to support traditional face-to-face pedagogies, but to enable new types of learning to take place over the Internet (Laurillard 2002; Pittinsky 2002; Salmon 2004).

Laurillard (2002) champions an approach to learning technology which begins with a consideration of how students learn best. A 'conversational framework' should be constructed using the technology to support learning. Laurillard (2002) also considers the different media of teaching, which she calls narrative, interactive, adaptive, communicative and productive. In this way, the process of learning becomes central to the potential for the use of technology in the classroom.

In *The Wired Tower*, Pittinsky (2002) sets out the pedagogical, theoretical and economic case for the use of VLEs in higher education. In particular, the idea that higher education can be delivered effectively by the 'brick and click' method advocated by Levine (Pittinsky 2002) supports the theory that virtual learning can be delivered alongside traditional learning in a single programme.

Salmon (2004) developed an approach to teaching and learning using VLEs which she calls e-moderating. This is the bringing together of online databases (such as library catalogues and archives), digital teaching materials and communications – either synchronous (happening at the same time) or asynchronous (happening in the same place but at different times). Her ideas about how these elements might combine is exemplified as a '5-stage model' (Figure 10.1), where access to and motivation to use a VLE lead to online socialisation and information exchange and then to knowledge construction and further development. In this way Salmon shows how online learning can be truly interactive and enable learners to develop new ideas.

The enthusiasm of these leaders in virtual learning is grounded firmly in pedagogical thinking. Rather than developing technology for its own sake, such thinking is likely to continue to have the greatest impact on development of VLE usage in higher, secondary and primary education.

Exploiting the potential of virtual learning environments

A recent literature review (Becta 2003a) outlines the stage of VLE development throughout education. In the higher education sector VLEs are

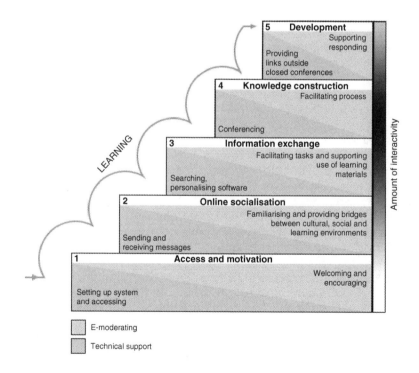

Figure 10.1 Salmon's 5-stage model of e-moderating (Salmon 2004).

now reasonably common. Some institutions have bought 'off the peg' solutions and others have introduced their own in-house solutions. JISC is funding developments in this area, transferring from VLEs to MLEs (managed learning environments), where the VLE works in a connected way with other in-house data and systems such as the student database and library services, through a single portal. In a study of MLE activity in further and higher education (JISC 2002) significant levels of MLE development activity were evident in all the institutions in the survey, with four in five further and higher education institutions using a VLE.

The Becta report on VLEs describes the VLEs in the schools sector as both immature and volatile (Becta 2003a). At the time of writing, research about VLEs at the level of compulsory education was very limited and inconsistent, but recent developments in the broadband network, along with the work of Regional Broadband Consortia, have begun to give UK schools access to VLE technology. However, there is still some work to be done before the available technology becomes successfully embedded in the pedagogical approach of schools. This is most likely to happen in the secondary sector first, where the text-based nature of VLEs is likely to be more appropriate for learners and where schools have the dedicated ICT personnel necessary to drive forward such innovations.

By contrast, VLEs are rare in primary schools. However, many primary schools are beginning to use the *constituents* of a VLE separately in various ways. E-mail communication and online discussion can be restricted by problems with connectivity, but has substantial potential; the Internet is often used as a source of information and as more primary schools have effective connections to a broadband network more children can access quality online learning resources (Murphy 2003).

Use of a VLE is beneficial for communication, databases and the delivery of resources because the teacher can present, edit and shape the learning tools and resources to suit their purposes. Imagine a teacher takes her Year 4 class to the school resources centre, gives the children a topic to research, then asks them to talk together about what they have learned and produce a presentation as an outcome. Under these circumstances the learners are being offered the resources and given the task but it is difficult for the teacher to monitor and intervene in the learning at each step.

A VLE enables the teacher to have far more control of the task through the creation of 'learning units'. Teachers can select the web resources pupils will use and enable and monitor communication about what has been found out. When presentations are made they can be shared electronically. Thus a VLE can 'repackage' the learning experience to make it more focused and enable the teacher to monitor and intervene far more effectively than if the resources are used separately. The relevant potential uses of a VLE are shown in Tables 10.1, 10.2, 10.3 and 10.4 below.

Table 10.1 Effective usage of a VLE to facilitate communication (a) through e-mail, and (b) through a discussion board

(a)

Aim	To e-mail.
What can VLEs do?	Provide secure e-mail facilities to communicate as individuals or groups.
Examples of effective use	E-mail can be used in a variety of ways to support learning. Not just for communication between learners and their teacher, but also to communicate more widely with the science communities, perhaps even globally (Murphy 2003)

(b)

Aim	To use a discussion board.
What can VLEs do?	Provide a space for learners and/or teachers to discuss the topic at hand.
Examples of effective use	This has the advantage over a 'face-to-face' discussion in that it can be reread and added to, therefore deepening the level of reflection. A teacher might ask learners to use a particular set of resources as part of their project and learners might post messages to the discussions about the usefulness of the resources.
	In this way a range of learning styles and skills are supported by the use of the discussion board enabling the learners to develop a range of deeper and strategic learning styles (Gibbs 1999)

Table 10.2 Effective usage of a VLE to access databases and other resources (a) to access a library catalogue, and (b) to work with computer simulations

(a)

Aim	To access a library catalogue, and other online resources.
What can VLEs do?	Provide direct access to the relevant part of online databases and resources, 'packaged' with tasks and selected by teachers, to meet groups' and individual needs, into 'learning objects'.
Examples of effective use	Online databases of such things as History resources can be used in research projects. Teachers can direct learners to the relevant parts of the database, rather than have them sort through layers. This saves time and reduces the possibility of learners going 'off task'. Increasingly this idea of construction of 'learning objects' is progressing and more complex packages of learning materials are being developed, including those which can be tailored to individual needs.

(b)

Aim	Work with computer simulations.
What can VLEs do?	Provide access to teacher-created simulations.
Examples of effective use	Teachers and pupils can create simulations of events in video or animation software, which can allow learners to experiment with ideas and 'walk through' situations and work creatively. The National Endowment for Science, Technology and the Arts (NESTA) have funded a range of projects in the field of ICT and learning, including Sodaplay (NESTA 2005) which is designed to allow pupils in primary and secondary schools to create simulations as a design tool or to create science scenarios through modelling using virtual springs.

Table 10.3 Effective usage of a VLE in using presentation technology

Aim	Present findings and share outcomes with others.
What can VLEs do?	Provide file exchange and viewing systems for work to be transferred between teachers and learners.
Examples of effective use	VLEs provide tools for teachers and learners to communicate about work produced in flexible ways. Using e-learning portfolios, learners can construct their own areas to display written, pictorial and multimedia work. Teachers can access these when learners need support and comment on work in progress. This method of teaching, which is supported with formative assessment, is useful to support learners' individual needs. The Becta quality framework for e-learning resources (Becta 2005) endorses this approach.

Table 10.4 Effective usage of a VLE to facilitate assessment

Aim	To assess learners' understanding and knowledge.
What can VLEs do?	Provide teachers with tools to assess pupils learning through tests and quizzes, which can give immediate and formative feedback, or serve as end of unit assessments.
Examples of effective use	E-assessment is a rapidly growing field. Well-constructed e-assessment can support and augment effective practice (Becta 2005). There are some straightforward ways in which VLEs can be used to deliver tests made up of multiple choice, ordering or matching exercises. However there are also some challenges for e-assessment, where it might be developed to assess metacognition and thinking styles via simulated group work.

How can VLEs support effective learning in primary science?

The vast majority of teaching of science in UK schools is reported as being satisfactory or better (96 per cent at Key Stage 1 and 95 per cent at Key Stage 2 – OFSTED 2004), yet much of it is very tightly focused and doesn't allow for links to be made from one part of the science curriculum to another or to other subjects. According to OFSTED, effective teaching and learning in science is characterised by pupils being actively involved in thinking and carrying out scientific enquiry. Perhaps this priority might best be achieved as the creative potential and possibilities of practical cross-curricular working (inspired by the Primary Strategy in UK schools – DfES 2004) are explored. Flexibility in teaching approaches seems central here (OFSTED 2004).

Clearly, it is not just OFSTED that asserts that good teaching in primary science is closely built around the investigative process. The Association for Science Education's (ASE) journal *Primary Science Review* (PSR) reflects good practice associated with an active approach to learning in the primary classroom. A good example is presented in Robertson's review of the 'Let's Think' programme (Robertson 2004). Here, the theme of a practical approach to science closely allied to the investigative process is evident, with a focus on children's ability to hypothesise, discuss and draw conclusions about scientific ideas (Rowell 2004).

Accepting that practical investigative skills really should be at the centre of teaching and learning of primary science means teachers must try to give pupils opportunities to do the following, as set out in the National Curriculum (DfEE/QCA 1999):

- Ask and answer questions
- Observe and measure
- Recognise a fair test
- Follow instructions to control risks
- Explore
- Compare and consider
- Communicate

The crucial question is how the use of virtual learning might support this. To provide an answer we must first consider the prerequisite hardware required to access VLEs in the primary classroom. Having done so, we can examine ways in which VLEs might be used in the primary classroom to facilitate investigative learning.

Using hardware to access virtual learning

Interactive whiteboards (IWBs) have become an ICT 'essential' in a very short time. With funding dedicated to installing them in classrooms in every school in the UK through the Standards Fund (DfES 2003) and research showing that they can have benefits for both pupil motivation and teaching strategies (Becta 2003b), it may not be long before most teachers have access to this type of technology. In addition, with the increase in wireless technology, portable ICT devices such as laptops and – crucially for the primary school sector – tablet PCs are increasingly common. Used together, these devices allow primary aged pupils to see and interact with online resources in ways that were not previously possible. The ability to use 'touch screen technology', both in groups using an IWB and as individuals using a tablet PC, means that children need not wrestle with input devices such as mice or keyboards which are designed for adults. Small keyboards with fewer keys and pen technology mean that even when text is needed, there are fewer possibilities for mistakes to be made. In short, tablets, laptops and IWBs have made virtual learning more accessible.

In addition, increasing bandwidth has led to the development of web resources that are more suited to primary aged children because they utilise still and moving images to support learning. More imaginative website design that exploits words and icons also means that web-based resources are less reliant on text. Thus the multi-modal nature of the tools that might engage children's learning is strongly emphasised. These developments, coupled with the developments in hardware, mean that the technology is well suited to the introduction of more virtual learning in the primary classroom.

As with all technological developments, teachers will only really integrate them into their practice where real benefits for teaching and learning can be seen. The answer lies in the link between the underlying themes of this chapter: the uses of VLEs, the developments in hardware provision and good practice in science teaching pedagogy, based around the practical skills of scientific enquiry.

A VLE in primary science

The push to get broadband into UK schools has led the Regional Broadband Consortia to investigate what kind of teaching and learning tools can utilise the power of broadband, not simply by allowing schools to access the Internet quickly but also by making full use of the available bandwidth. Most of these consortia are now providing a VLE with a range of content, such as access to video and audio resources, as well as the opportunity to create individually tailored learning units and objects for pupils.

Like websites, VLEs have a homepage which children would see when they log on. This interface can be easily changed to reflect the users, using text or pictures to indicate links and adding or reducing the tools available to users as required. In addition to the notice board, pupils' files, a calendar, students' folios and text and image files known as 'learning objects' are shown in Figure 10.2.

Indeed, one of the most powerful tools of the VLE is the ability to create learning objects. In essence, this is a way to package up information, images and web links so pupils can access all they need from one page. This has advantages in that it saves time and keeps pupils on task. In its simplest form, the learning object would contain a question or task and a link or picture to use in answering the question (for example, Figure 10.3). Pupils can then either use paper or digital media to record their findings and post their responses to the teacher via the VLE.

E-mail is a powerful tool for communication and it is also one of the

Figure 10.2   Example homepage for a VLE (created via the Netmedia Virtual
Learning Environment).

simplest and most useful tools available via the VLE. This is a simple way
for pupils and teachers to communicate with one another in a secure
environment. A teacher might e-mail the class to remind them of homework or an assignment. Alternatively, imagine a teacher is working with
a mixed Key Stage 2 class on a project about animals and finds a useful
website. She e-mails this site as a link to all pupils (Figure 10.4), who are
able to log on to the VLE at home. This supports their homework for the
week, where they are collecting animal names and trying to classify them.
    Let us consider some other simple examples of how a VLE could be
further used to support primary science, with an emphasis on aspects of
the science enquiry process. They do not represent complex or apparently
'advanced' use of ICT. In fact, using a VLE should make incorporation of
ICT into primary science much more straightforward. However, they do
show the range of opportunities that could be provided by a VLE.

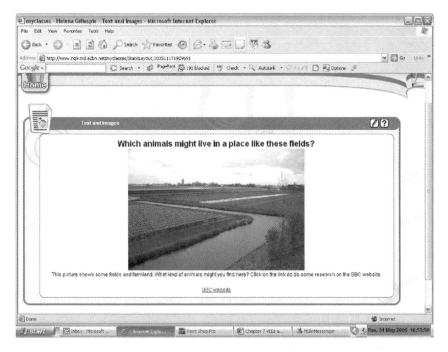

Figure 10.3 A simple learning object containing a question, a link and a picture (created via the Netmedia Virtual Learning Environment).

Ask and answer questions

Year 1 pupils follow a link in the VLE to the BBC website where they play on a science simulation game about forces. After a 20-minute session, working in pairs, their teacher asks them to work with their talking partners to come up with a question about what they have seen and learned on the site. They share these questions with the rest of the class and then decide on a question they can investigate as part of their practical science.

Observe and measure

Each week, children in Year 3 who are investigating the growth of plants over a period of time photograph a bean plant, a sunflower and some cress, all grown from seed. These photos are then put into three separate Power-Point presentations and the children observe the changes which happen over the weeks by viewing the presentations via the VLE.

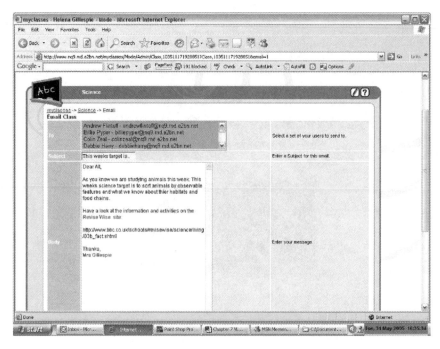

Figure 10.4   E-mail to pupils (created via the Netmedia Virtual Learning environment).

Recognise a fair test

Year 2 and Year 3 pupils are working on devising a fair test. The teacher constructs a simple investigation, rolling some cars down a ramp to see which car goes furthest. He videos the investigation three times, once as a fair test, once where he changes the height of the ramp and once where he varies the covering on the ramp as well as the car. The children view these videos via the VLE and are asked to say which is the fair test and why. After this activity the teacher introduces the children to the idea of simple variables and the children watch the videos again, this time naming the variables in each investigation and saying which have changed.

Follow instructions to control risks

Year 6 pupils are planning investigations into their topic on micro-organisms. Using links on the VLE they research practical investigations that could otherwise be harmful. They then share their findings on a discussion board and the teacher uses excerpts from this discussion to construct a list of 'dos' and don'ts' for the topic.

Explore

Year 4 pupils in an urban school could investigate their school habitat. They use a link on the VLE which takes them directly to the BBC's 'bird-cam' where they can compare the birds they see in their school with the countryside-based birdcam. They keep records of what they see via the VLE's discussion board, which logs dates and times birds are spotted as messages are posted.

Compare and consider

Groups of Year 5 pupils construct a simple, one-page PowerPoint presentation about what they have found out about the effects of the sun and moon on the Earth. They use links to the Science Museum website via the VLE as a starting point for their research. Their finished slides are linked together into a presentation by the teacher, who then posts this to the VLE. In a subsequent lesson, the pupils are asked to summarise in pairs what groups have found out.

Communicate

A teacher of a Year 4 class is working on a project about habitats around the world. Linking up with teachers in different countries, the class exchange e-mails and information about plants and animals in their local area.

Conclusions

Good use of information and communications technology in teaching does not have to use complex hardware or applications. Virtual learning environments *can* make using technology in primary science teaching simpler, by collecting together resources to support research, by enabling pupils to interact with resources such as moving images and by encouraging communication and collaboration. But the question is *will* they and, if so, *when*?

The answers are not straightforward. Barriers to successful integration of virtual learning still exist in areas like the professional development of teachers, availability of suitable hardware and even in the curriculum and assessment systems currently in place. However, if these barriers can be overcome the possibilities for virtual learning in primary science are diverse and numerous and could help to support the best practical science teaching, which focuses on the investigative process and on children engaging in real science learning.

References

Becta (2003a) *A Review of the Research Literature on the Use of Managed Learning Environments and Virtual Learning Environments in Education, and a Consideration of the Implications for Schools in the United Kingdom*. Coventry: Becta.

Becta (2003b) *What the Research Says about Interactive Whiteboards*. Coventry: Becta.

Becta (2005) *BECTA's View. A Quality Framework for E-learning Resources*. Coventry: Becta.

Department for Education and Employment (DfEE)/Qualifications and Curriculum Authority (QCA) (1999) *Science – The National Curriculum for England: Key Stages 1–4*. Norwich: HMSO.

Department for Education and Skills (DfES) (2003) *Fulfilling the Potential*. London: DfES.

DfES (2004) *The National Primary Strategy. Transforming Teaching and Learning through ICT in Schools*. London: DfES.

Gibbs, G.R. (1999) 'Learning how to learn using a virtual learning environment for philosophy', *Journal of Computer Assisted Learning*, 15: 221–231.

Harrison, C., Comber, C., Fisher, T. *et al.* (2002) *ImpaCT2: The Impact of Information and Communication on Pupil Learning and Attainment. Strand 1 Report*. London: DfES.

JISC (2002) *Managed Learning Environment Activity in Further and Higher Education in the UK*. London: JISC.

JISC (2005) 'Virtual and managed learning environments', http://www.jisc.ac.uk/index.cfm?name=issue_vle_mle (June 2005).

Laurillard, D. (2002) *Rethinking University Teaching: A Framework for the Effective Use of Learning Technologies*. London: Routledge Falmer.

Murphy, C. (2003) *Literature Review in Primary Science and ICT*. Bristol: NESTA Futurelab.

National Endowment for Science, Technology and the Arts (NESTA) (2005) 'News releases', http://www.nesta.org.uk/mediaroom/newsreleases/3863/index.html (June 2005).

Office for Standards in Education (OFSTED) (2004) *OFSTED Subject Reports 2002/03: Science in Primary Schools*. London: HMSO.

Pittinsky, M. (2002) *The Wired Tower: Perspectives on the Impact of the Internet on Higher Education*. Upper Saddle River, NJ: Financial Times/Prentice Hall.

Robertson, A. (2004) 'Let's Think! Two years on', *Primary Science Review*, 82: 4–7.

Rowell, P.M. (2004) 'Why do bees sting? Reflecting on talk in science lessons', *Primary Science Review*, 82: 15–17.

Salmon, G. (2004) *E-moderating: The Key to Teaching and Learning Online (Second Edition)*. London: Kogan Page.

ICT AND PRIMARY SCIENCE – WHERE ARE WE GOING?

Angela McFarlane

From the richness and complexity of human endeavour three domains of knowledge have been selected and privileged above all others to form the core of the UK education system. The study of language, mathematics and science are legally compulsory for all students from the ages of 5 to 16 and attainment in these subjects is the sole measurement by which the success or failure of our primary education system is judged. For this reason, it is worth considering why it is science rather than, say, humanities, creative arts, philosophy or any other field that has been chosen for such special investment, what we as a society hope to achieve through this focus and the extent to which we are indeed doing so.

Castells (1996) in his definitive trilogy *The Information Age: Economy, Society and Culture* sets out his analysis of the rise and implications of 'the network society'. In the early twenty-first century it seems we inhabit a world where economic prosperity and all that depends on this – democracy, health, our very survival and that of the species with which we share our planet and on which we in turn depend – rests on the ability to benefit from the effects of globalisation. According to the 'Globalisation Guide' website (www.globalisationguide.org), 'Globalisation is the rapid increase in cross-border economic, social, technological exchange under conditions of capitalism'. Globalisation is both fuelled by and fuels the unparalleled ability we now have to share information across boundaries of time and geography as a result of the web of communications technologies we share. There is a network of social and economic inter-dependencies which criss-cross our world and depend on communications technologies and a level of connectivity that is unprecedented in human history. The technologies we use to support this web of communication are the result of over a hundred years of development that started, according to the Smithsonian Institute, with Morse's invention of the telegraph in 1837. His was the first machine to transmit information over long distances almost instantaneously. Today this simple notion has developed into the enabler of the so-called 'knowledge economies' where wealth creation depends on an ability to innovate. As a result our technology and information-rich era is also known as the 'knowledge age', where knowledge creation is or is predicted to be the basis of wealth and economic growth in developed nations for the first half of the twenty-first century. As a consequence of this vision there are widespread calls for our education systems to change in order to prepare learners to take their place in a knowledge economy.

It seems that to take a place in this connected world we have decided young people need to be able to use language, work with number and shape, and know about science (OECD 2001). It is interesting to note that the curriculum requires only one language in the case of England – the other global languages including Spanish and Chinese, which are used by as many people as English, being almost entirely ignored in primary education in the UK. It may be that the Anglophone dominance of communications technologies has reinforced this linguistic isolationism, although change is on the horizon as we wake up to the need to work in languages other than English. Currently, however, we are not only limiting the scope of languages our young people experience, we are also taking a very partial view of the necessary competences they need in English. It is the use of the written form of language that has dominated schooling, with other forms of communication such as film or multimedia almost ignored and even speaking and listening skills being seemingly relegated to a poor second place despite their central importance in everyday life. Nowhere is this more evident than in the explosion in the use of voice-dependent technologies such as telephony. Consider how many everyday business exchanges that used to be undertaken in writing are now dealt with entirely by telephone – albeit often through interaction with a semi-automated system. Indeed, the use of voice-based technology is set to undergo a further expansion with the use of voiceover Internet protocols (VIP) making voice communication worldwide cheap and accessible to a much wider user base. Even as I am writing this chapter the Internet search engine company Google has announced their venture into the Internet voice communications arena.

This, then, is something of the background of worldwide development against which structures and developments within the education system might be held to account. Against this background, the following remarks consider the place of science education and, in particular, some of the issues associated with an examination of some of the more anachronistic features of the UK school science curriculum.

The role of science in the curriculum

It seems that science is seen as a necessary preparatory experience for life in a technologically framed world, where innovation and knowledge creation are seen as key to economic success for the individual and the nation. So what is it about science that could have led policy makers to this conclusion?

If the study of science is meant to underpin a technology dependent culture, why is science and not technology itself the core subject? Is it because the 'pure' sciences underpin subjects such as engineering and computer science, biomedical sciences and material science? Will knowledge of the sciences aid an understanding of these more applied fields? The domains of science chosen for the school curriculum, especially that of the primary key stages, do not obviously suggest this. Indeed, it is difficult to infer the logic behind the selection of content in the school science curriculum beyond a clear desire to represent the three traditional school subjects of biology, physics and chemistry. These selections may reflect the personal histories and allegiances of those who wrote the curriculum since they do not necessarily map onto any significant practice of science beyond school. After that an air of stamp collecting invades the UK science curricula, with a smattering of pretty examples from a range of countries in the album. Clear linking themes, or big ideas, or even progression of understanding across the elements or the key stages are, however, sometimes hard to discern.

The identified skill sets behind the curriculum show a welcome coherence in contrast, since they are present in all four key stages. These skills are set out in detail in Chapter 2 and I will not repeat them here. The key skills are predicated on an experimental model of science, where hypotheses are tested through investigation and observation and conclusions drawn based on the evidence accumulated. However, this model of science, and the so-called scientific method, is only one approach to the development of scientific understanding. The use of models in science, as discussed in Chapter 6, is just one alternative. It is not clear why we devote 11 years of schooling to one experimental method, or why even then we do not apparently teach this very well, hypothesising being particularly poorly developed (see the House of Lords 2001 and House of Commons 2002 reports on this topic). As pointed out in Chapter 3, the competencies credited in tests of science learning used in England can equally well be acquired through drill and practice as through an experimental approach. Moreover, there is very good evidence that if it is understanding of scientific content that is the objective, the experimental approach leaves much to be desired (see McFarlane and Sakellariou 2002 for a discussion of this).

A skill set vital to science that is not even mentioned within the defined curriculum is the ability to recognise and take part in reasoned, evidence-based discussion. This may be because this skill set is not unique to science, but is central to an active intellectual life in any knowledge domain in western society. However, there is little evidence of reasoned discussion elsewhere in the curriculum, even as a desired cross-curricular aspiration in the introductory parts of the curriculum orders (which encapsulate many worthy aims but rarely seem to influence practice in teaching or assessment). Fortunately, despite this absence in the curriculum orders, debate and argumentation have been the subject of a small number of highly important research and development projects and are certainly achieving prominence in post–16 science courses, particularly those dealing with bioethics, such as the Salters-Nuffield A-level biology course.

The ability to recognise and distinguish between ideas and beliefs is at the heart of this process and in an ever more complex world is a vital skill set for everyone who ever has to make a choice about the use of technology – either for themselves, a dependant or society at large. We are faced daily with questions about our own behaviours that affect others directly through the process of globalisation – from which brand of coffee to buy, to vaccinating our children, to who we should vote for if we care about climate change policy. All of these issues have at their heart a need to understand and respond to a range of views, arguments and counter-arguments in order to make an informed personal choice. We also need to be able to recognise when we and others make decisions from the head or the heart, using ideas or beliefs, evidence or instinct. This is not about making the right choice, it is about making informed choice; not about being told what to think or do, but to understand how and why we think and act and to take responsibility for the consequences. And all the while to recognise that there will always be a degree of uncertainty, and that there is almost never an entirely risk-free answer.

It will be clear from the above that there is much debate concerning the nature and purpose of school science (see House of Lords 2000). If we consider purpose, is the main purpose of school science to winnow out what will inevitably be a minority for a science-related career, or to prepare all for active participation in a scientifically based culture? Arguably, at the moment the school science curriculum in the UK does neither well and in fact needs to do both, with scientific literacy a requisite for all. We have only to look at the level of science discourse in the popular media to realise that whatever else science education has achieved in the last 100 years, general scientific literacy is not among the accolades we can boast. We have, however, been good in the past at educating science specialists. The

UK leads the world in a range of scientific and technology enterprises as a result, and we must not forget this in the gloom that tends to attach to policy debates around science education. However, even here there is no room for complacency; we have lost ground as undergraduate recruitment stagnates and the numbers taking any science post-16 are not growing.[1]

As I have suggested, it is easy to find evidence of our poor scientific literacy in the popular media, where even on otherwise intellectually robust platforms we daily hear such remarks as 'we need to be able to buy our vitamins free of chemicals' (as in a piece on threatened EU legislation on dietary supplements on the *Today* programme on BBC Radio 4). A recent exchange in a weekend broadsheet was more thought provoking. A short and admittedly light-hearted piece advised readers not to look up information on their health worries on the Internet on a Friday as the result would be a certainty that they did indeed have a terminal complaint. The weekend would then be ruined as they waited and worried until Monday to get the reassurance they needed from a doctor that this was not in fact the case. The following week saw a response from a reader who had secured the treatment she needed for her daughter and avoided the loss of sight in one of her eyes with the aid of information and support she had accessed through the Internet. A rare condition – unlikely to be seen by an individual GP – was diagnosed, a worldwide community of sufferers and their parents joined and consulted, and a child's life changed immeasurably through the use of communications technology.

Surely an objective of good scientific education in the information age should be to equip learners with the skill sets they need to deal with either of the situations described? Indeed, patients turning up with printouts from the Internet is now commonplace for primary healthcare professionals and the 'expert patient' initiative is a web-based project backed by the National Health Service to encourage patients with chronic conditions such as diabetes and arthritis to share information and experience in order to make living with their condition as easy as possible.

ICT and scientific reasoning

The sheer amount of information now available to any individual is enormous and pupils need to be equipped to evaluate it and build personal knowledge. They need to know how to distinguish a statement which may be true (e.g. our sun is 4.5 billion years old) from a fact (e.g. the earth moves around the sun) and how to distinguish the knowledge produced by pseudo-science (e.g. astrology) from science (e.g. astronomy). Moreover, modern society requires citizens to make decisions on many issues related to the cultural implications of scientific achievements (e.g. cloning). For these reasons public understanding of science necessitates that pupils understand not only the content of science, but also its methods. It is argued (Driver *et al.* 1996) that emphasising scientific knowledge is not enough for pupils to be scientifically literate. They need to be introduced to the ways that scientists came to these conclusions.

But the way that scientists come to conclusions is not entirely straight-forward or uniform. Helms (1998: 128) identifies scientific method as all the skills and processes, technologies and tools employed by scientists to gather valid and reliable data in order to verify, falsify or formulate a theory. This is very similar to the model that the UK National Curriculum identifies above others. Other authors (Hodson 1985; Driver *et al.* 1996; Leach 1998) argue that epistemology shows that there is not a single method or an 'algorithm' (Millar 1996: 15) that scientists follow in order to solve a scientific problem. Some scientists perform experiments whereas others do not. For instance, astronomers cannot intervene to conduct an experiment since they are only able to see what happened in the past (sometimes millions of years ago) in systems they cannot possibly influence. In addition, while some scientists develop a theory after experimentation, sometimes theories come first and experimentation supports or disproves the theory later.

The above examples illustrate the diversity of strategies that real scientists employ and also that scientific method cannot be templated. Therefore, if there is not any simple algorithm which describes sufficiently the ways that scientists work, can the scientific method be taught? It is difficult, if not impossible, for pupils to learn all scientific strategies through school investigations. Nevertheless, it might be realistic to introduce pupils to at least some of them. Here I want to concentrate on the understanding of the relationship between *evidence*, the *conclusions* based on that evidence and the development of rational-calculative approaches to this relationship which can be termed 'scientific reasoning'. A very simple proposition can usefully illustrate the underlying objective of a science curriculum aimed at developing scientific reasoning. A student who has successfully completed such a curriculum, when faced with the report of a scientific investigation in the popular media, would automatically ask the questions 'How do they know that?', 'Who is writing this?', and perhaps 'Who is paying for this work?'. Whilst the non-expert cannot be expected to interpret the raw data, or even perhaps the arguments put in full in the original source, the scientifically literate will have the requisite skills to interpret the more popular reports and make a valid judgement as to the likely validity or otherwise of their claims, as well as any likely bias in interpretation based on its provenance and the credibility of its sources. In particular, it should be possible to question whether the logical deductions in the argument are sound and if the data offered does indeed support the conclusions drawn. This will involve the understanding of and ability to apply such concepts as probability, risk and certainty[2] which allow us to make judgements as to the likely validity of such reports, and the personal and social consequences associated with related behaviours or policy decisions. These skills have always been important to an individual who wishes to play an active role in any democracy with a culture underpinned by science and technology. Arguably, in this era of information overload they are essential. How else are we to avoid intellectual paralysis as we are bombarded with information and mis-information, claim and counter-claim on such important topics as food safety, genetic manipulation, nuclear power, climate change and environmental pollution? Anyone who takes any interest in these issues can easily discover an overwhelming range of sources of conflicting information through print and electronic media, some original research reports as well as critiques and analyses based on them which may be interpreted from very particular positive or negative perspectives.

Home access to the Internet is growing and access through libraries and other public facilities such as learning centres mean anyone who wants access to the World Wide Web in the developed world can have it pretty much irrespective of income or age. The skills needed to turn this overwhelming sea of information into authentic knowledge include an ability to search vast multimedia sources, identify and interpret relevant information, critique sources in terms of provenance including source, accuracy, validity and reliability, weigh evidence which may be conflicting, and finally collect and synthesise sources into an authentic representation of personal knowledge. These are important elements of ICT literacy which are relevant to scientific literacy and to the development of scientific reasoning.

Extensive discussion of scientific literacy and the relevance of such literacy to science education has, of course, taken place elsewhere (see Osborne 2002). Here I wish only to flag the importance of the role of the Internet and the World Wide Web as contexts for the development of these important skills sets. This is particularly so when the experience of access to information sources, including broadcast and Internet media in the wider community, is growing so rapidly and is such a central part of young people's experience of the world beyond school (Buckingham and McFarlane 2001).

Electronic communications

Much prominence is given to the facility that electronic communications affords educational users to access vast quantities of information from an ever-expanding range of sources. Indeed, scientific sources are at the forefront of this trend as the speed of discovery and dissemination of findings outstrips the rate at which print sources can support the culture of scientific research. It is well known that the original protocols for communicating information over what has become the Internet were devised to support sharing of data between physicists working at laboratories in Switzerland, Italy and England (CERN 2001).

Unfortunately, education policy in the UK has tended to focus on the ability of these networks to disseminate information rather than to support communication. Whilst brief mention is given to student involvement in production and publication, the model implicit in the 'Curriculum on-line' consultation paper produced by the Department for Education and Employment (now Education and Skills) is firmly one of broadcast of digital content to a receptive audience (DfEE 2001). Much of the discourse still tends to assume a view of education – including science – as a process of passing on a discrete body of knowledge to the learner. This is to miss an opportunity to use the developing ICT infrastructure as a means of developing students' ability to be critically informed users and producers of information and, in the case of science, to develop the skills needed to apply scientific reasoning skills to the analysis and critique of related information sources. There is an important role for the active learner here in the manipulation and production of multimedia sources (Bonnett *et al.* 1999). Thus the model of science education which fully exploits electronic media should incorporate both the location and analysis of scientific information and the publishing of the resulting critique as part of an active electronic community of learners. In this way school pupils can expose their interpretations of science to peer review and truly experience the way research proceeds in an authentic fashion.

Reasoned argument in the primary classroom

So if this degree of scientific reasoning is a key objective of science education, how might the foundations be prepared in the primary curriculum? Work with philosophy in the primary curriculum shows that even young children are capable of engaging with debate and reasoning (Lipman 1988). We know that Key Stage 2 children use the Internet regularly and have a worrying degree of confidence in what they find there (McFarlane and Roche 2003). We also know that children are very aware of politicised scientific issues such as conservation and climate change and that they can be left feeling disturbed and disempowered as a result (Chapter 2 in this volume). The context for work on authentic consideration of scientific issues is set and indeed there is a real need to support children's engagement with issues that they find troubling.

To map how such issues might be tackled in the classroom it is useful to point to much relevant and important work in this area that is already ongoing. Of particular concern is how pupils might be brought to a critical awareness of (and engagement with) the nature and methods of science (Warwick and Stephenson 2002). Put another way, the challenge is to 'design instructional sequences and learning environment conditions that help pupils become members of epistemic communities' (Duschl 2000: 188). This is the primary concern of the ongoing EPSE project (see Chapter 1 in this volume), whilst the ASE and King's College Science Investigations in Schools project (AKSIS – Goldsworthy *et al.* 2000: 4) has 'explored the effects of Sc1 in the National Curriculum on current practice and made recommendations for its future development'. AKSIS has had, as a central concern, the exemplification of different types of scientific enquiry and the production of materials to support pupil thinking in relation to the processes of scientific enquiry. A substantive part of this work has been predicated on the notion that if procedural understanding and a wider understanding of the nature of science are to be developed, a vital element of the process is necessarily the extent to which evidence is questioned. It could be argued that the interpretation of evidence is the activity around which all the understandings in science, and of science, pivot. With reference to science education, Duschl (2000: 189) cites Driver *et al.* (1996) in stating that the evaluation of evidence is one of three strands of curriculum emphasis that 'explicitly establish an epistemological basis for scientific knowledge claims'. Thus, research into the uses and interpretations of all forms of evidence is central to elucidating pupils' developing understandings of the personal relevance of science. Warwick and Siraj-Blatchford (in press) recognise that 'the development of a science education that includes a focus upon the nature of science suggests the need for "pedagogic tools" that can be used to engage children with the procedural understandings that are central to a scientific approach to enquiry'. Amongst these tools they report that the use of secondary data for comparative analysis of secondary and investigative data can provide a basis for such engagement. However, they note that 'such comparative analysis will only mirror the collaborative nature of the scientific enterprise where children have guided opportunities to discuss their understanding of the issues revealed by the comparisons . . . (and where) . . . the data is contextualised through connection with the knowledge claims made in science'.

But it seems that in some cases the curriculum is still a long way from even recognising the importance of teaching such critical engagement, whilst the uses of information technologies do not seem to be strongly allied to this purpose. In recent work with post-16 teachers it was surprising to find frustration with students' rather unthinking use of electronic sources, with claims that students tend to use cut and paste uncritically rather than engage with the sources. Yet even though these same teachers and students had been in the same schools for some six years, there was no recognition that this inability to make meaningful use of electronic sources might highlight a deficit in the study skills developed while in the school. Science teachers, it seems, are commonly ill-equipped to teach science in a way that prepares students for citizenship and decision-making (Levinson and Turner 2001; House of Commons 2002). Children, however, do want to know about contemporary science and to engage meaningfully with investigations (Osborne and Collins 2002).

Given the level of use of the Internet even in Key Stage 2 we cannot wait until secondary school to begin to teach children how to make meaningful, critical use of information sources. By the age of 12 bad habits may already be well established. Rather, we need to develop good questioning skills from the earliest stages, and where better to begin with the development of these skills than in science? Science, after all, is all about asking questions and the best scientists ask the best questions. Yet all too often the questions we explore in science are not particularly good or inspiring, and they are certainly not the questions the children would ask. In many lessons we set up contexts that are full of pitfalls for anyone who diverges from the set path as the science around them is complex and hard if not impossible to demonstrate in the classroom. Whoever decided the physics of running cars down a ramp was easy?

However, by talking about systems we are examining and facing up to what we can and cannot deduce about them; we can learn as much, if not more, about both the system and the processes of scientific reasoning as we can through manipulating apparatus in search of an answer. In science, knowing what we cannot know is as important as knowing what we can know. Pretending that science has all the answers is perhaps the greatest disservice we can do, to pupils and to science. And all too easily this can be the impression gained by young investigators, who have to leave an 'experiment' with an answer. In fact, all too often their observations are not adequate to get to an answer. For example, you may have seen that large sugar crystals take longer to dissolve than small ones, but can you be sure why this is just by observing them? One memorable training video showed a group left firmly convinced this is because the large crystals had an invisible coating on them. This conclusion had their teacher stumped and with no time to challenge this view as the class had to move on to another topic. Yet it is perhaps one of the commonest failings of the trainee experimental scientist, and social scientist, to extrapolate their conclusions beyond anything the data can support.

Conclusion

To speak of the role of ICT in science education it is necessary first to identify the objectives of that education and then disaggregate the various forms of ICT in order to discuss the potential relevance or otherwise of each. Where investigative science plays a central part, there are applications of ICT which can both support 'live' investigation and some which can replace it, providing a virtual system to investigate using the same principles as in the laboratory. Moreover, models of the idealised system can be animated alongside a simulation of the real system to reinforce the relationship between practice and theory.

A second and complementary method can be to adopt an analytical

approach to scientific information found in popular and scientific literature, especially the wealth of each available on the Internet. Where an
understanding of various scientific methods and the relationship between
evidence and conclusions are required this can be a more potent experience, dealing as it does with science that cannot be replicated in school
and topical subjects of greater inherent interest to pupils than much of the
rather stodgy content still found in the school curriculum.

In following either of these approaches exclusively there may be a danger of creating a social divide in school science, where perhaps the more
able follow an empirical science curriculum and the less able the more
populist model. In order to avoid such a potentially divisive curriculum, it
might be better to model a curriculum for all which has an equitable
balance between investigative empirical science, supported with ICT so
that it is more effective, and investigative critical science which is supported through access to scientific sources and published analysis shared
and discussed with peers. In this way pupils will experience a range of
approaches to science which will be more likely to enthuse them to follow
a career in science, and ensure they become scientifically literate citizens.
This process cannot begin too early.[3]

Notes

1 A recent report by the higher education funding council into the state of vulner-
able subjects at university level concluded that the closure of university physics
departments per se was not a cause for concern since the number of students
studying the more contemporary but related branches of physical science was
compensating for the decline. Time will tell if this interpretation of the situation
prevails.
2 Probability and risk remain poorly understood concepts as illustrated by a personal favourite, when media reports put the odds of winning the lottery at less
than those of contracting new variant CJD (Creutzfeldt–Jakob disease). Meanwhile government sources were encouraging the population in the one hand to
buy lottery tickets and on the other to continue to eat beef.
3 Some parts of this text appeared in an earlier paper written with Silvestra
Sakellariou and published in 2002.

References

Bonnett, M.R., McFarlane, A.E. and Williams, J. (1999) 'ICT in subject teaching
– an opportunity for curriculum renewal?' *The Curriculum Journal*, 10 (3):
345–359.
Buckingham, D. and McFarlane, A.E. (2001) *A Digitally Driven Curriculum?* London:
Institute for Public Policy Research.
Castells, M. (1996) *The Information Age: Economy, Society and Culture – Volume 1: The
Rise of the Network Society*. Oxford: Blackwell.

CERN (European Organisation for Nuclear Research) (2001), http:// public.web.cern.ch/Public/ACHIEVEMENTS/web.html

Department for Education and Employment (OfEE) (2001) 'Curriculum on-line – a consultation paper'. London: DfEE.

Driver, R., Leach, J., Millar, R. and Scott, P. (1996) *Young People's Images of Science*. Buckingham: Open University Press.

Duschl, R. (2000) 'Making the nature of science explicit', in Millar, R., Leach, J. and Osborne, J. (eds) *Improving Science Education: The Contribution of Research*. Buckingham: Open University Press.

Goldsworthy, A., Watson, R. and Wood-Robinson, V. (2000) *AKSIS Investigations: Developing Understanding*. Hatfield/Cambridge: ASE/Black Bear.

Helms, J.V. (1998) 'Learning about the dimensions of science through authentic tasks', in Wellington, J. (ed.) *Practical Work in School Science – Which Way Now?* London: Routledge.

Hodson, D. (1985) 'Philosophy of science, science and science education', *Studies in Science Education*, 12: 25–57.

House of Commons Science and Technology Committee (2002) *Science Education from 14 to 19*. London: The House of Commons Stationery Office, http:// www.rsc.org/pdf/education/scied1419.pdf

House of Lords (2000) *Science and Society: 'The Jenkin Report'*. Select Committee on Science and Technology, Third Report, 23 February 2000 by the Select Committee appointed to consider Science and Technology.

House of Lords (2001) *Select Committee on Science and Technology First Report*. London: United Kingdom Parliament, http://www.parliament. the-stationery-office.co.uk/pa/ld200001/ldselect/ldsctech/49/4901.htm

Leach, J. (1998) 'Teaching about the world of science in the laboratory: the influence of students' ideas', in Wellington, J. (ed.) *Practical Work in School Science – Which Way Now?* London: Routledge.

Lipman, M. (1988) *Philosophy Goes to School*. Philadelphia: Temple University Press.

McFarlane, A. and Sakellariou, S. (2002) 'The role of ICT in science education', *Cambridge Journal of Education*, 32 (2): 219–232.

McFarlane, A. and Roche, E. (2003) 'Kids and the net: constructing a view of the world', *Education, Communications and Information*, 3 (1).

Millar, R. (1996) 'Towards a science curriculum for public understanding', *School Science Review*, 77 (280): 7–18.

Organisation for Economic Co-operation and Development (OECD) (2001) *Schooling for Tomorrow Series – Learning to Change: ICT in Schools*. Paris: Organisation for Economic Co-operation and Development.

Osborne, J. (2002) 'Science without literacy: a ship without a sail?', *Cambridge Journal of Education*, 32 (2): 203–218.

Warwick, P. and Stephenson, P. (2002) 'Reconstructing science in education: insights and strategies for making it more meaningful', *Cambridge Journal of Education*, 32 (2): 143–151.

Warwick, P. and Siraj-Blatchford, J. (in press) 'Using data comparison and interpretation to develop procedural understandings in the primary classroom: case study evidence from action research', *International Journal of Science Education*.

INDEX

6455444244444445

Joint Information Systems Committee (JISC), and VLEs, 162, 164

Key Stage, 1
National Curriculum, 128–9, 131
science teaching at, 167
use of ICT at, 47, 48–9
and visual literacy, 81–2
Key Stage, 2
and the Internet, 182, 184
National Curriculum, 128
science teaching at, 167
teaching of 'Force' at, 46–7
Key Stage test results, 3, 4
Kidspiration software, 109, 114–23
kinaesthetic learning styles, 150
King's Sutton School, Reception class, 135
Knight, P., 130
knowledge
home knowledge/school knowledge distinction, 114, 124
and scientific literacy, 71, 81, 178–9, 180–1
theories of, 150
knowledge age, 176
preparing learnings, 10
knowledge economies, 176
Kolb, D.A., 149–50
Koopman, Ryan, 97–8, 100
Kress, G., 81, 124
Kutnick, P., 124

language
and inclusive learning, 62
and the National Curriculum, 176
laptop computers, 37
and mind mapping, 109, 114, 115–23
and virtual learning, 168, 169
Laurillard, D., 10, 162
'Learning at the Vets' software, 135–6
learning difficulties, children with, 7, 54
contribution of science to the education of, 55
increasing engagement, 58–60
see also inclusion
learning styles, and museum learning, 149–51
Lemke, J., 81
length, concept of, 15

lesson planning, integrating ICT into science, 36–9
'Let's Think' programme, 168
Lias, S., 25–6
life processes
and CD-ROMs, 39
and practical science, 35
light gates, for data loggers, 48
Linn, M.C., 5
Logo programming language, 97
London, Science Museum, 152, 154, 157, 173
Loveless, A., 72, 74

McFarlane, A., 5, 18–19, 21, 28, 73
McIntyre, D., 2
Manson, I., 124
media
and scientific literacy, 179
and scientific reasoning, 180
metaphors, visual, 73–4
microscopes see digital microscopes
microworlds, and analogical modelling, 96–7, 98
mind mapping, 8–9, 109, 111, 113–14
and Kidspiration software, 109, 114–23
MLEs (managed learning environments), 164
mobile phones, and museum learning, 157
modelling, 93–104
analogical, 8, 93, 95–6
and authenticity, 94–5
children thinking and practising science with ICT, 97–8
ecological simulations, 100–1
and emergent science education, 142–5
interaction analogies and the 'real world', 101–3
model-building and model-use, 104
role of ICT, 96–7
and science in the curriculum, 177
virtual pets, 98–100
Montgomery, D., 54
Moose Crossing, 94
Moovl software, 8, 70, 75–81, 89–90
and concept development, 86–7, 88–9
and visual literacy, 82–4
Morse, S., 176

Related books from Open University Press

Purchase from www.openup.co.uk or order through your local bookseller

---

EARLY EXPLORATIONS IN SCIENCE
SECOND EDITION
Jane Johnston

Reviewers' comments on the first edition:

> Jane Johnston communicates a sense of effervescent enthusiasm for teaching and science, and her treatment is comprehensive.
>
> *TES*

> At last! A serious attempt to explore the scientific potential of infant and pre-school children . . . The author explains how scientific skills can be developed at an early stage, stimulating the natural inquisitive streak in children. This book . . . will start you thinking about science in a much more positive light.
>
> *Child Education*

This accessible and practical book supports good scientific practice in the early years. It helps practitioners to be creative providers, and shows them how to develop awe and wonder of the world in the children they teach. The book highlights the importance of a motivating learning environment and skilled interaction with well-trained adults. In addition, fundamental issues are explored such as the range, nature and philosophical underpinning of early years experiences and the development of emergent scientific skills, understandings and attitudes.

New features for this edition include:

* An extended age range encompassing early learning from 0–8
* Updated material for the Foundation Stage Curriculum for 3–5-year-olds and the National Curriculum 2000 for 5–8-year-olds
* A new chapter focusing on conceptual understanding and thinking skills in the early years
* An emphasis on the importance of informal learning and play in early development

The book introduces and discusses new research and thinking in early years and science education throughout, making it relevant for current practice. This is an indispensable resource for all trainee and practising primary school teachers and early years practitioners.

c.208pp   0 335 21472 X (Paperback)

SUPPORTING SCIENCE, DESIGN AND TECHNOLOGY IN THE EARLY YEARS

John Siraj-Blatchford and Iain MacLeod-Brudenell

- How do young children learn science, design and technology?
- How can we support young children and help them to develop scientific, design and technology skills?

This practical and accessible text answers these questions and provides guidance for adults working with young children in a variety of formal and informal settings. Concrete advice is given to show how parents, carers, teachers and other professionals can provide a rich learning environment and support children in this important area of the curriculum. The differing needs of both adults and children are recognized and a variety of stimulating activities is illustrated. A clear and helpful discussion of a developmental framework enables readers to strengthen their own practice and understanding. The book will be of value to all early childhood professionals as well as being of great interest to parents and carers.

Contents
Acknowledgements – Preface – Introduction – Science, design and technology in the home and local environment – Responding to the differing needs of children – Developing good practice in early years settings – An integrated approach to science design and technology education – Making science, design and technology more relevant to the child – Ensuring progression and continuity – Conclusion – The way forward – Appendices – References – Index.

160pp   0 335 19942 9 (EAN 9 780335 199429) (Paperback)

ENQUIRING CHILDREN, CHALLENGING TEACHING
INVESTIGATING SCIENCE PROCESSES

Max De Boo

This book describes the development of children's enquiry skills offering a rationale and theoretical basis for teaching and learning using this approach and showing its particular relevance to scientific enquiries. The teacher's role is discussed and practical suggestions are given to stimulate effective classroom practice. The author shows how children's ideas can be supported, challenged and assessed, and considers how to resource enquiries and expand these within the school and local environment. The nature of knowledge is explored, with a focus on scientific knowledge about our world. Communication and language skills are discussed, emphasising effective questioning and ways to encourage children's questions. Guidance is given as to how to promote and integrate problem-solving skills into class teaching, particularly in practical cross-curricular and technological projects.

The book will be of great value to both student and practising primary school teachers as well as providing informed support for parents and governors.

Contents

Introduction – 'Look what I've found!': Exploring our world – 'Do stones float?': Investigating our environment – 'Babies don't eat real food!': Thinking, reason and creativity – 'I know why bees are furry. Do you?': Knowledge and understanding – 'Do you mean what I think you mean?': Communication skills – How to cross the river without getting wet: Problem solving – Conclusion – Appendix: Basic resources for science enquiries – Bibliography – Index.

192pp   0 335 20096 6 (EAN 9 780335 200962) (Paperback)